Precise GNSS Positioning and Navigation: Methods, Challenges, and Applications

Precise GNSS Positioning and Navigation: Methods, Challenges, and Applications

Editors

Zhetao Zhang
Wenkun Yu
Giuseppe Casula

MDPI • Basel • Beijing • Wuhan • Barcelona • Belgrade • Manchester • Tokyo • Cluj • Tianjin

MDPI

Editors
Zhetao Zhang
Hohai University
China

Wenkun Yu
Central South University
China

Giuseppe Casula
Istituto Nazionale di
Geofifisica e Vulcanologia
(INGV)
Italy

Editorial Office
MDPI
St. Alban-Anlage 66
4052 Basel, Switzerland

This is a reprint of articles from the Special Issue published online in the open access journal *Remote Sensing* (ISSN 2072-4292) (available at: https://www.mdpi.com/journal/remotesensing/special_issues/39A72305U7).

For citation purposes, cite each article independently as indicated on the article page online and as indicated below:

LastName, A.A.; LastName, B.B.; LastName, C.C. Article Title. *Journal Name* **Year**, *Volume Number*, Page Range.

ISBN 978-3-0365-7622-0 (Hbk)
ISBN 978-3-0365-7623-7 (PDF)

Contents

About the Editors

Zhetao Zhang

Dr. Zhetao Zhang is currently an Associate Professor with the School of Earth Sciences and Engineering, Hohai University, Nanjing, China. He received his Ph.D. degree with distinction in Geodesy from Tongji University, Shanghai, China, in 2019. He was a Postdoctoral Fellow and Visiting Ph.D. student at the Hong Kong Polytechnic University and the University of Calgary, respectively. His primary research focuses on precise positioning and navigation with GNSS and multi-sensor systems under complex conditions, including challenging environments, low-cost devices, multi-source data, etc. He has presided over more than 20 projects, including the National Natural Science Foundation of China, and published more than 70 academic papers, including in the Journal of Geodesy, GPS Solutions, and IEEE Transactions on Aerospace and Electronic Systems. In addition, his interests include the research and development of core technologies and the transformation of corresponding achievements, including new technologies and software.

Wenkun Yu

Dr. Wenkun Yu is currently an Assistant Professor with the School of Geosciences and Info-physics, Central South University, Changsha, China (since 2019). He received his Ph.D. (2019) from the Hong Kong Polytechnic University. His scientific interests include high-precision GNSS positioning and innovative applications for deformation monitoring. He has been engaged in the development of GNSS deformation monitoring algorithms and software systems, and he is now studying the integration of multiple sensors to enhance monitoring performance.

Giuseppe Casula

Dr. Giuseppe Casula has held a permanent position as a senior technologist at the Istituto Nazionale di Geofisica e Vulcanologia (INGV), Unit of Bologna, Italy, since 2003. He studied applied physics with a focus on geophysics at the University of Bologna, where he graduated in 1990 with a thesis on gravimetry. His scientific interests are centered around the use of satellite positioning geodetic techniques, such as GNSS, and advanced differential interferometric synthetic aperture radar, for the study of crustal deformation in seismic and volcanic areas and areas of hydro-geological instability, such as those subject to landslides and subsidence. Recently, he has also studied the use of terrestrial laser scanners and close-range photogrammetry technologies applied to the study of cultural heritage.

remote sensing

Editorial

Editorial for Special Issue "Precise GNSS Positioning and Navigation: Methods, Challenges, and Applications"

Zhetao Zhang [1,*], Wenkun Yu [2] and Giuseppe Casula [3]

[1] School of Earth Sciences and Engineering, Hohai University, Nanjing 211100, China
[2] School of Geosciences and Info-Physics, Central South University, Changsha 410083, China; geowkyu@csu.edu.cn
[3] Istituto Nazionale di Geofisica e Vulcanologia (INGV)—Sezione di Bologna, Viale Berti Pichat 6/2, 40127 Bologna, Italy; giuseppe.casula@ingv.it
* Correspondence: ztzhang@hhu.edu.cn; Tel.: +86-188-8877-0616

Citation: Zhang, Z.; Yu, W.; Casula, G. Editorial for Special Issue "Precise GNSS Positioning and Navigation: Methods, Challenges, and Applications". *Remote Sens.* **2023**, *15*, 2271. https://doi.org/10.3390/rs15092271

Received: 17 April 2023
Accepted: 23 April 2023
Published: 25 April 2023

1. Introduction

The Global Navigation Satellite System (GNSS) can provide users with high-precision positioning information continuously and benefits all walks of life, e.g., unmanned driving, urban navigation, deformation monitoring, etc. The important scientific research and application value of GNSSs have prompted many countries and regions to develop GNSS technologies. GNSS core positioning technologies, such as Precise Point Positioning (PPP) and Real-Time Kinematic positioning (RTK), can provide decimeter-level or even centimeter-level positioning accuracy in open environments. However, active GNSS positioning technologies are susceptible to complex conditions, including canyon environments, low-cost receivers, and multi-GNSS situations, and, on occasion, cannot provide accurate, continuous, and reliable positioning information. The diversification of GNSS systems and constellations, receiver types, and observation environments puts forward higher requirements for technology and algorithms to maintain high-precision positioning and navigation services. Advanced algorithms are key to solving GNSS practical application problems and expanding the scope of GNSS applications.

This Special Issue aims at studies covering improved methods and the latest challenges in precise GNSS positioning and navigation, especially under complex conditions for various research investigations as well as a range of practical applications. Both theoretical and applied research contributions to the GNSS high-precision technology in all disciplines are considered. Topics may cover anything from precise muti-GNSS positioning algorithms and GNSS data processing to more comprehensive targets and scales. Therefore, new algorithms for high-precision positioning and navigation, GNSS receivers, software development for data collection and processing, and their applications in various fields are all included.

2. Overview of Contributions

The following is the synthesis of results obtained in each paper published in the Special Issue "Precise GNSS Positioning and Navigation: Methods, Challenges, and Applications".

Wang et al. [1] evaluated both a multi-baseline solution (MBS) model and a constrained-MBS (CMBS) model that had prior constraints of the spatial-correlated tropospheric delay in deformation monitoring. The reliability and validity of the MBS model was verified using the GPS/BDS data set from ground-based settlement deformation monitoring, with a baseline length of about 20 km and a height difference of about 200 m. They reported that the MBS model reduced the positioning standard deviation (STD) and root-mean-squared (RMS) errors by up to (47.4/51.3/66.2%) and (56.9/60.4/58.4%) in the north/east/up components compared with the single-baseline solution (SBS) model, respectively. Moreover, the combined GPS/BDS localization performance of the MBS model outperformed the

GPS-only and BDS-only localization models by an average of about 13.8 and 25.8 percent, respectively, with highest accuracy improvements of about 41.6 and 43.8 percent. With the additional tropospheric delay constraint, the CMBS model improved monitoring precision in the up direction by about 45.0%.

Guo et al. [2] verified the impact of random interference from walkers on positioning signals in an indoor environment. Based on this phenomenon, the authors proposed a novel real-time dynamic Beacons selection method (RD) in the field of indoor positioning. First, the authors introduced a machine learning algorithm for the real-time anomaly detection of received signals from different Beacons. Then, the Beacon selection was completed based on the real-time anomaly detection results and RSSI. In an indoor scene, the authors verified the positioning accuracy of three other methods when selecting various numbers of Beacons. Then, the authors used the best selection strategies to compare with the RD method. Experiments showed that the RD method can use the least Beacons to obtain higher accuracy and stable positioning results.

Liu et al. [3] verified that the broadcast clock offset had smaller and more stable biases in the long term to compare with the post-processing clock offset and proposed a regional clock offset estimation strategy using broadcast clock offsets for a priori constraints. The results showed that the new algorithm could effectively reduce the biases in PPP-B2b clock offsets. The new clock offset product could improve convergence speeds by 25% and 10% in the horizontal and vertical directions, respectively. For positioning accuracy, the improvements were 22% and 17%, respectively. The absolute error of timing could also be reduced by 60%.

Yang et al. [4] proposed a new strategy to real-time estimate high-accuracy satellite clock offsets. The clock offsets estimated by the new strategy showed good consistencies with the GBM clock offsets. The averaged STD of their differences in the MEO was 0.14 ns, and the clock offsets estimated by the new strategy presented less fluctuations in the 1-day fitting residuals. Applying the new clock offsets to prediction, BDS-3 could reduce its clock offset errors from 1.05 ns to 0.29 ns (RMS), which was a value of about 72%. The above results indicated that the new estimated strategy for clock offsets could improve the accuracy of the clock offset parameters of BDS-3 effectively.

Liu et al. [5] deduced a strategy based on the "density" of common satellites (OBS-DEN) to guarantee baselines of the highest accuracy to be selected. It takes the number of co-viewing satellites per unit distance between stations as the criterion. This method ensured that the independent baseline network had both sufficient observations and short baselines. With single-day solutions and annual statistics computed with parallel processing, the method demonstrated that it had the ability to obtain comparable or even higher positioning accuracies than the conventional methods. With a clearer meaning, OBS-DEN could be an option alongside the previous methods in an independent baseline search.

Qu et al. [6] carried out a series of experiments with a shaking table to assess the structural health monitoring performances of a set of 100 Hz GNSS equipment and three commonly used GNSS positioning techniques: PPP (precise point positioning), PPP-AR (precise point positioning with ambiguity resolution), and RTK (real-time kinematic). They found that the standard deviations of the 100 Hz GNSS displacement solutions derived from PPP, PPP-AR, and RTK techniques were 5.5 mm, 3.6 mm, and 0.8 mm, respectively, when the antenna was in quasi-static motion, and about 9.2 mm, 6.2 mm, and 3.5 mm, respectively, when the antenna was vibrating (up to about 0.7 Hz), under typical urban observational conditions in Hong Kong. They also found that the higher the sampling rate of a, the lower the accuracy of a measured displacement series. On average, the 10 Hz and 100 Hz results were 5.5% and 10.3%, respectively, which were noisier than the 1 Hz results.

Lan et al. [7] provided a comprehensive evaluation of the accuracies of the satellite's precise real-time orbit and clock products, including BDS-3 PPP-B2b precise products and the precise real-time products provided by four IGS centers (CAS, DLR, GFZ, and WHU). In addition, the influences of these real-time precise satellite products on the PPP positioning accuracies with single-frequencies and dual-frequencies were also studied.

Furthermore, the accuracies of the broadcast ephemeris and IGS ultra-rapid products were studied, as well as their impacts on the PPP accuracies. The results illustrated that the orbit accuracies of the PPP-B2b orbits were 9.42 cm, 21.26 cm, and 28.65 cm in the radial, along-track, and cross-track components, which were slightly lower than those of the real-time orbits provided by the four IGS centers. However, the accuracy of the PPP-B2b clock biases was 0.18 ns, which was higher than those provided by IGS Real-Time Service (RTS). In the static positioning test, the 3D positioning accuracy of the B1I+B3I dual-frequencies PPP and the B1C single-frequency PPP were centimeter-level while using the PPP-B2b service, which were slightly lower in the horizontal components compared to those obtained based on the IGS RTS products. The results of the dynamic vehicle test indicated that the positioning accuracies of the B1I+B2b dual-frequency PPP were about 50 cm and 120 cm in the horizontal and vertical components, respectively, which were close to those of the B2b single-frequency PPP using the PPP-B2b service.

Wang et al. [8] proposed a new empirical PWV grid model (called ASV-PWV) using the zenith wet delay from the Askne model, improved by the spherical harmonic function and vertical correction. The proposed method was convenient and enabled the user to gain PWV data with only four input parameters (e.g., the longitude and latitude, time, and atmospheric pressure of the desired position). The profiles of 20 radiosonde stations in Qinghai Tibet Plateau, China, along with the latest publicly available C-PWVC2 model, were used to validate local performance. The PWV data from ASV-PWV and C-PWVC2 were generally consistent with the radiosonde (the average annual bias was −0.44 mm for ASV-PWV and −1.36 mm for C-PWVC2; the root mean square error (RMSE) was 3.44 mm for ASV-PWV and 2.51 mm for C-PWVC2). The proposed ASV-PWV performed better than C-PWVC2 in terms of seasonal characteristics. In general, a sound consistency existed between PWV values of ASV-PWV and the fifth generation of the European Centre for Medium-Range Weather Forecasts Atmospheric Reanalysis (ERA5) (a total of 7381 grid points in 2020). The average annual bias and RMSE were −0.73 mm and 4.28 mm, respectively. The ASV-PWV had a similar performance as the ERA5 reanalysis products, indicating that ASV-PWV was a potentially alternative option for rapidly gaining PWV.

Min et al. [9] carried out a model by integrating between-satellite single-differenced (BSSD) PPP, a low-cost Inertial Navigation System (INS), and an odometer via an extended Kalman filter. The performance of this integration model was assessed with vehicle-borne data. The results demonstrated that the position RMS (Root Mean Square) values of the BSSD PPP were 64.33 cm, 53.47 cm, and 154.11 cm. Compared with BSSD PPP, position improvements of about 31.2%, 23.3%, and 27.3% could be achieved by using INS. Further enhancements of the RMS positions benefiting from the odometer were 1.34%, 1.41%, and 1.73% in the three directions. The accuracy of the BSSD PPP/INS/Odometer tightly coupled integration was slightly higher than that of the undifferenced PPP/INS/Odometer integration, with average improvement percentages of 7.71%, 3.09%, and 0.27%. Meanwhile, the performance of the BSSD PPP/INS/Odometer integration during the periods with satellite outages was better than the undifferenced PPP-based solutions. The improvements in attitudes from an odometer were more significant on heading angles than the other two attitudes, with percentages of 25.00% each. During frequent GNSS outage periods, the reduction in average maximum position drifts provided by the INS were 18.01%, 8.95%, and 20.74%. After integrating with an odometer, the drifts could be decreased further by 25.11%, 15.96%, and 20.69%. For attitude, an about 41.67% reduction in the average maximum drifts of the heading angles was obtained.

Zhang et al. [10] comprehensively assessed the BDS-2/BDS-3 final (ISC), rapid (ISR), and ultra-rapid (ISU) products based on B1I/B3I and B1C/B2a frequencies from the international GNSS Monitoring and Assessment System (iGMAS). Specifically, at first, the precise orbits from iGMAS were compared with the ones from the IGS ACs. Based on this, the Satellite Laser Ranging inspected the precise orbits from iGMAS. Finally, the orbit errors were discussed systematically by considering the Beta and Elongation angles. Using one year of data, the orbit accuracies of geostationary orbit, inclined geosynchronous orbit, and medium earth orbit (MEO) satellites could reach an almost meter to decimeter level, a decimeter to sub-decimeter level, and a centimeter level, respectively, where the ISC products were the best. The ISC, ISR, and ISU products based on B1I/B3I frequencies were generally better than the ones based on B1C/B2a frequencies. Additionally, according to the SLR data, the results showed that the accuracy of the precise orbits of the BDS-3 was better than that of the BDS-2. The mean values of orbit biases of the BDS-3 MEO satellites were approximately 2.88 cm. In addition, the orbit errors were related to the beta angle and elongation angle to some extent, and the manufacturers may also have had an influence on the orbit errors.

Zhou and Wang. [11] provided a comprehensive analysis of pseudorange-based/single point positioning (SPP) among GPS, BDS-3, and Galileo on a global scale. First, the positioning accuracy distribution of adding IGSO and GEO to the MEO of BDS-3 was analyzed. The results showed that the accuracy of the third dimension in the Asia–Pacific region was significantly improved after adding IGSO and GEO. Then, the positioning accuracies of the single-system and single-frequency SPPs were validated and compared. The experimental results showed that the median RMS values for the GPS, Galileo, and BDS-3 were 1.10/1.10/1.30 m and 2.57/2.69/2.71 m in the horizontal and vertical components, respectively. For the horizontal component, the GPS and Galileo had better positioning accuracy in the middle- and high-latitude regions, while the BDS-3 had better positioning accuracy in the Asia–Pacific region. For the vertical component, poorer positioning accuracy could be seen near the North Pole and the equator for all three systems. Meanwhile, in comparison with the single-system and single-frequency SPPs, the contribution of adding pseudorange observations from other satellite systems and frequency bands was analyzed fully. Overall, the positioning accuracy could be improved to varying degrees.

Viler et al. [12] focused on the quality of 2D and 3D kinematic positionings of different geodetic and low-cost GNSS devices, using the professional mobile mapping system (MMS) as a reference. Kinematic positionings were performed simultaneously with a geodetic Septentrio AsteRx-U receiver, two u-blox receivers—ZED-F9P and ZED-F9R—and a Xiaomi Mi 8 smartphone, which were then compared with Applanix Corporation GPS/INS MMS reference trajectories. As expected, some results in the GNSS positionings were subject to position losses, large outliers, and multipath effects; however, after removing them, they were quite promising, even for the Xiaomi Mi8 smartphone. From the comparison of the GPS and GNSS solutions, as expected, the GNSS processing achieved many more solutions for position determination and allowed a relevant higher number of fixed ambiguities, even if this was not true, in general, for the Septentrio AsteRx-U, in particular, in a surveyed non-urban area with curves and serpentines characterized by a reduced signal acquisition. In the GNSS mode, the Xiaomi Mi8 smartphone performed well in situations with thresholds less than 1 m, with the percentages varying from 50% for the urban areas to 80% for the non-urban areas, which offered potential in view of future improvements for applications in terrestrial navigation.

Shang et al. [13] investigated the temporal behaviors of differential inter-system bias (DISB) and implemented an inter-system model for smartphones. They accessed the data from a Huawei P40 (HP40) smartphone and reported: (1) For the HP40, the frequencies of code-division-multiple-access systems were free of receiver channel-dependent phase bias, which provided chances for additional interoperability among these systems. However, the code observations of HP40 were affected by the receiver channel-dependent code bias. Therefore, it was suggested to set a large initial STD value for code observations in the positioning. (2) GPS L1/ QZSS L1 and BDS-2 B1I/BDS-3 B1I were free of phase DISB, and there were evident phases DISB between GPS L1 and Galileo E1. Even then, the valuations were sufficiently stable, with an STD close to 0.005 cycles. However, the GPS L1/BDS B1I phase DISB was unstable. (3) For kinematic positioning, when the stable phase DISB was introduced, a 3–38.9% improvement in the N/E/U directions of the positioning accuracies in the inter-system differencing was achieved compared with the intra-system differencing.

Author Contributions: Conceptualization, Z.Z.; writing—original draft preparation, Z.Z., W.Y. and G.C.; writing—review and editing, Z.Z., W.Y. and G.C. All authors have read and agreed to the published version of the manuscript.

Conflicts of Interest: The authors declare no conflict of interest.

References

1. Wang, H.; Dai, W.; Yu, W. BDS/GPS Multi-Baseline Relative Positioning for Deformation Monitoring. *Remote Sens.* **2022**, *14*, 3884. [CrossRef]
2. Guo, Y.; Zheng, J.; Di, S.; Xiang, G.; Guo, F. A Beacons Selection Method under Random Interference for Indoor Positioning. *Remote Sens.* **2022**, *14*, 4323. [CrossRef]
3. Liu, J.; Tang, C.; Zhou, S.; Hu, X.; Yang, Y.; Yang, J.; Liu, Y. The Bias in PPP-B2b Real-Time Clock Offset and the Strategy to Reduce It. *Remote Sens.* **2022**, *14*, 4569. [CrossRef]
4. Yang, J.; Tang, C.; Zhou, S.; Song, Y.; Liu, J.; Xiang, Y.; Liu, Y.; Tian, Q.; Yang, Y.; Yang, Z.; et al. High-Accuracy Clock Offsets Estimation Strategy of BDS-3 Using Multi-Source Observations. *Remote Sens.* **2022**, *14*, 4674. [CrossRef]
5. Liu, T.; Du, Y.; Nie, W.; Liu, J.; Ma, Y.; Xu, G. An Observation Density Based Method for Independent Baseline Searching in GNSS Network Solution. *Remote Sens.* **2022**, *14*, 4717. [CrossRef]
6. Qu, X.; Shu, B.; Ding, X.; Lu, Y.; Li, G.; Wang, L. Experimental Study of Accuracy of High-Rate GNSS in Context of Structural Health Monitoring. *Remote Sens.* **2022**, *14*, 4989. [CrossRef]
7. Lan, R.; Yang, C.; Zheng, Y.; Xu, Q.; Lv, J.; Gao, Z. Evaluation of BDS-3 B1C/B2b Single/Dual-Frequency PPP Using PPP-B2b and RTS SSR Products in Both Static and Dynamic Applications. *Remote Sens.* **2022**, *14*, 5835. [CrossRef]
8. Wang, X.; Chen, F.; Ke, F.; Xu, C. An Empirical Grid Model for Precipitable Water Vapor. *Remote Sens.* **2022**, *14*, 6174. [CrossRef]
9. Min, Y.; Gao, Z.; Lv, J.; Lan, R.; Xu, Q.; Yang, C. Low-Cost IMU and Odometer Tightly Augmented PPP-B2b-Based Inter-Satellite Differenced PPP in Urban Environments. *Remote Sens.* **2023**, *15*, 199. [CrossRef]
10. Zhang, Z.; Zeng, P.; Wen, Y.; He, L.; He, X. Comprehensive assessment of BDS-2 and BDS-3 precise orbits based on B1I/B3I and B1C/B2a frequencies from iGMAS. *Remote Sens.* **2023**, *15*, 582. [CrossRef]
11. Zhou, F.; Wang, X. Some Key Issues on Pseudorange-Based Point Positioning with GPS, BDS-3, and Galileo Observations. *Remote Sens.* **2023**, *15*, 797. [CrossRef]
12. Viler, F.; Cefalo, R.; Sluga, T.; Snider, P.; Pavlovčič-Prešeren, P. The Efficiency of Geodetic and Low-Cost GNSS Devices in Urban Kinematic Terrestrial Positioning in Terms of the Trajectory Generated by MMS. *Remote Sens.* **2023**, *15*, 957. [CrossRef]
13. Shang, R.; Gao, C.; Gan, L.; Zhang, R.; Gao, W.; Meng, X. Multi-GNSS Differential Inter-System Bias Estimation for Smartphone RTK Positioning: Feasibility Analysis and Performance. *Remote Sens.* **2023**, *15*, 1476. [CrossRef]

remote sensing

Article

Comprehensive Assessment of BDS-2 and BDS-3 Precise Orbits Based on B1I/B3I and B1C/B2a Frequencies from iGMAS

Zhetao Zhang, Ping Zeng, Yuanlan Wen *, Lina He and Xiufeng He

School of Earth Sciences and Engineering, Hohai University, Nanjing 211100, China
* Correspondence: wwwgwyl@126.com; Tel.: +86-135-4968-7989

Abstract: The BeiDou Global Navigation Satellite System (BDS), including the second generation (BDS-2) and the third generation (BDS-3), has been widely used in areas of positioning, navigation, and timing (PNT). One of the essential prerequisites for accurate PNT service is the precise satellite orbits of multi-frequency and multi-constellation BDS-2 and BDS-3 satellites. As usual, the precise orbit products can be obtained from analysis centers (ACs) of the international GNSS Service (IGS). The precise orbits can also be downloaded from the international GNSS Monitoring and Assessment System (iGMAS). Compared with the IGS ACs, the iGMAS can provide featured services such as satellite orbits based on the new B1C/B2a BDS signals. Considering the indispensability of the new signals, the performance of all BDS precise orbits from iGMAS needs to be known. However, there is no comprehensive assessment of BDS-2 and BDS-3 precise orbits based on B1I/B3II and B1C/B2a frequencies from iGMAS, especially for the period after the BDS entered the stable operation stage. In this paper, BDS-2/BDS-3 final (ISC), rapid (ISR), and ultra-rapid (ISU) products based on B1I/B3I and B1C/B2a frequencies from iGMAS are all assessed comprehensively. Specifically, at first, the precise orbits from iGMAS are compared with the ones from the IGS ACs. Based on this, the satellite laser ranging inspects the precise orbits from iGMAS. Finally, the orbit errors are discussed systematically by considering the beta and elongation angles. Using one year of data, the orbit accuracy of geostationary orbit, inclined geosynchronous orbit, and medium earth orbit (MEO) satellites can almost reach meter to decimeter level, decimeter to sub-decimeter level, and centimeter level, respectively, where the ISC products are the best. The ISC, ISR, and ISU products based on B1I/B3I frequencies are generally better than the ones based on B1C/B2a frequencies. Additionally, according to the SLR data, the results show that the accuracy of precise orbits of BDS-3 is better than that of BDS-2. The mean values of orbit biases of BDS-3 MEO satellites are approximately 2.88 cm. In addition, the orbit errors are related to the beta angle and elongation angle to some extent, and the manufacturers may also have an influence on the orbit errors.

Keywords: BDS; B1I/B3I frequencies; B1C/B2a frequencies; iGMAS; precise orbits

Citation: Zhang, Z.; Zeng, P.; Wen, Y.; He, L.; He, X. Comprehensive Assessment of BDS-2 and BDS-3 Precise Orbits Based on B1I/B3I and B1C/B2a Frequencies from iGMAS. *Remote Sens.* **2023**, *15*, 582. https://doi.org/10.3390/rs15030582

Academic Editor: Andrzej Stateczny

Received: 10 December 2022
Revised: 14 January 2023
Accepted: 14 January 2023
Published: 18 January 2023

1. Introduction

Up to now, the BeiDou Navigation Satellite System (BDS) has been completed. It has been put into use in areas of positioning, navigation, and timing (PNT), which includes the second-generation one (BDS-2) and third-generation one (BDS-3) [1,2]. The BDS is a multi-source heterogeneous constellation with three types of satellites and five frequencies in total. Specifically, according to the China Satellite Navigation Office (CSNO), the three types of satellites are geostationary orbit (GEO), inclined geosynchronous orbit (IGSO), and medium earth orbit (MEO) satellites, respectively. The five available frequencies are B1C (1575.420 MHz), B1I (1561.098 MHz), B3I (1268.520 MHz), B2b/B2I (1207.140 MHz), and B2a (1176.450 MHz) signals, respectively [3–7]. As of December 2022, there are 15 BDS-2 satellites in orbit, including five GEO, seven IGSO, and three MEO satellites, respectively. For BDS-3 satellites, there are 29 BDS-3 satellites in total, consisting of two GEO, three IGSO, and 24 MEO satellites in orbit [8]. Moreover, the BDS-2 GEO/IGSO/MEO and

BDS-3 GEO/IGSO satellites are manufactured by the China Academy of Space Technology (CAST), and BDS-3 MEO satellites are built by the CAST or Shanghai Engineering Center for Microsatellites (SECM). Apparently, obtaining the precise satellite orbits of BDS-2 and BDS-3 satellites is the prerequisite of high-precision and high-reliability PNT service [9].

As usual, global navigation satellite system (GNSS)-related products can be obtained from the international GNSS Service (IGS), including satellite orbits, clock corrections, Earth rotation parameters, geocentric coordinates of IGS tracking stations, and tropospheric and ionospheric parameters [10]. Currently, twelve IGS analysis centers (ACs) can provide related precise products [11]. Specifically, they are Natural Resources Canada (EMR), Wuhan University (WHU), Geodetic Observatory Pecny (GOP), Space geodesy team of the CNES (GRG), European Space Agency (ESA), GeoForschungsZentrum (GFZ), Center for Orbit Determination in Europe (CODE), Jet Propulsion Laboratory (JPL), Massachusetts Institute of Technology (MIT), National Geodetic Survey (NGS), Scripps Institution of Oceanography (SIO), and U.S. Naval Observatory (USNO). Recently, the Multi-GNSS Experiment (MGEX) has been set up and can provide multi-GNSS, including global positioning system (GPS), GLONASS (GLObal'naya NAvigatsionnaya Sputnikovaya Sistema), Galileo, and BDS products [12,13], etc. Many scholars have analyzed the four-system orbit and clock products provided by MGEX ACs [14–17]. However, it is worth noting that one may combine the above products from different ACs to improve reliability in real applications [18].

To further monitor and assess the performance and status of BDS-2 and BDS-3, the international GNSS Monitoring and Assessment System (iGMAS) has been established [19,20]. As of December 2022, the iGMAS has 30 tracking stations, three data centers, and 12 ACs. The 12 ACs include Xi'an Satellite Control Center, Shanghai Astronomical Observatory, Information Engineering University, Beijing Aerospace Control Center, Chinese Academy of Surveying and Mapping, Innovation Academy for Precision Measurement Science and Technology, Chang'an University, Wuhan University, National Time Service Center, China University of Mining and Technology, Xi'an Research Institute of Surveying and Mapping, and Tongji University [21]. Based on the observed multi-frequency and multi-constellation GNSS observations, including BDS, GPS, GLONASS, and Galileo collected by the tracking stations, the ACs obtain the multi-GNSS data from data centers and generate the precise products, including satellite orbits, clock corrections, Earth rotation parameters, geocentric coordinates of iGMAS tracking stations, and tropospheric and ionospheric parameters, etc. Based on the products from different ACs, the iGMAS will combine and generate the final (ISC), rapid (ISR), and ultra-rapid (ISU) orbit products. Therefore, the iGMAS can promote compatibility and interoperability among different GNSS constellations [16,22]. One of the most notable features of iGMAS is that it supports new BDS signals B1C/B2a and related products.

Since the orbit products will directly affect the PNT performance and other related applications [23–25], assessing the precision and reliability of the iGMAS orbit products at hand is crucial. At earlier times, compared with the IGS products, the precision of BDS-2 GEO/IGSO and MEO satellites from iGMAS is better than 10 cm and approximately 1 m, respectively [26]. Later, several studies show that by using the iGMAS products, the precision of BDS-3 IGSO and MEO satellites is approximately 10 cm and 40 cm [27], which somewhat shows improvements. To further evaluate the BDS products from iGMAS, the satellite-induced code pseudorange variation [28], differential code bias [29], and signal quality [30] are studied. In addition, the satellite laser ranging (SLR) data are applied to validate the BDS-2 precise orbits [31]. Actually, since the middle of 2019, the iGMAS has been the only center that can provide the orbit products based on B1I/B3I and B1C/B2a signals simultaneously. In contrast, the others can only offer the ones of B1I/B3II signals. Recently, the latest research found the potential advantage of B1C/B2a signals compared with the B1I/B3I signals, such as in better signal-in-space range error [32]. In theory, the B1C/B2a-based orbit products are advised when using the B1C/B2a signals for better consistency. Therefore, it is indispensable to systematically evaluate all kinds of BDS products from iGMAS.

Unfortunately, to our best knowledge, there is no comprehensive assessment of BDS-2 and BDS-3 precise orbits based on B1I/B3II and B1C/B2a frequencies from iGMAS, especially for the period after the BDS entered the stable operation stage. In this paper, all kinds of BDS products are systematically studied, where the BDS-2/BDS-3 ISC, ISR, and ISU precise orbit products based on B1I/B3I and B1C/B2a frequencies from iGMAS are all included. Specifically, to comprehensively evaluate the different orbits of iGMAS, the MGEX precise orbits from four primary ACs that can provide BDS orbits, particularly the BDS-3, are used for orbit comparison. Secondly, the optical technique of SLR is used as an external independent validation for the radial components of iGMAS products. Thirdly, the orbit errors from iGMAS are discussed in depth, where the influential factors including beta angle and elongation angle are considered.

2. Methodology

To evaluate the precise orbits of iGMAS, the first strategy is to compare the products of iGMAS with those of other ACs, such as GFZ, ESA, CODE and WHU, etc. Whereas the product difference can only depict the consistency between two arbitrary agencies, new products, such as precise orbits based on B1C/B2a frequencies, cannot be assessed with high accuracy and high reliability. Therefore, the SLR residuals are further applied to inspect the precise orbits of iGMAS, where the B1I/B3I and B1C/B2a are included.

The inspection with SLR residuals is actually the differences between the computed satellite-to-receiver range with precise products and the observed SLR observations. The SLR observation equation can be expressed as follows [33]:

$$\rho_{SLR} = \frac{1}{2}c\tau + d_t + d_{sr} + d_{td} + d_{pm} + \epsilon \tag{1}$$

where ρ_{SLR} denotes the SLR observation; c and τ denotes the light speed in a vacuum and the time interval from laser emission to reception; the d_t, d_{sr}, d_{td}, d_{pm} and ϵ denote the time offset, station-specific range bias, tidal displacement correction of site position, plate motion correction of the station, and other observation error, respectively. Here, the ϵ mainly includes the center-of-mass correction, tropospheric delay, space-time curvature correction, and random error, etc. In addition, according to the satellite precise orbit, the computed satellite-to-receiver range reads:

$$\rho_{sr} = \rho + d_t + d_{sr} + d_{td} + d_{pm} + \varepsilon \tag{2}$$

where ρ_{sr} denotes the computed satellite-to-receiver range; $\rho = \sqrt{(x^s - x_r)^2 + (y^s - y_r)^2 + (z^s - z_r)^2}$ with coordinate components of satellite (x^s, y^s, z^s) and receiver (x_r, y_r, z_r) denotes the satellite-to-receiver range; and ε mainly contains the residual systematic error [34,35] and random error, which is similar to the observation error in Equation (1). Therefore, the SLR residuals can be obtained as follows:

$$v_{SLR} = \rho_{SLR} - \rho_{sr} \tag{3}$$

where v_{SLR} denotes the SLR residuals to be used in this study.

To make a more comprehensive study on the performance of precise products in this study. The relations between the orbit error and beta angle/elongation angle are analyzed in this study. Figure 1 depicts the geometric space relations among the sun, Earth, and GNSS satellite, where the beta angle and elongation angle are marked. Here, the beta angle represents the elevation angle of the sun above the orbital plane, and the angle between the sun, the satellite, and the Earth is elongation angle [22].

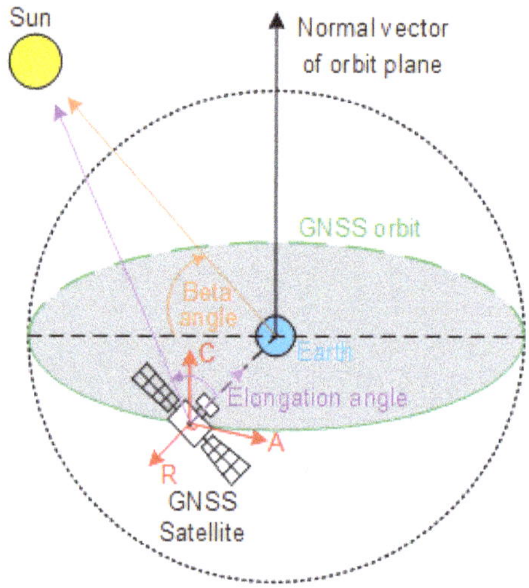

Figure 1. Space geometric angle and position among the sun, Earth, and GNSS satellite.

3. Data and Experiment

In this study, we have compared iGMAS orbits with four MGEX ACs from 10 July 2021 to 10 July 2022 in 366 days. Table 1 summarizes the details of precise orbit products provided by iGMAS, GFZ, ESA, CODE, and WHU. It should be noted that these four ACs were chosen because their products are widely used and include the BDS-3 satellites. The institution, ID, orbit latency, orbit update, orbit sampling, and frequency are all included. It can be seen that the iGMAS has the most extensive range of products.

Table 1. Overview of the orbit products from iGMAS and four MGEX ACs.

Institution	ID	Orbit Latency	Orbit Update	Orbit Sampling	Frequency
iGMAS	ISC	12 days	weekly	15 min	B1I/B3I; B1C/B2a
	ISR	17 h	daily	15 min	B1I/B3I; B1C/B2a
	ISU	3 h	6 h	15 min	B1I/B3I; B1C/B2a
GFZ	GFZ	daily	daily	5 min	B1I/B3I
ESA	ESA	daily	daily	5 min	B1I/B3I
CODE	COD	daily	daily	5 min	B1I/B3I
WHU	WUM	daily	daily	15 min	B1I/B3I

As of July 2022, GFZ, ESA, and CODE are provided BDS orbit with 5-min sampling based B1I/B3I, while WHU and iGMAS orbit products are supplied at 15 min intervals. Figure 2 shows the availability of orbit products from iGMAS and four MGEX ACs. The availability of GFZ, ESA, and COD can reach 100%, while WUM is approximately 92.4%. The ISC, ISR, and ISU products based on B1I/B3I also with high availability, approximately 100%, 99.7%, and 100%, respectively. For the B1C/B2a-based orbit products of iGMAS, the ISC can maintain long-term availability with approximately 91.8%, although interruptions have occurred at some time. In contrast, the availability of ISR and ISU products is much lower than ISC, approximately 14.4%, and 14.5%, respectively. It seems that iGMAS stopped to provide these two products after 27 November 2021. We noticed that iGMAS combined products included all BDS-2 satellites and 27 BDS-3 satellites. Specifically, there are 10 BDS-3 MEO satellites built by SECM, 15 BDS-3 MEO, and 3 BDS-3 IGSO satellites built by CAST.

The PRNs are from C19 to C46, except for the experiment satellite of C31. The satellite products of GFZ and WHU are similar to iGMAS, while BDS-2 GEO satellites are excluded from ESA and CODE products. Note that the time system used in iGMAS products is BDS Time, while MGEX ACs is GPS time (GPST). Therefore, the iGMAS products are converted to GPST compared with the four MGEX ACs. In order to validate the iGMAS product, SLR validation is used in this paper. The SLR observations were collected from the International Laser Ranging Service (ILRS). The other details of the processing strategies are shown in Table 2.

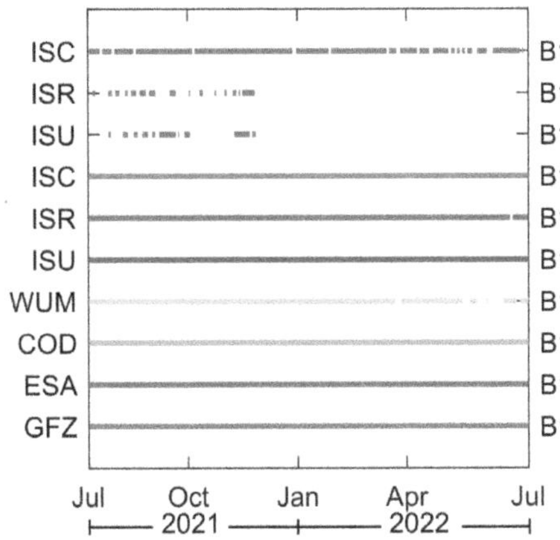

Figure 2. Availability of orbit products from iGMAS and four MGEX ACs.

Table 2. Processing strategies of orbit products.

Item	Strategy
Time Span	DOY 190 in 2021 to DOY 190 in 2022
Orbit compare	iGMAS orbit products convert to GPST and are compared with MGEX ACs BDS-2: all of the GEO, IGSO, and MEO satellites BDS-3:C19-C49 (except C31)
SLR validation	BDS-2 GEO: C01 BDS-2 IGSO: C08, C10 BDS-2 MEO: C11 BDS-3 CAST MEO: C20, C21 BDS-3 SCEM MEO: C29, C30

4. Analysis of Orbit Products

4.1. ISC Products

Figure 3 shows the root mean squares (RMSs) for BDS-2 satellites in along-track, cross-track, and radial directions between iGMAS and MGEX ACs based on B1I/B3I. The BDS-2 GEO satellites are not provided by ESA and CODE. Thus, this paper compares the BDS-2 GEO satellites with GFZ and WHU. As seen in Figure 3, GEO satellites show the most significant RMSs, especially in the along-track direction. Taking the C01 satellite as an example, we can find that the RMS of cross-track between ISC and WHU is much smaller than that of ISC and GFZ. It indicates that the processing strategies of ISC and WHU for GEO satellites may be more similar. For the IGSO satellite, the RMS is at the sub-decimeter to decimeter level between ISC and four MGEX ACs in three directions. For MEO satellites, the RMS is better than GEO and IGSO satellites, especially for the radial direction, which is at the centimeter level. Moreover, the radial accuracy of MEO satellites is much better than both along-track and cross-track directions. This is reasonable because the observations are oriented around the radial direction rather than the along-track and cross-track directions. Overall, the consistency of the BDS-2 GEO orbits between ISC and WHU products is better than that between ISC and GFZ. For IGSO satellites, CODE showed the best agreement regarding ISC; the RMSs of along-track, cross-track, and radial directions are 0.120 m, 0.113 m, and 0.102 m, respectively. For BDS-2 MEO satellites, ESA showed the best agreement concerning ISC, and the RMSs of along-track, cross-track, and radial directions are 0.102 m, 0.063 m, and 0.051 m, respectively.

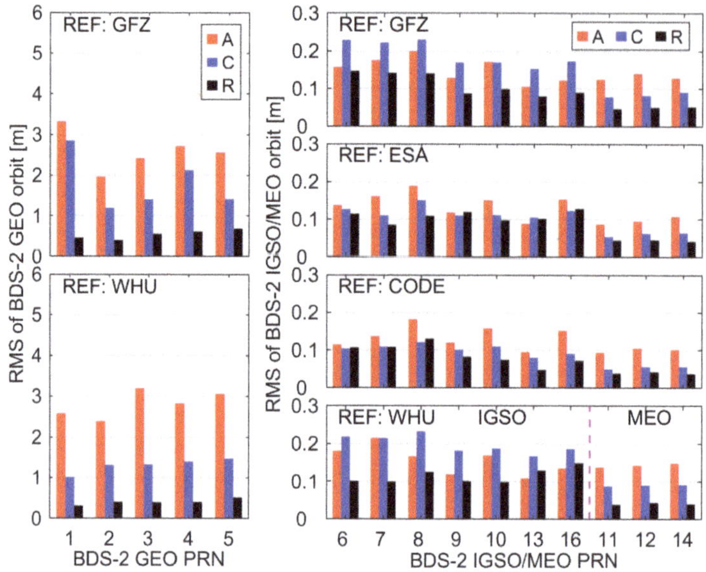

Figure 3. RMS of BDS-2 ISC orbits based on B1I/B3I compared with different MGEX ACs in the along-track (A), cross-track (C), and radial (R) directions, respectively.

Figure 4 shows the RMSs of BDS-3 ISC orbits based on B1I/B3I and B1C/B2a compared with MGEX ACs precise orbit products. The horizontal axis is the RMSs of B1I/B3I and the vertical axis is the RMSs of B1C/B2a. The gray dashed line indicates that the RMSs of B1I/B3I and B1C/B2a are the same. Therefore, the upper part of the dashed line means that B1I/B3I is better than B1C/B2a and vice versa. The circle represents MEO satellites, and the triangle represents IGSO satellites. In addition, different colors are used to distinguish the different directions. That is, the green, blue, and red colors indicate the along-track, cross-track, and radial directions, respectively. As shown in Figure 4, we can see that the comparison results of along-track, cross-track, and radial directions of MEO are basically above the dashed line. This indicates that the consistency of B1C/B2a orbit products is not as good as B1I/B3I. The main reason may be that the phase center offsets and variations corrections for the B1C/B2a orbit are not accurate enough.

Figure 5 shows the box plot of RMSs of ISC orbits based on B1I/B3I and B1C/B2a in along-track, cross-track, and radial directions for SECM MEO, CAST MEO and CAST IGSO satellites. The five short horizontal lines from the top to the bottom of the box chart represent the 100%, 75%, 50%, 25%, and 0% quantiles, respectively. It can be seen that the median values of RMSs of the along-track based on B1I/B3I for SECM MEO, CAST MEO, and CAST IGSO satellites are approximately 0.055 m, 0.050 m, and 0.150 m, respectively. This indicates that the accuracy of SECM MEO satellites is similar to CAST MEO satellites, and both are better than CAST IGSO satellites. However, an extensive variation range of RMS of 0.045 m to 0.072 m can be seen for CAST MEO satellites. This means that the consistency of CAST MEO satellites is not better than SECM MEO satellites. For the cross-track, the median values of RMSs of B1I/B3I for SECM MEO, CAST MEO, and CAST IGSO satellites are approximately 0.042 m, 0.043 m, and 0.137 m, respectively. For the radial directions, the median values of RMSs of B1I/B3I for SECM MEO, CAST MEO, and CAST IGSO satellites are approximately 0.036 m, 0.038 m, and 0.141 m, respectively.

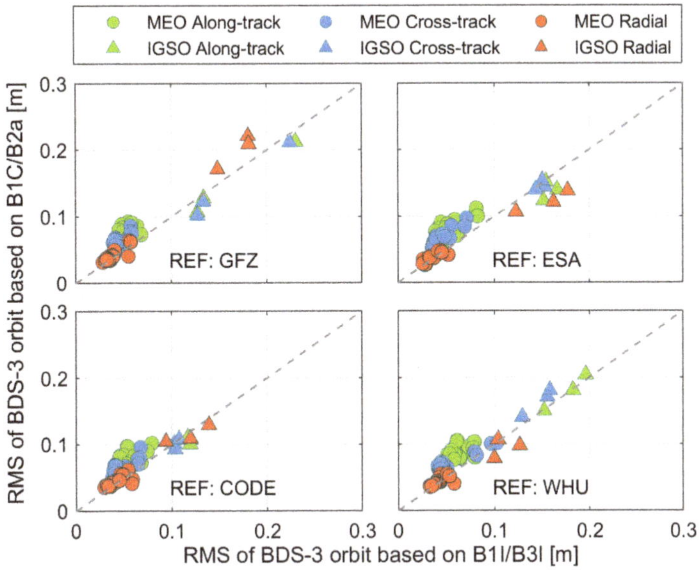

Figure 4. RMS of BDS-3 ISC orbits based on B1I/B3I and B1C/B2a compared with different MGEX ACs in the along-track (A), cross-track (C), and radial (R) directions, respectively.

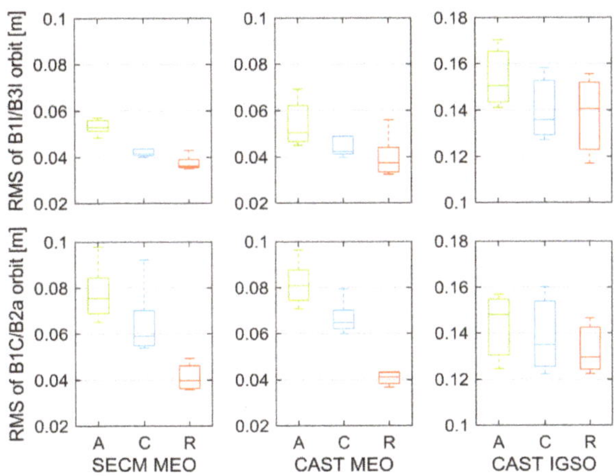

Figure 5. Box plot of RMSs of ISC orbits based on B1I/B3I and B1C/B2a in along-track, cross-track, and radial directions for SECM MEO, CAST MEO, and CAST IGSO satellites, respectively.

4.2. ISR Products

For ISR products, Figure 6 illustrates the RMSs for BDS-2 satellites in along-track, cross-track, and radial directions between ISR B1I/B3I-based and MGEX ACs. We can find that GEO satellites show the largest RMSs in the along-track, cross-track, and radial directions, of which the values are approximately 3.0 m, 1.5 m, and 0.5 m, respectively. For IGSO satellites, the consistency of ISR and MGEX ACs is different. Specifically, the consistency of the cross-track between ISR and ESA/CODE is better than ISR and GFZ/WHU, while the along-track is in contrast. It implies that the different ACs may apply different dynamic models. For MEO satellites, the consistency of ISR and MGEX ACs is similar. On the other hand, by comparing Figures 3 and 6, we can find that the accuracy of ISR B1I/B3I-based orbits is slightly lower than that of ISC, which is reasonable since the ISC products are the final combined ones.

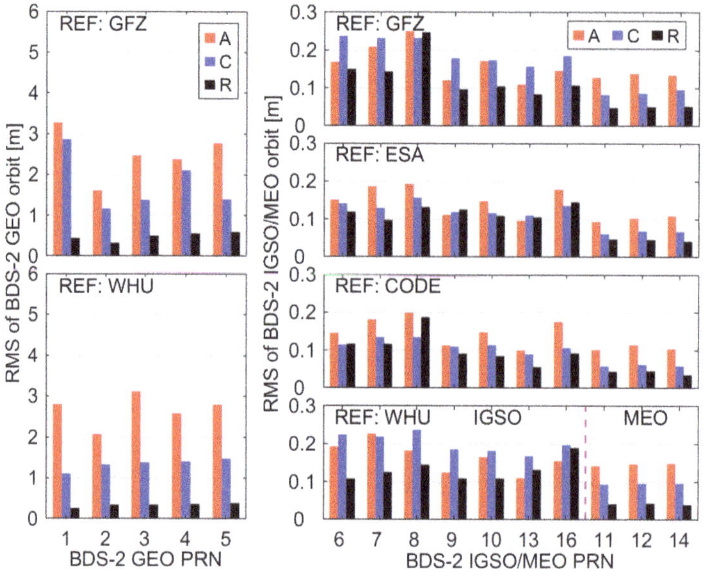

Figure 6. RMS of BDS-2 ISR orbits based on B1I/B3I compared with different MGEX ACs in the along-track (A), cross-track (C), and radial (R) directions, respectively.

To comprehensively assess the difference between ISR B1I/B3I-based and B1C/B2a-based orbit products, Figure 7 shows the comparison results of ISR with four MGEX ACs. Once again, the gray dashed line indicates that the accuracy of ISR B1I/B3I-based and B1C/B2a-based are the same. Thus, the upper part of the dashed line indicates that the accuracy of ISR B1I/B3I-based is better than B1C/B2a-based, and vice versa. We can find that the consistency of ISR B1C/B2a-based orbit products for BDS-3 MEO satellites is worse than B1I/B3I-based. In addition, by comparing Figures 4 and 7, we can find that the orbit accuracy of ISR B1C/B2a-based has decreased compared to ISC, especially for MEO satellites. For the radial directions of BDS-3 IGSO satellites, we can see that the consistency of B1I/B3I-based and B1C/B2a-based for different MGEX ACs is different, and the CODE shows the best consistency with ISR. The reason may be that the processing strategies of ISC and CODE are more consistent than other ACs.

Figure 8 shows the box plot of RMSs of ISR orbits based on B1I/B3I and B1C/B2a in along-track, cross-track, and radial directions for SECM MEO, CAST MEO, and CAST IGSO satellites. As can be seen in Figure 8, the median values of RMSs of the along-track based on B1I/B3I for SECM MEO, CAST MEO, and CAST IGSO satellites are approximately 0.055 m, 0.058 m, 0.175 m, respectively. This demonstrates that the accuracy of SECM MEO satellites is better than CAST MEO satellites in along-track direction. In addition, SECM and CAST MEO satellites are better than CAST IGSO satellites. For the B1C/B2a-based, the median values of RMSs of the along-track for SECM MEO, CAST MEO, and CAST IGSO satellites are approximately 0.085 m, 0.148 m, and 0.164 m, respectively. This means that the consistency of CAST MEO satellites is not better than SECM MEO satellites in along-track direction, and B1C/B2a is significantly different from B1I/B3I. For the cross-track, the median values of RMSs of B1I/B3I for SECM MEO, CAST MEO, and CAST IGSO satellites are approximately 0.042 m, 0.053 m, and 0.153 m, respectively. For the radial directions, the median values of RMSs of B1I/B3I for SECM MEO, CAST MEO, and CAST IGSO satellites are approximately 0.032 m, 0.037 m, and 0.147 m, respectively.

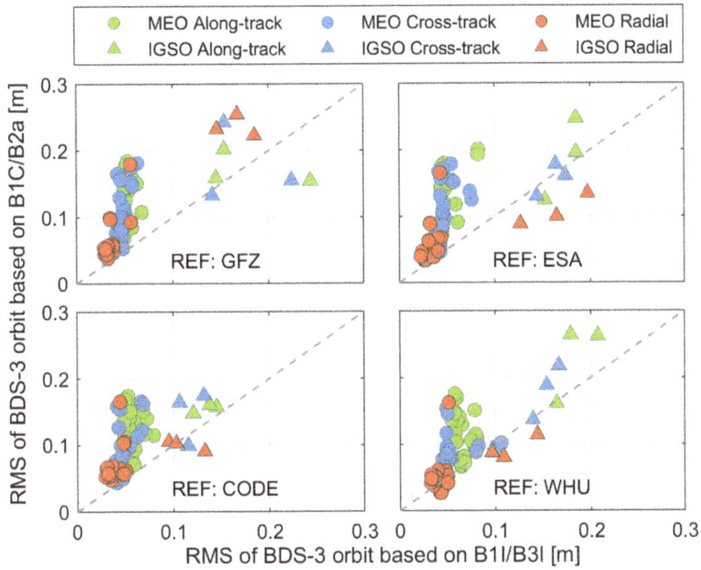

Figure 7. RMS of BDS-3 ISR orbits based on B1I/B3I and B1C/B2a compared with different MGEX ACs in the along-track (A), cross-track (C), and radial (R) directions, respectively.

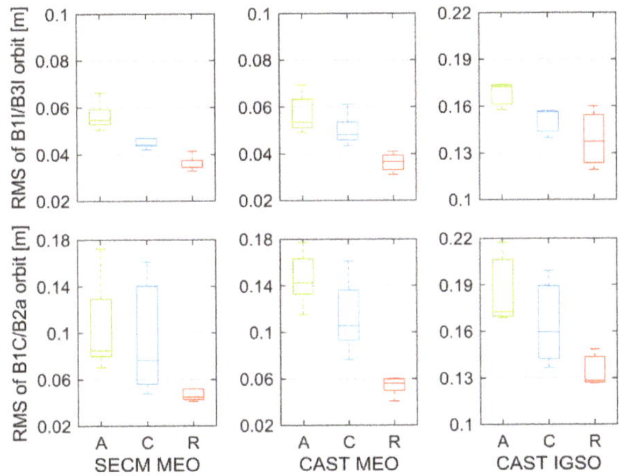

Figure 8. Box plot of RMSs of ISR orbits based on B1I/B3I and B1C/B2a in along-track, cross-track, and radial directions for SECM MEO, CAST MEO, and CAST IGSO satellites, respectively.

4.3. ISU Products

Figure 9 illustrates the RMSs for BDS-2 satellites in along-track, cross-track, and radial directions between ISU B1I/B3I-based and MGEX ACs. It can be clearly seen that the accuracy of ISU becomes much worse, even when compared to WHU. For the along-track direction, the consistency between ISU and GFZ is better than ISU and WHU. However, for the cross-track and radial directions, the consistency of ISU and WHU is better than ISU and GFZ. For IGSO satellites, the consistency of ISU and MGEX ACs is different. That is, the consistency of the cross-track between ISU and ESA/CODE is better than ISU and

GFZ/WHU, while the along-track is in contrast. For MEO satellites, the consistency of ISU and MGEX ACs is similar. Compared with the ISC and ISR products, we can find that the accuracy of ISU products is not as good as others, especially for the BDS-2 GEO satellites.

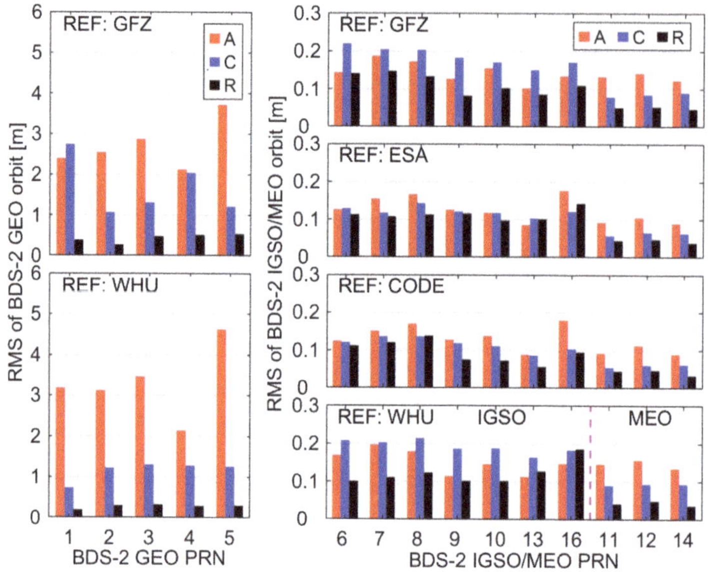

Figure 9. RMS of BDS-2 ISU orbits based on B1I/B3I compared with different MGEX ACs in the along-track (A), cross-track (C), and radial (R) directions, respectively.

Figure 10 shows the RMSs of ISU orbits based on B1I/B3I and B1C/B2a in along-track, cross-track, and radial directions for SECM MEO, CAST MEO, and CAST IGSO satellites, where the gray dashed line indicates that the accuracy of ISU B1I/B3I-based and B1C/B2a-based are the same. We can find that the consistency of ISU B1C/B2a-based orbit products for BDS-3 MEO satellites is worse than B1I/B3I-based. For the radial directions of BDS-3 IGSO satellites, we can see that the consistency of B1I/B3I-based and B1C/B2a-based for different MGEX ACs are different, and the GFZ shows the best consistency with ISU.

Figure 11 shows the box plot of RMSs of ISU orbits based on B1I/B3I and B1C/B2a in along-track, cross-track, and radial directions for SECM MEO, CAST MEO, and CAST IGSO satellites. The five short horizontal lines from the top to the bottom of the box chart represent the 100%, 75%, 50%, 25%, and 0% quantiles, respectively. It can be seen that the median values of RMSs of the along-track based on B1I/B3I for SECM MEO, CAST MEO, and CAST IGSO satellites are approximately 0.061 m, 0.056 m, and 0.146 m, respectively. It indicates that the accuracy of CAST MEO satellites is better than SECM MEO satellites, and both are better than CAST IGSO satellites. However, an extensive variation range of RMS of 0.052 m to 0.075 m can be seen for CAST MEO satellites. This means that the consistency of some CAST MEO satellites is not better than SECM MEO satellites in ISU products. For the B1C/B2a-based, the median values of RMSs of the along-track for SECM MEO, CAST MEO, and CAST IGSO satellites are approximately 0.076 m, 0.136 m, and 0.124 m, respectively. This means that the consistency of CAST MEO satellites is not better than SECM MEO satellites, and B1C/B2a is significantly different from B1I/B3I. For the cross-track, the median values of RMSs of B1I/B3I for SECM MEO, CAST MEO, and CAST IGSO satellites are approximately 0.045 m, 0.096 m, and 0.140 m, respectively. For the radial directions, the median values of RMSs of B1I/B3I for SECM MEO, CAST MEO, and CAST IGSO satellites are approximately 0.041, 0.052, and 0.104 m, respectively.

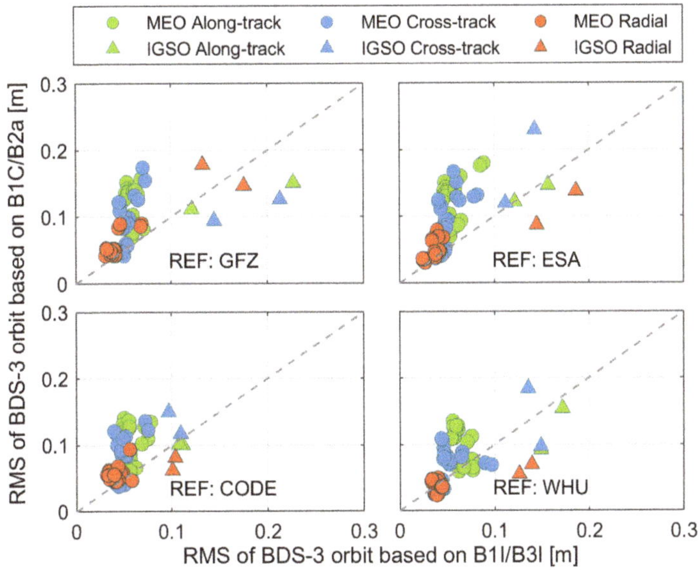

Figure 10. RMS of BDS-3 ISU orbits based on B1I/B3I and B1C/B2a compared with different MGEX ACs in the along-track (A), cross-track (C), and radial (R) directions, respectively.

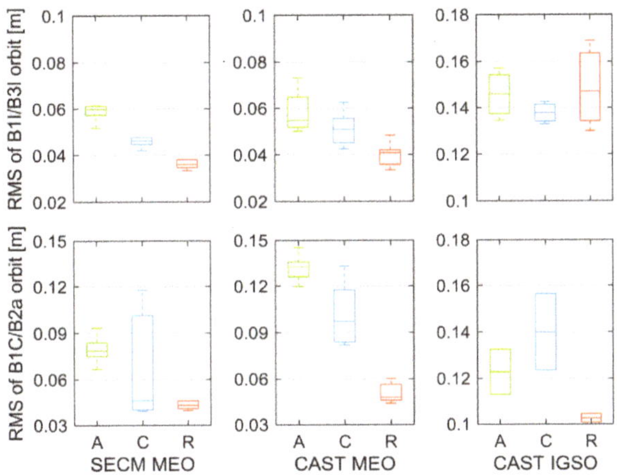

Figure 11. Box plot of RMSs of ISU orbits based on B1I/B3I and B1C/B2a in along-track, cross-track, and radial directions for SECM MEO, CAST MEO, and CAST IGSO satellites, respectively.

5. Inspection with SLR Residuals

5.1. B1I/B3I Frequencies

The SLR technique measures the distance from the SLR station to the laser retroreflector arrays of the satellite. This distance can also be calculated using satellite and station coordinates. The SLR residuals are the difference between the observed and computed distance and are traditionally used as an external inspection of GNSS satellite orbits. Figure 12 shows the SLR residuals for the BDS-2 GEO (C01), IGSO (C08 and C10), and MEO (C11) satellites in the iGMAS ISC, ISR, and ISU B1I/B3I-based orbit products from 10 July 2021 to 10 July 2022. The panels from top to bottom denote the SLR residuals of different satellites. It is worth noting that the residuals with absolute values larger than

1.5 m, 0.5 m, and 0.5 m were removed as outliers for the GEO, IGSO, and MEO satellites in this study. We can find that the GEO and IGSO satellites almost have a systematic bias of approximately 0.3 m. In addition, the performance of C11 is better than that of C01, C08, and C10, which is possibly attributed to the better geometry condition of MEO satellites. The panels from left to right denote the SLR residuals of ISC, ISR, and ISU products. It can be found that for the same satellite, the SLR residuals have similar behaviors to some extent, but some differences can also be found. Specifically, the magnitude of the fluctuations of the SLR residuals is different for ISC, ISR, and ISU products, where the ISC product is the best.

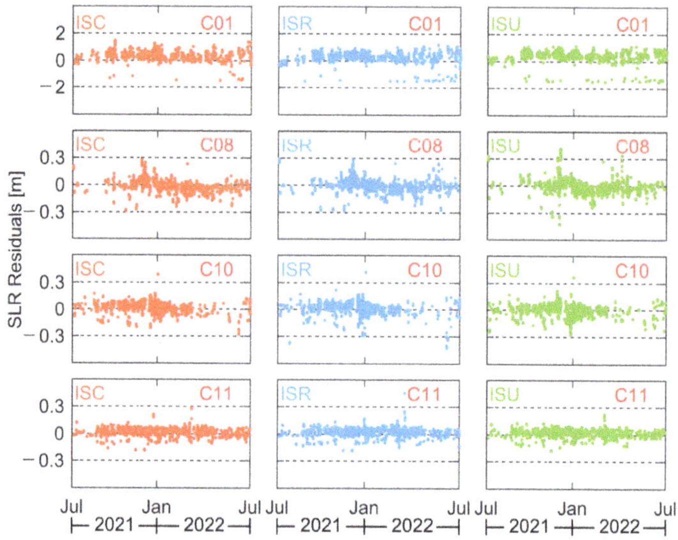

Figure 12. SLR residuals of BDS-2 satellites based on three types of iGMAS B1I/B3I orbits.

Figure 13 shows the SLR residuals for the BDS-3 CAST MEO (C20 and C21) and SECM MEO (C29 and C30) satellites in the iGMAS ISC, ISR, and ISU B1I/B3I-based orbit products from 10 July 2021 to 10 July 2022. The SLR residuals show significant changes in these periods. After the detection and removal of outliers in the SLR observations, there are 2795 normal points available for C20, 2702 for C21, 2483 for C29, and 2528 for C30. First of all, compared with Figure 12, the amplitudes of residuals in Figure 13 are smaller than the ones in Figure 12, thus indicating that the accuracy of precise orbits of BDS-3 is better than that of BDS-2. The mean values are 3.49 cm, 3.36 cm, −2.52 cm, and −2.16 cm for C20, C21, C29, and C30, respectively. It indicates a systematic error in the SLR residuals of the CAST MEO and SECM MEO satellites; one is generally positive, and the other is negative. This phenomenon may be related to the manufacturer, which can be considered for modeling in the future. In addition, the SLR residuals are obviously abnormal over a period of time. The main reason is that the force models are not accurate enough when satellites enter the eclipse period.

5.2. B1C/B2a Frequencies

Figure 14 shows the SLR residuals for the BDS-3 CAST MEO (C20 and C21) and SECM MEO (C29 and C30) satellites in the iGMAS ISC, ISR, and ISU B1C/B2a-based orbit products from 10 July 2021 to 10 July 2022. It can be found that the performance of C20 and C21 is better than that of C29 and C30. The panels from left to right denote the SLR residuals of ISC, ISR, and ISU products. It can be seen that for the same satellite, the SLR residuals have similar behaviors to some extent, but some differences can also be found. Specifically, the missing SLR residual data for ISR and ISU are more significant mainly because there is no corresponding orbit product. Taking a closer look at Figure 14, the accuracy of precise orbits based on B1C/B2a frequencies is generally worse than that of B1I/B3I frequencies. Specifically, the STD values of ISC/ISR/ISU for CAST MEO (C20 and C21) and SECM MEO (C29 and C30) satellites are approximately 0.058/0.088/0.089 m, and 0.065/0.067/0.051 m, respectively.

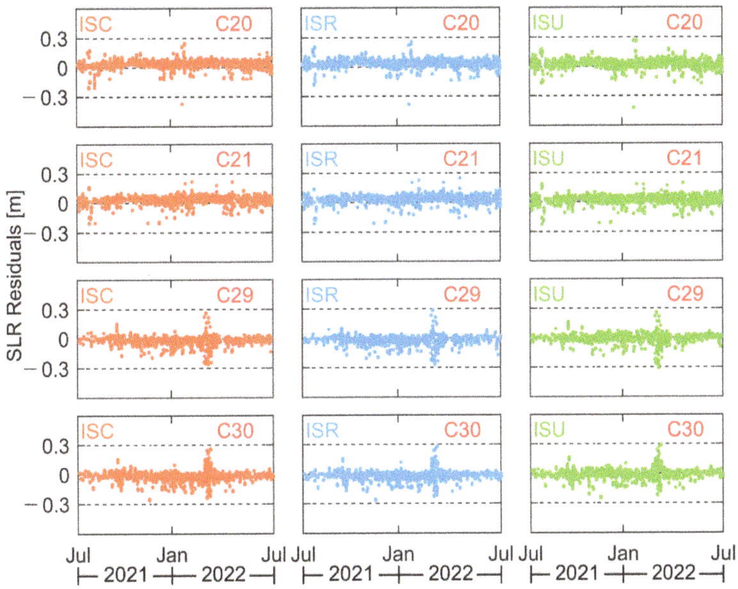

Figure 13. SLR residuals of BDS-3 satellites based on three types of iGMAS B1I/B3I orbits.

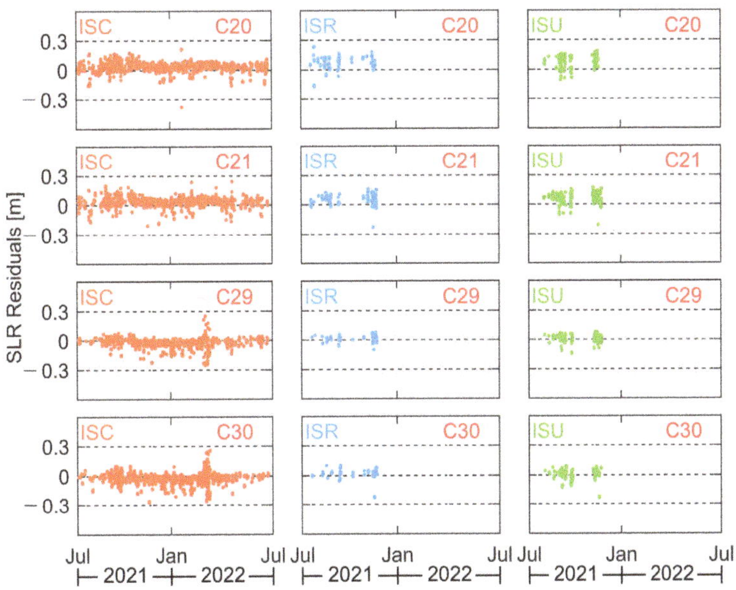

Figure 14. SLR residuals of BDS-3 satellites based on three types of iGMAS B1C/B2a orbits.

6. Discussion of Orbit Errors

6.1. Beta Angle

Figure 15 illustrates the relationship between SLR residuals and beta angle. The panels from top to bottom are the results of BDS-2 GEO (C01), IGSO (C08 and C10), and MEO (C11) satellites. The orange, blue, and green scatters represent the SLR residuals of ISC, ISR, and ISU orbits, respectively. The grey-shaded area in Figure 15 represents the eclipse period, and the rest are non-eclipse periods. As shown in Figure 15, we can find that the beta angle ranges approximately from −49 to +49 degrees for IGSO, −33 to 33 degrees for MEO, and from −23 to +23 degrees for GEO, respectively. The RMS of SLR residuals for C01 of ISC, ISR, and ISU orbits in the non-eclipse/eclipse periods are 0.473/0.455 m, 0.480/0.458 m, and 0.597/0.533 m, respectively. One of the main reasons for the poor accuracy of GEO satellites is that there is little change in the geometry of the satellite and the station. In addition, it is noteworthy that the SLR residuals of the GEO satellite in eclipse period are better than non-eclipse period, which indicates the apparent deficiency of the non-conservative force model. For the IGSO satellites, The RMS of SLR residuals in the non-eclipse/eclipse periods are approximately 0.056~0.066/0.084~0.112 m; the accuracy decreased by approximately 30% in the eclipse period. It implies that IGSO behaves inconsistently with GEO after entering the eclipse periods. For the BDS-2 MEO satellite, the SLR residuals in the non-eclipse/eclipse periods are approximately 0.044~0.049/0.050~0.056 m; the accuracy decreased by approximately 12% in the eclipse period. It is clear that IGSO/MEO has a similar phenomenon when entering the eclipse periods.

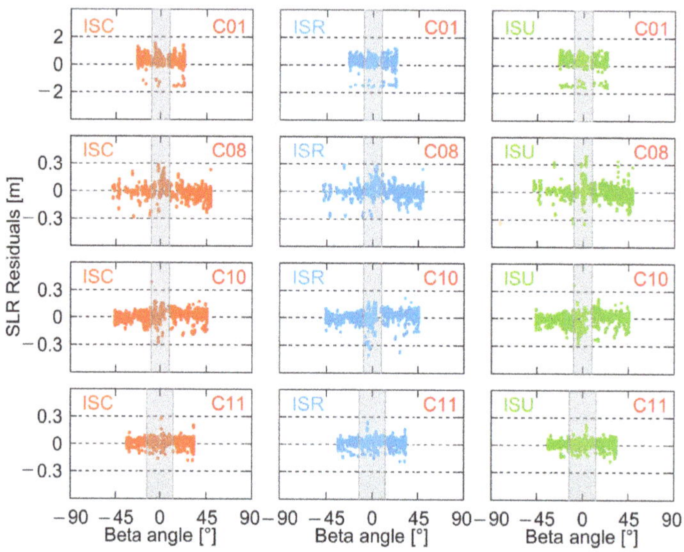

Figure 15. SLR residuals of BDS-2 satellites based on three types of iGMAS B1I/B3I orbits versus beta angle.

Figures 16 and 17 show the relationship between SLR residuals and beta angle for B1I/B3I-based and B1C/B2a-based orbits, respectively. From Figures 16 and 17, we can find that the characteristics of satellites produced by the same manufacturer are more similar. The beta angle ranges from −68 to +68 degrees for CAST MEO satellites (C20 and C21) and −32 to +32 degrees for SECM MEO satellites (C29 and C30). Figure 16 shows that the RMS of SLR residuals for CAST MEO satellites in the non-eclipse /eclipse periods are approximately 0.049/0.068 m, while the SECM MEO satellites are approximately 0.045/0.068 m. Obviously, the accuracy of both BDS-3 CAST MEO and SECM MEO satellites is similar, and both decrease in accuracy after entering the eclipse periods by approximately

33%. Due to the low availability of ISR and ISU products and the low volume of data during the eclipse period, only the ISC products for B1C/B2a were compared.

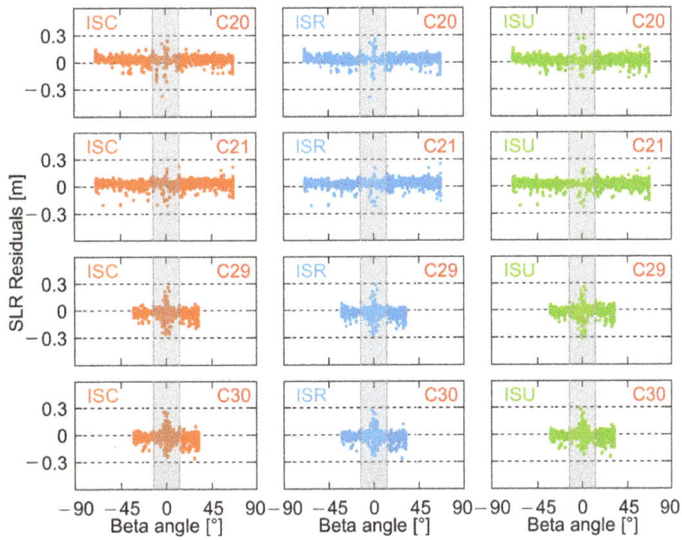

Figure 16. SLR residuals of BDS-3 satellites based on three types of iGMAS B1I/B3I orbits versus beta angle.

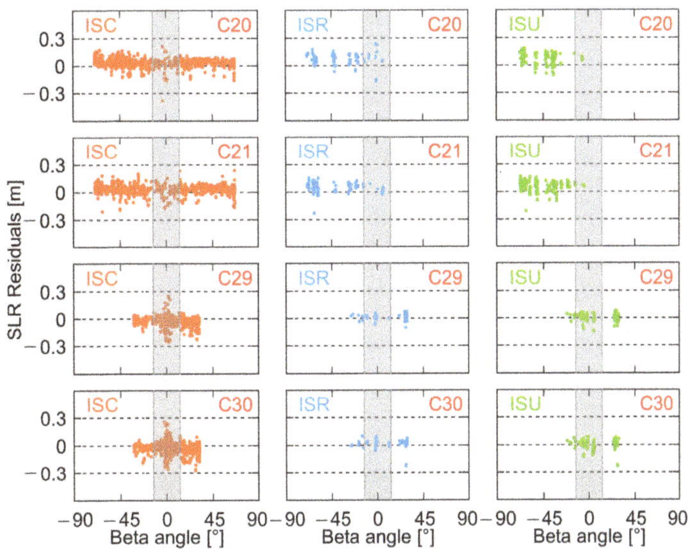

Figure 17. SLR residuals of BDS-3 satellites based on three types of iGMAS B1C/B2a orbits versus beta angle.

6.2. Elongation Angle

Figure 18 shows the relationship between SLR residuals for ISC, ISR, and ISU orbits of BDS-2 GEO (C01), IGSO (C08 and C10), MEO (C11) satellites, and the satellite–sun elongation angle. Once again, the orange, blue, and green scatters represent the SLR residuals of ISC, ISR, and ISU orbits, respectively. The purple line represents the linear function of the satellite laser residuals concerning the elongation angle. We can find that the elongation angle range for C01 is approximately from 1 to 117 degrees and shows a strong correlation with the SLR residuals as the elongation angle increases. The slopes of ISC/ISR/ISU orbit are approximately 0.1627/0.1982/0.0843 cm/°. In addition, We can find that C01 in ISU obviously has unmodeled error. The reason may be that the GEO observation conditions are poor and the force models are not accurate enough. Moreover, the ISU orbits are generated without a higher precision initial orbit, which is generally worse than ISR and ISC orbits. For the IGSO satellites, C08 and C10 have different characteristics from C01. Specifically, the range of elongation angle for IGSO is approximately 1 to 143 degrees, which is more significant than C01. In addition, their SLR residuals are negatively correlated with the elongation angle. The SLR residuals of IGSO satellites are mostly negative when the elongation angle is close to 140° and positive when the elongation angle is close to 0°. This systematic error may be due to the fact that the surface area of the Z surface of the satellite is much smaller than the X surface area and the higher-order solar radiation pressure (SRP) uptake term is not yet well modeled. For the MEO satellite, C11 has the most extensive elongation angle range, approximately from 2 to 163 degrees, and has the most minor linear correlation with SLR residuals. Thus, this indicates that the SLR residuals exhibit systematic error characteristics related to the type of satellite. The non-conservative force models such as SRP may not be suitable for all satellites. Therefore, building a more accurate non-conservative force model may be one of the ways to weaken the systematic errors of SLR for iGMAS orbit products.

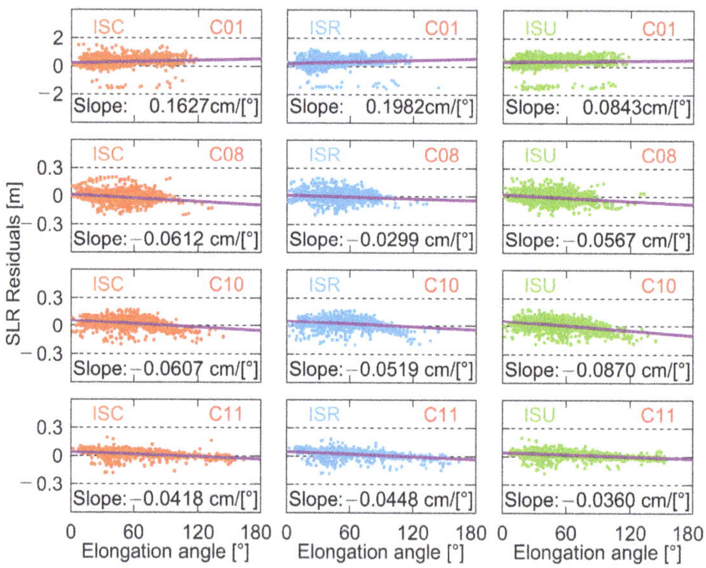

Figure 18. SLR residuals of BDS-2 satellites based on three types of iGMAS B1I/B3I orbits versus elongation angle.

Figures 19 and 20 show the relationship between SLR residuals and elongation angle for B1I/B3I-based and B1C/B2a-based orbits, respectively. Figures 19 and 20 show that the elongation angle range for CAST MEO and SECM MEO satellites is similar to BDS-2 MEO satellites, approximately from 2 to 163 degrees. We note that all elongation angles are more significant than 0 degrees. This is because the sun–satellite–Earth is co-linear, and SLR observations are unavailable. Judging from Figure 19, we can find the slope for C20 of ISC/ISR/ISU orbit is approximately −0.0156/−0.0131/0.011 cm/°, and for the other CAST MEO satellite C21, the slope has similar characteristics, with positive ISC/ISR and negative ISU. The main reasons are the insignificant degree of linear correlation, and the data period is not very long. For SECM MEO satellites, the slopes of C29 and C30 are both positive and more significant than CAST MEO satellites. Figure 20 shows that the SLR residuals of the ISC B1C/B2a-based orbits are similar to the B1I/B3I-based orbits, with no significant systematic errors. However, the SLR residuals of the ISR/ISU B1C/B2a-based orbits show a strong linear relationship with the elongation angle, mainly due to the large fitting error caused by insufficient data.

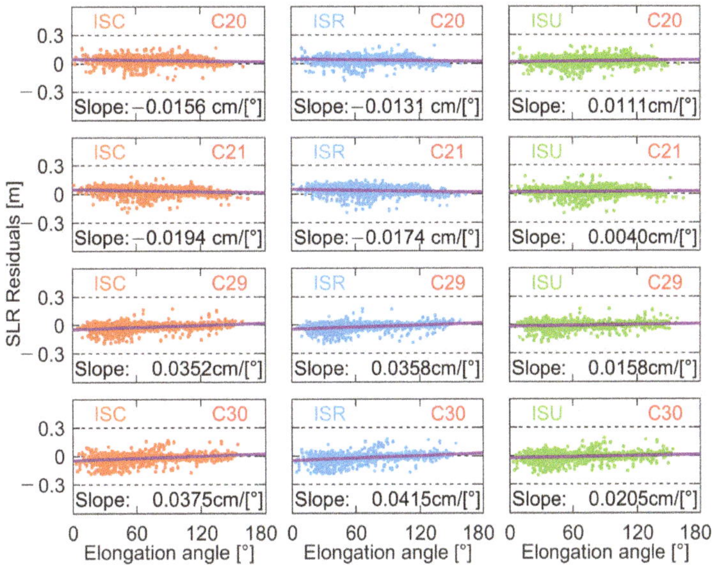

Figure 19. SLR residuals of BDS-3 satellites based on three types of iGMAS B1I/B3I orbits versus elongation angle.

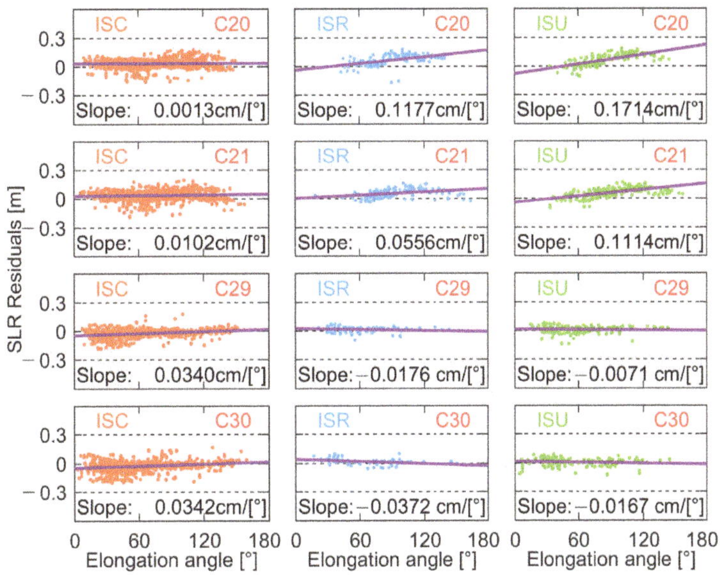

Figure 20. SLR residuals of BDS-3 satellites based on three types of iGMAS B1C/B2a orbits versus elongation angle.

7. Conclusions

In this paper, three kinds of precise orbit, i.e., ISC, ISR, and ISU products, are comprehensively assessed, where the BDS-2/BDS-3 based on B1I/B3I and B1C/B2a frequencies are all considered. The main conclusions can be made based on one year of iGMAS data and other supplementary data, including four precise orbits of four ACs and SLR data. Several main conclusions can be made.

Considering the orbit accuracy according to the four ACs, for the ISC products based on B1I/B3I frequencies, GEO satellites show the most significant biases. The accuracy of IGSO satellites is sub-decimeter level to decimeter level. Then, the accuracy for MEO satellites in three directions can reach the centimeter level, especially the radial direction. Additionally, the consistency of B1C/B2a orbit products is not as good as B1I/B3I. In the future, there is room for improvement in B1C/B2a-based products. Similar conclusions can be found for the ISR products, and the accuracy of B1I/B3I-based ISR products is slightly lower than that of ISC, which is reasonable since the ISC products are the final combined ones. For the ISU products, the accuracy becomes much worse than the other two products, especially for the BDS-2 GEO satellites.

By using the SLR technique, the precise orbits are inspected. For the B1I/B3I frequencies, the BDS-2 GEO and IGSO satellites have a systematic bias of approximately 0.3 m, whereas the BDS-2 MEO satellites do not display this. The accuracy of precise orbits of BDS-3 are better than those of BDS-2. Specifically, the mean values are 3.49 cm, 3.36 cm, −2.52 cm, and −2.16 cm for C20, C21, C29, and C30, respectively. Comparing three types of orbit products, the ISC product has the best performance. Additionally, we find that the orbit errors are related to the manufacturer, which can be considered for modeling in the future. For the B1C/B2a frequencies, similar conclusions can be found. Then, the accuracy of precise orbits based on B1C/B2a frequencies is generally worse than that of B1I/B3I frequencies. Specifically, the mean STD values of ISC/ISR/ISU for CAST MEO and SECM MEO satellites are approximately 0.078 m and 0.061 m, respectively.

When analyzing the impact factors of orbit errors, the results show that they are related to the beta angle and elongation angle. Specifically, for the beta angle, the BDS-2 GEO satellites are more accurate in the eclipse period, mainly due to the deficiency of the non-conservative force model. At the same time, the accuracy of the precise orbits in the non-eclipse period is better in BDS-2/BDS-3 IGSO and MEO satellites. For the elongation angle, there is a strong positive correlation with the SLR residuals for BDS-2 GEO satellites, whereas negative correlations can be found for BDS-2 IGSO and MEO satellites. This indicates that the non-conservative force models such as SRP may not be suitable for all satellites. Then, for the BDS-3 MEO satellites, the behaviors in terms of elongation angle are similar to the BDS-2 MEO satellites. It is worth noting that, compared with the behaviors of B1I/B3I frequencies, the B1C/B2a frequencies are different, especially for the ISR/ISU products, where a strong linear relationship exists with the elongation angle. Unlike the beta angle, the impacts of the elongation angle are related to the manufacturer.

Author Contributions: Conceptualization, Z.Z. and Y.W.; methodology, Z.Z. and P.Z.; software, L.H. and P.Z.; validation, Y.W., L.H. and X.H.; data Curation, Z.Z. and P.Z.; writing—original draft preparation, Z.Z. and P.Z.; writing—review and editing, Z.Z. and P.Z. All authors have read and agreed to the published version of the manuscript.

Funding: This study is partly supported by the National Natural Science Foundation of China (42004014), the Natural Science Foundation of Jiangsu Province (BK20200530), and the National Natural Science Foundation of China (41974001).

Data Availability Statement: The datasets analyzed in this study are managed by the School of Earth Sciences and Engineering, Hohai University, and can be available upon reasonable request from the corresponding author.

Acknowledgments: The authors thank the iGMAS, IGS, MGEX, and ILRS authorities for providing the data for this study.

Conflicts of Interest: The authors have no relevant financial or non-financial interest to disclose.

References

1. Yang, Y.; Mao, Y.; Sun, B. Basic Performance and Future Developments of BeiDou Global Navigation Satellite System. *Satell. Navig.* **2020**, *1*, 1. [CrossRef]
2. Yang, Y.; Gao, W.; Guo, S.; Mao, Y.; Yang, Y. Introduction to BeiDou-3 Navigation Satellite System. *Navigation* **2019**, *66*, 7–18. [CrossRef]
3. *CSNO BeiDou Navigation Satellite System Signal in Space Interface Control Document Open Service Signal B1C (Version 1.0)*; China Satellite Navigation Office: Beijing, China, 2017.
4. *CSNO BeiDou Navigation Satellite System Signal in Space Interface Control Document Open Service Signal B2a (Version 1.0)*; China Satellite Navigation Office: Beijing, China, 2017.
5. *CSNO BeiDou Navigation Satellite System Signal in Space Interface Control Document Open Service Signal B3I (Version 1.0)*; China Satellite Navigation Office: Beijing, China, 2018.
6. *CSNO BeiDou Navigation Satellite System Signal in Space Interface Control Document Open Service Signal B1I (Version 3.0)*; China Satellite Navigation Office: Beijing, China, 2019.
7. *CSNO BeiDou Navigation Satellite System Signal in Space Interface Control Document Open Service Signal B2b (Version 1.0)*; China Satellite Navigation Office: Beijing, China, 2020.
8. CSNO. Fundamental PNT Service. 2022. Available online: http://www.csno-tarc.cn/en/system/constellation (accessed on 1 December 2022).
9. Zhao, Q.; Guo, J.; Wang, C.; Lyu, Y.; Xu, X.; Yang, C.; Li, J. Precise Orbit Determination for BDS Satellites. *Satell. Navig.* **2022**, *3*, 2. [CrossRef]
10. Kouba, J. *A Guide to Using International GNSS Service (IGS) Products*; Maryland Biological Stream Survey; Versar, Inc.: Springfield, VA, USA, 2009; Volume 4, p. 106.
11. IGS. IGS Analysis Centers. 2022. Available online: https://igs.org/acc/#ac-list (accessed on 1 December 2022).
12. Montenbruck, O.; Steigenberger, P.; Prange, L.; Deng, Z.; Zhao, Q.; Perosanz, F.; Romero, I.; Noll, C.; Stürze, A.; Weber, G.; et al. The Multi-GNSS Experiment (MGEX) of the International GNSS Service (IGS)–Achievements, Prospects and Challenges. *Adv. Space Res.* **2017**, *59*, 1671–1697. [CrossRef]
13. Montenbruck, O.; Steigenberger, P.; Khachikyan, R.; Weber, G.; Langley, R.; Mervart, L.; Hugentobler, U. IGS-MGEX: Preparing the Ground for Multi-Constellation GNSS Science. *Inside GNSS* **2014**, *9*, 42–49.
14. Ai, Q.; Maciuk, K.; Lewinska, P.; Borowski, L. Characteristics of Onefold Clocks of GPS, Galileo, BeiDou and GLONASS Systems. *Sensors* **2021**, *21*, 2396. [CrossRef] [PubMed]
15. Jaduszliwer, B.; Camparo, J. Past, Present and Future of Atomic Clocks for GNSS. *GPS Solut.* **2021**, *25*, 27. [CrossRef]
16. Chen, Q.; Song, S.; Zhou, W. Accuracy Analysis of GNSS Hourly Ultra-Rapid Orbit and Clock Products from SHAO AC of IGMAS. *Remote Sens.* **2021**, *13*, 1022. [CrossRef]
17. Kazmierski, K.; Zajdel, R.; Sośnica, K. Evolution of Orbit and Clock Quality for Real-Time Multi-GNSS Solutions. *GPS Solut.* **2020**, *24*, 111. [CrossRef]
18. Zhou, W.; Cai, H.; Chen, G.; Jiao, W.; He, Q.; Yang, Y. Multi-GNSS Combined Orbit and Clock Solutions at IGMAS. *Sensors* **2022**, *22*, 457. [CrossRef]
19. Jiao, W.; Ding, Q.; Li, J.; Lu, X.; Feng, L.; Ma, J.; Chen, G. Monitoring and Assessment of GNSS Open Services. *J. Navig.* **2011**, *64*, S19–S29. [CrossRef]
20. Cai, H.; Chen, K.; Xu, T.; Chen, G. The IGMAS Combined Products and the Analysis of Their Consistency. In Proceedings of the China Satellite Navigation Conference (CSNC) 2015 Proceedings, Xi'an, China, 13–15 May 2015; Springer: Berlin/Heidelberg, Germany, 2015; Volume III, pp. 213–226.
21. iGMAS. Analysis Centers. 2022. Available online: http://en.igmas.org/About/Showigmas/detail/nav_id/6/cate_id/62.html (accessed on 1 December 2022).
22. Ye, F.; Yuan, Y.; Yang, Z. Validation and Evaluation on B1IB3I-Based and B1CB2a-Based BDS-3 Precise Orbits from IGMAS. *Adv. Space Res.* **2022**, *70*, 2167–2177. [CrossRef]
23. Ye, F.; Yuan, Y.; Deng, Z. Improved Ultra-Rapid UT1-UTC Determination and Its Preliminary Impact on GNSS Satellite Ultra-Rapid Orbit Determination. *Remote Sens.* **2020**, *12*, 3584. [CrossRef]
24. Li, Z.; Wang, N.; Wang, L.; Liu, A.; Yuan, H.; Zhang, K. Regional Ionospheric TEC Modeling Based on a Two-Layer Spherical Harmonic Approximation for Real-Time Single-Frequency PPP. *J. Geod.* **2019**, *93*, 1659–1671. [CrossRef]
25. Li, X.; Chen, X.; Ge, M.; Schuh, H. Improving Multi-GNSS Ultra-Rapid Orbit Determination for Real-Time Precise Point Positioning. *J. Geod.* **2019**, *93*, 45–64. [CrossRef]
26. Chen, K.; Xu, T.; Chen, G.; Li, J.; Yu, S. The Orbit and Clock Combination of IGMAS Analysis Centers and the Analysis of Their Precision. In Proceedings of the China Satellite Navigation Conference (CSNC) 2015 Proceedings, Xi'an, China, 13–15 May 2015; Springer: Berlin/Heidelberg, Germany, 2015; Volume II, pp. 421–438.
27. Tan, B.; Yuan, Y.; Wen, M.; Ning, Y.; Liu, X. Initial Results of the Precise Orbit Determination for the New-Generation BeiDou Satellites (BeiDou-3) Based on the IGMAS Network. *ISPRS Int. J. Geo-Inf.* **2016**, *5*, 196. [CrossRef]

28. Zhang, X.; Li, X.; Lu, C.; Wu, M.; Pan, L. A Comprehensive Analysis of Satellite-Induced Code Bias for BDS-3 Satellites and Signals. *Adv. Space Res.* **2019**, *63*, 2822–2835. [CrossRef]
29. Li, X.; Xie, W.; Huang, J.; Ma, T.; Zhang, X.; Yuan, Y. Estimation and Analysis of Differential Code Biases for BDS3/BDS2 Using IGMAS and MGEX Observations. *J. Geod.* **2019**, *93*, 419–435. [CrossRef]
30. Xie, X.; Fang, R.; Geng, T.; Wang, G.; Zhao, Q.; Liu, J. Characterization of GNSS Signals Tracked by the IGMAS Network Considering Recent BDS-3 Satellites. *Remote Sens.* **2018**, *10*, 1736. [CrossRef]
31. Yang, H.; Xu, T.; Nie, W.; Gao, F.; Guan, M. SLR Validation and Evaluation on BDS Precise Orbits from 2013 to 2018. *Adv. Space Res.* **2019**, *64*, 475–490. [CrossRef]
32. Montenbruck, O.; Steigenberger, P.; Wang, N.; Hauschild, A. Characterization and Performance Assessment of BeiDou-2 and BeiDou-3 Satellite Group Delays. *NAVIGATION J. Inst. Navig.* **2022**, *69*, navi.526. [CrossRef]
33. Arnold, D.; Montenbruck, O.; Hackel, S.; Sośnica, K. Satellite Laser Ranging to Low Earth Orbiters: Orbit and Network Validation. *J. Geod.* **2019**, *93*, 2315–2334. [CrossRef]
34. Zhang, Z.; Li, B.; Shen, Y.; Gao, Y.; Wang, M. Site-Specific Unmodeled Error Mitigation for GNSS Positioning in Urban Environments Using a Real-Time Adaptive Weighting Model. *Remote Sens.* **2018**, *10*, 1157. [CrossRef]
35. Zhang, Z.; Li, B. Unmodeled Error Mitigation for Single-Frequency Multi-GNSS Precise Positioning Based on Multi-Epoch Partial Parameterization. *Meas. Sci. Technol.* **2020**, *31*, 025008. [CrossRef]

remote sensing

MDPI

Article

BDS/GPS Multi-Baseline Relative Positioning for Deformation Monitoring

Haonan Wang, Wujiao Dai * and Wenkun Yu

School of Geosciences and Info-Physics, Central South University, Changsha 410083, China
* Correspondence: wjdai@csu.edu.cn

Abstract: The single-baseline solution (SBS) model has been widely adopted by the existing global navigation satellite system (GNSS) deformation monitoring systems due to its theoretical simplicity and ease of implementation. However, the SBS model neglects the mathematical correlation between baselines, and the accuracy and reliability can be degraded for baselines with long length, large height difference or frequent satellite signal occlusion. When monitoring large-area ground settlement or long-spanned linear objects such as bridges and railroads, multiple reference stations are frequently utilized, which can be exploited to improve the monitoring performance. Therefore, this paper evaluates the multi-baseline solution (MBS) model, and constrained-MBS (CMBS) model that has a prior constraint of the spatial-correlated tropospheric delay. The reliability and validity of the MBS model are verified using GPS/BDS datasets from ground settlement deformation monitoring with a baseline length of about 20 km and a height difference of about 200 m. Numerical results show that, compared with the SBS model, the MBS model can reduce the positioning standard deviation (STD) and root-mean-squared (RMS) errors by up to (47.4/51.3/66.2%) and (56.9/60.4/58.4%) in the north/east/up components, respectively. Moreover, the combined GPS/BDS positioning performance for the MBS model outperforms the GPS-only and BDS-only positioning models, with an average accuracy improvement of about 13.8 and 25.8%, with the highest accuracy improvement of about 41.6 and 43.8%, respectively. With the additional tropospheric delay constraint, the CMBS model improves the monitoring precision in the up direction by about 45.0%.

Keywords: relative positioning; GPS/BDS; multi-baseline solution; deformation monitoring

Citation: Wang, H.; Dai, W.; Yu, W. BDS/GPS Multi-Baseline Relative Positioning for Deformation Monitoring. *Remote Sens.* **2022**, *14*, 3884. https://doi.org/10.3390/rs14163884

Academic Editor: Xiaogong Hu

Received: 27 July 2022
Accepted: 8 August 2022
Published: 11 August 2022

1. Introduction

The global navigation satellite system (GNSS) is an all-weather, real-time system with high-accuracy and high-automation; therefore, it has been widely used in the field of deformation monitoring in recent decades [1–3]. A GNSS deformation monitoring system, in general, employs relative positioning that can achieve millimeter accuracy since errors such as clock error, orbital error, and atmospheric delay are eliminated or attenuated [4,5]. The relative positioning is commonly implemented in simple single-baseline solution (SBS) models. However, satellite signal occlusion and low-quality satellite observations frequently present in the complex environments of deformation monitoring, result in limited positioning accuracy and reliability [6]. In recent years, the rapid development of multiple GNSSs has obviously increased the available satellite numbers and GNSS observations for deformation monitoring [7,8]. For example, in experiments with simulated signal oscillation, Roberts et al. [9] found that combining GPS and BDS observations could improve the positioning accuracy especially for the height component. In experiments using real data from a large loess landslide and a simulated slow deformation test, Huang et al. [10] verified that the GPS/BDS solution could significantly enhance positioning reliability and accuracy. Xi et al. [11] found that, in bridge deformation monitoring, the accuracy of combined GPS/BDS positioning was higher than that of a single GNSS system by 20–30%.

Another issue of the traditional SBS model is the residual tropospheric delay. Tropospheric delays, especially in baselines with large height differences or long lengths, cannot be eliminated by operations such as differencing and simple parameter estimation, leading to serious degradation of the monitoring accuracy [12–20]. Many data processing strategies have been proposed to reduce the influence of the residual troposphere delay in relative positioning. For example, Li et al. [21] introduced the relative zenith tropospheric delay (RZTD) parameter into the ambiguity function method (AFM) equation to improve the accuracy and reliability of tropospheric parameter estimation for deformation monitoring of super high-rise buildings. Jacek [22] proposed a multi-antenna atmospheric error constraint algorithm for multiple baselines to improve the ambiguity success rate of medium-long baselines in RTK. Although multiple antennas provide more observations, the geometric strength of the constraint is limited since the distance between antennas is short.

Multiple reference stations are widely used in GNSS deformation monitoring projects especially for those with a large monitoring area [23]. As a result, the multi-baseline solution (MBS) model is sometimes more practical. The MBS model is theoretically more rigorous than the SBS model since it considers inter-baseline correlation. More importantly, MBS enhances the satellite geometry and increases observation redundancy [24]. For example, Amin et al. [25] employed the MBS model to increase the reliability of GPS monitoring networks by adopting a uniform datum and eliminating synchronous baseline closure errors in the optimal design of large GNSS control networks. Wang et al. [26] enhanced the availability of RTK positioning by roughly 10% using a multi-reference station constraint algorithm based on the equivalence principle in autonomous vehicle navigation. Fan et al. [27] proposed a dual-antenna RTK algorithm with baseline vector constraint to improve the system ambiguity resolution (AR) success rate from 48 to 85%. The MBS model has been less explored in deformation monitoring than the SBS model. One important reason is that in MBS the stochastic model for differential observations becomes more sophisticated [28]. Most of the present commercial software is developed based on the SBS model [29–31]. To solve this problem, some researchers have proposed the equivalent double-differenced (DD) observation equations [32–34], in which a transformation matrix is generated to eliminate receiver clock errors and satellite clock errors. Although the equivalent DD equations can preserve the diagonality of the stochastic model, the solution is still inefficient due to the transformation matrix. Previous research of the MBS mainly focused on the solutions of large control networks, the improvement of the ambiguity fixing rate for dynamic positioning, and the complexity reduction of the solutions of traditional DD equations.

Section 2 gives the data processing strategy of the MBS model. We first derive a multi-baseline mathematical model based on the traditional DD observation equation, and then constrain the MBS model on the tropospheric parameters between multiple baselines based on their spatial correlation. In Section 3, a comparative analysis with the SBS model was performed using actual observation datasets to verify and evaluate the performance of the proposed method. The datasets contained baselines with medium-long lengths and large height differences. The experiments were conducted to analyze the performance of the MBS model under different combinations of reference stations, satellite systems, and height differences. Finally, the conclusions are given in Section 4.

2. Methods

2.1. MBS Mathematical Model

2.1.1. Function Model

As shown in Figure 1, there are m reference stations and one rover station tracking n satellites synchronously, forming m independent baselines. Taking the single-GNSS double-baseline situation as an example, there are n_j and n_k common satellites between the rover station and the j-th and k-th reference stations, respectively. The carrier phase and code DD observation equations of the epoch t_i can be respectively expressed as:

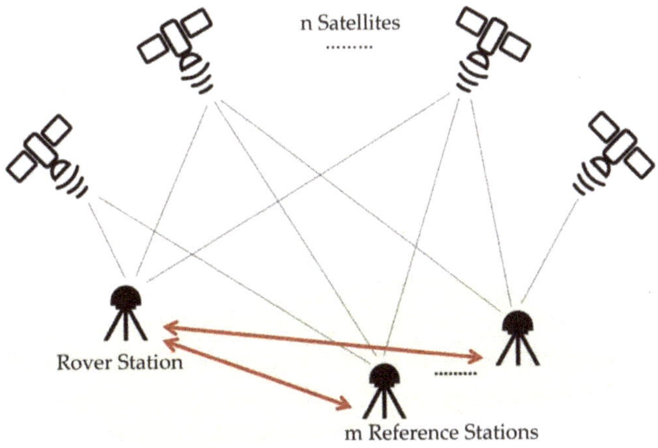

Figure 1. Schematic graph of relative positioning with one rover station and multiple reference stations.

$$\begin{cases} \lambda\nabla\Delta L_{rb_j}^{l_jq_j}(t_i) = \nabla\Delta\rho_{rb_j}^{l_jq_j}(t_i) - \nabla\Delta I_{rb_j}^{l_jq_j}(t_i) + \nabla\Delta T_{rb_j}^{l_jq_j}(t_i) + \lambda\nabla\Delta N_{rb_j}^{l_jq_j} + \nabla\Delta\varepsilon_{rb_j}^{l_jq_j}(t_i) \\ \nabla\Delta P_{rb_j}^{l_jq_j}(t_i) = \nabla\Delta\rho_{rb_j}^{l_jq_j}(t_i) + \nabla\Delta I_{rb_j}^{l_jq_j}(t_i) + \nabla\Delta T_{rb_j}^{l_jjq_j}(t_i) + \nabla\Delta\delta_{rb_j}^{l_jq_j}(t_i) \\ \lambda\nabla\Delta L_{rb_k}^{l_kq_k}(t_i) = \nabla\Delta\rho_{rb_k}^{l_kq_k}(t_i) - \nabla\Delta I_{rb_k}^{l_kq_k}(t_i) + \nabla\Delta T_{rb_k}^{l_kq_k}(t_i) + \lambda\nabla\Delta N_{rb_k}^{l_kq_k} + \nabla\Delta\varepsilon_{rb_k}^{l_kq_k}(t_i) \\ \nabla\Delta P_{rb_k}^{l_kq_k}(t_i) = \nabla\Delta\rho_{rb_k}^{l_kq_k}(t_i) + \nabla\Delta I_{rb_k}^{l_kq_k}(t_i) + \nabla\Delta T_{rb_k}^{l_kq_k}(t_i) + \nabla\Delta\delta_{rb_k}^{l_kq_k}(t_i) \end{cases} \quad (1)$$

where $\nabla\Delta$ is the DD operator. L and P represent the carrier phase and code observations, respectively. l_j and l_k are the reference satellites of the j-th and k-th baselines. q_j and q_k are the common satellites of the j-th and k-th baselines. r denotes the rover station. b_j and b_k denote the j-th and k-th reference stations, and ρ denotes the geometric distance between the receiver and the satellite. I denotes ionospheric delay, which is deemed minor enough to be ignored after DD for short baselines, and can be eliminated using a dual-frequency ionosphere-free (IF) linear combination for medium-long baselines. T denotes tropospheric delay, which can be neglected for short baselines, and can be reduced by random walk parameter estimation in the case of medium-long baselines or large height differences. λ denotes the wavelength of the carrier phase. N is the phase ambiguity. ε and δ represent the measurement noise of the carrier phase and code, respectively.

2.1.2. Stochastic Model

The following stochastic model based on the elevation angle is utilized in this study [35]. The variance of the undifferenced (UD) observations can be expressed as:

$$\sigma^2 = a^2 + \frac{b^2}{\sin^2(Ele)} \quad (2)$$

where represents the variance of UD observations. is the satellite elevation angle. It is assumed that the standard deviation of the code observations is 100 times that of the carrier phase, with for the carrier phase and for the code. The variance-covariance matrices of the

UD carrier phase observations of the -th, -th reference station, and the rover station can be expressed as Equation (3).

$$\begin{cases} R_{b_j,\varphi} = diag(\sigma^2_{\varphi_{b_j}^{1_j}}, \ldots, \sigma^2_{\varphi_{b_j}^{q_j}})_{n_j} \\ R_{b_k,\varphi} = diag(\sigma^2_{\varphi_{b_k}^{1_k}}, \ldots, \sigma^2_{\varphi_{b_k}^{q_k}})_{n_k} \\ R_{r,\varphi} = diag(\sigma^2_{\varphi_r^{1_j}}, \sigma^2_{\varphi_r^{1_k}}, \ldots, \sigma^2_{\varphi_r^{qmax}})_{n_{max}} \end{cases} \quad (3)$$

where and are, respectively, the numbers of the common satellites of the rover station and the -th and -th reference stations. is the total number of common satellites in baselines.

According to the error propagation law, the variance-covariance matrix of the single-differenced (SD) carrier phase observations of baselines can be derived as:

$$R_{rb,\varphi}^{Single} = \begin{bmatrix} \widetilde{E}_{n_1 n_{\max}} R_{r,\varphi} \widetilde{E}^T_{n_{\max} n_1} + R_{b_1,\varphi} & \widetilde{E}_{n_1 n_{\max}} R_{r,\varphi} \widetilde{E}^T_{n_{\max} n_2} & \cdots & \widetilde{E}_{n_1 n_{\max}} R_{r,\varphi} \widetilde{E}^T_{n_{\max} n_m} \\ \widetilde{E}_{n_2 n_{\max}} R_{r,\varphi} \widetilde{E}^T_{n_{\max} n_1} & \widetilde{E}_{n_2 n_{\max}} R_{r,\varphi} \widetilde{E}^T_{n_{\max} n_2} + R_{b_2,\varphi} & \cdots & \widetilde{E}_{n_2 n_{\max}} R_{r,\varphi} \widetilde{E}^T_{n_{\max} n_m} \\ \vdots & \vdots & \ddots & \vdots \\ \widetilde{E}_{n_m n_{\max}} R_{r,\varphi} \widetilde{E}^T_{n_{\max} n_1} & \widetilde{E}_{n_m n_{\max}} R_{r,\varphi} \widetilde{E}^T_{n_{\max} n_2} & \cdots & \widetilde{E}_{n_m n_{\max}} R_{r,\varphi} \widetilde{E}^T_{n_{\max} n_m} + R_{b_m,\varphi} \end{bmatrix}_{m \times m} \quad (4)$$

where $\widetilde{E}_{n_i n_{max}}$ is the SD operator between stations, $i = 1, 2, \ldots, j, k, \ldots m$.

Equation (4) can be divided into diagonal submatrix $\widetilde{E}_{n_i n_{max}} R_{r,\varphi} \widetilde{E}^T_{n_{max} n_i} + R_{b_i,\varphi}$ and non-diagonal submatrix $\widetilde{E}_{n_i n_{max}} R_{r,\varphi} \widetilde{E}^T_{n_{max} n_{i+1}}$. The diagonal submatrix can be simplified and expressed as $diag(\sigma^2_{\varphi_r^{1_j}} + \sigma^2_{\varphi_{b_j}^{1_j}}, \ldots, \sigma^2_{\varphi_r^{q_j}} + \sigma^2_{\varphi_{b_j}^{q_j}})_{n_j}$, which is equal to the inter-station SD stochastic model of SBS. The non-diagonal submatrix is related to common satellites between baselines, which is essentially the difference between the MBS and SBS stochastic models.

Then the inter-station SD observations of m baselines are differentiated between satellites to obtain DD observations, which are arranged sequentially based on the order of stations. According to the variance-covariance error propagation law, the variance-covariance matrix of DD carrier phase observations can be derived as:

$$R_{rb,\varphi}^{Double} = D R_{rb,\varphi}^{Single} D^T \quad (5)$$

$$\begin{cases} D = diag(D_1, D_2, \ldots, D_m)_{m \times m} \\ D_i = \begin{bmatrix} -1 & 1 & 0 & \cdots & 0 \\ -1 & 0 & 1 & \cdots & 0 \\ \vdots & \vdots & \vdots & \ddots & \vdots \\ -1 & 0 & 0 & \cdots & 1 \end{bmatrix}_{(n_i - 1) \times n_i} \end{cases} \quad (6)$$

where D denotes the SD operator between satellites.

2.2. MBS Parameter Estimation

In this section, Kalman filtering (KF) was used for MBS parameter estimation. The main parameters to be estimated included ambiguities and three-dimensional coordinates. Additionally, the correlation between the atmospheric errors of the reference station and the monitoring station decreased as the baseline length or height difference increased. The DD ionospheric and tropospheric residual errors were no longer negligible for baselines with medium-long length or large height difference. Ionospheric delay can be eliminated by using the dual-frequency ionosphere-free (IF) linear combination. The tropospheric delay can be separated into hydrostatic and wet components, with the hydrostatic delay being corrected by a specific model at mm level accuracy [36], whereas the wet delay is more complex and related to the water vapor in the atmosphere, which is difficult to accurately model. Therefore, in this paper, the GPT2_1W surface meteorology model was used to interpolate the station meteorological elements [37], and the hydrostatic delay was corrected using the Saastamoinen model and the Vienna Mapping Function1 (VMF1) mapping function [38]. The relative zenith wet delay (RZWD) between the rover and reference stations has spatial and temporal correlation, thus is estimated as an unknown parameter of a random walk process.

The state vector of the rover station associated with m baselines is:

$$X = \left[X_r, Y_r, Z_r, RZWD_{rb_1}, \ldots, RZWD_{rb_m}, \Delta\nabla N_{rb_1}^{l_1 q_1}, \ldots,, \Delta\nabla N_{rb_m}^{l_m q_m}\right]^T \tag{7}$$

where X_r, Y_r, Z_r denote the three-dimensional coordinate vector of the rover station. The coordinate components of the rover stations in each baseline should be the same in the same epoch. The SBS model treats them as separate parameters, resulting in several baseline solutions for the same rover station. The MBS model unifies the coordinate parameters of the rover stations in each baseline in data processing and employs only one set of rover station coordinate parameters, considerably increasing the model strength.

2.3. A Priori Constraint on Tropospheric Delay

The atmospheric delay errors are characterized by spatial and temporal correlation. The relationship between the RZWD and the height difference in the j-th and k-th baselines is considered to be given by the following formula. The exponential function is used to describe the variation of tropospheric wet delay in the height direction [39–41]:

$$\frac{RZWD_{rb_j}}{e^{k*\Delta h_j}} = \frac{RZWD_{rb_k}}{e^{k*\Delta h_k}} \tag{8}$$

where the value of k is empirically set between $1e^{-5}$ and $1e^{-4}$; Δh_i denotes the height difference between the rover station and the i-th reference station.

This prior information can be expressed as virtual observations:

$$AX = M \tag{9}$$

where X is the unknown state vector.

Taking the 1st, 2nd, and 3rd baselines as examples, coefficient matrix A and constant matrix M can be expressed as:

$$A = \begin{bmatrix} 0 & 0 & 0 & 1/e^{k*\Delta h_1} & -1/e^{k*\Delta h_2} & 0 & 0 & \ldots & 0 \\ 0 & 0 & 0 & 1/e^{k*\Delta h_1} & 0 & -1/e^{k*\Delta h_3} & 0 & \ldots & 0 \end{bmatrix} \tag{10}$$

$$M = \begin{bmatrix} \left(1/e^{k*\Delta h_2} - 1/e^{k*\Delta h_1}\right) * ZWD_0(r) + 1/e^{k*\Delta h_1} * ZWD_0(b_1) - 1/e^{k*\Delta h_2} * ZWD_0(b_2) \\ \left(1/e^{k*\Delta h_3} - 1/e^{k*\Delta h_1}\right) * ZWD_0(r) + 1/e^{k*\Delta h_1} * ZWD_0(b_1) - 1/e^{k*\Delta h_3} * ZWD_0(b_3) \end{bmatrix} \tag{11}$$

where ZWD_0 denotes the wet delay a priori value calculated by the empirical model. The a priori standard deviation (STD) of the virtual observations is empirically set to 0.001 m.

2.4. Data Processing Strategy

Figure 2 depicts the flow chart of the MBS model. A three-step method was used to solve the IF combined AR problem [42]. Firstly, the wide-lane (WL) ambiguity was fixed by rounding, and multi-epoch smoothing was performed using HATCH filtering [43]. Next, the float solution of the IF ambiguity was solved by KF. Then, the L1 ambiguity float solution was extracted from the IF ambiguity using the WL ambiguity, and the L1 ambiguity fixing solution was searched using the well-known LAMBDA method. Finally, the IF ambiguity-fixed solution was obtained.

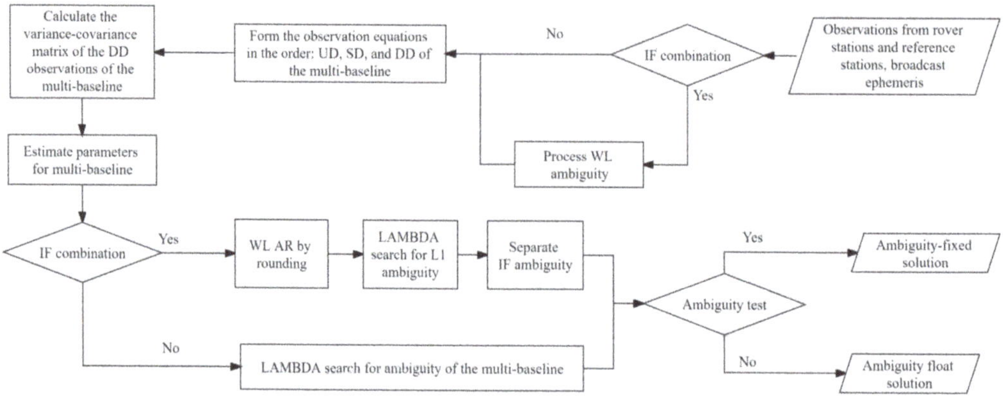

Figure 2. Flow chart of multi-baseline relative positioning.

The observation redundancy is considerably increased by the multi-GNSS MBS compared to the single-GNSS SBS [26]. Using more observations inevitably increases the AR dimension. As the number of AR dimensions increases, the difference in the size of the residual quadratic corresponding to the optimal candidate solution and the suboptimal candidate solution of ambiguity becomes less significant, and the ratio value gets close to 1.0, gradually [44]. Therefore, the commonly adopted empirical threshold of 3.0 is too strict for high-dimension AR, and it is easy to reject the ambiguity that can be fixed correctly. Empirically, the ratio test threshold for the MBS model should be set at 1.3 to 2.5.

3. Experiments and Analysis

In this section, the MBS model was evaluated by conducting experiments using baselines of medium-long lengths and large height differences. The datasets were obtained from an actual ground settlement monitoring project in Shanxi Province, China. Calculations were performed in static mode using a self-developed C++ program.

3.1. Medium-Long Baseline Experiment

3.1.1. Experimental Design

To compare the positioning performance of the MBS model and SBS model, an ideal monitoring environment was firstly considered. Four relatively stable stations with similar heights were selected. Figure 3 shows the distribution map. The dataset contained dual-frequency GPS observations of 60 days (from 15 February 2020 to 14 April 2020), with a sampling interval of 1 min and a cut-off elevation angle of 15°. Three independent baselines were formed, using XJ06 as the rover station and XJ12, WJ01, and XJ05 as the reference stations. The approximate lengths were 16, 20, and 23 km, respectively.

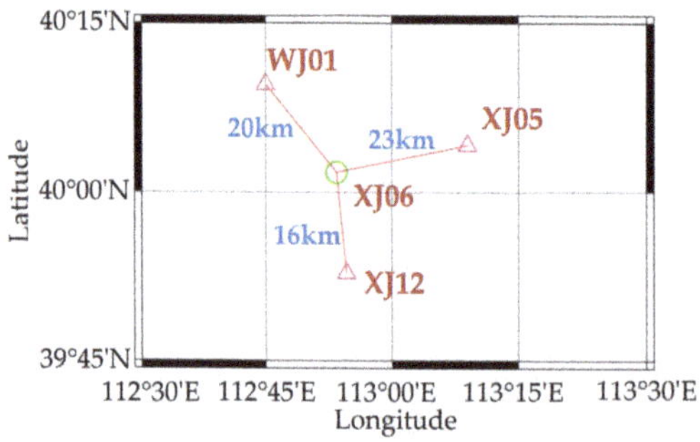

Figure 3. Distribution of medium and long baseline experimental stations.

Figure 4 shows the position dilution of precision (PDOP) and the number of GPS satellites tracked by the station XJ06 on 15 February 2020. Station XJ06 tracked 5 to 10 GPS satellites with PDOP values oscillating at about 2.3.

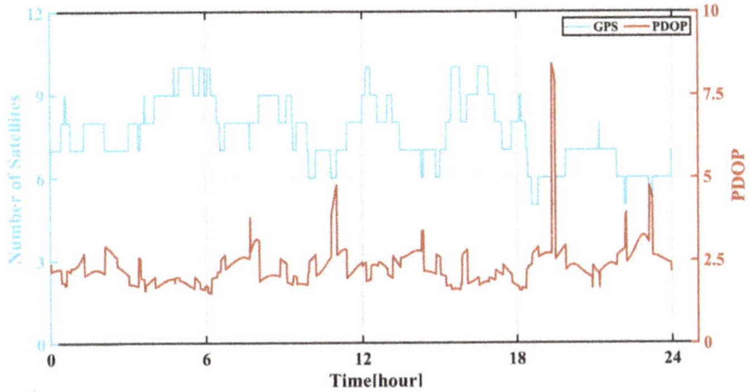

Figure 4. Number of tracked satellites (blue line) and PDOP values (red line) of station XJ06.

Seven data processing strategies with different reference stations were designed and are listed in Table 1. Strategies S_{XJ12}^{XJ06}, S_{WJ01}^{XJ06} and S_{XJ05}^{XJ06} used the SBS model, with the reference stations being XJ12, WJ01 and XJ05 respectively. Strategies $M_{WJ01+XJ12}^{XJ06}$, $M_{WJ01+XJ05}^{XJ06}$ and $M_{XJ05+XJ12}^{XJ06}$ adopted the MBS model with double reference stations (namely, WJ01 + XJ12, WJ01 + XJ05, and XJ05 + XJ12). Strategy $M_{WJ01+XJ05+XJ12}^{XJ06}$ adopted the MBS model using triple reference stations. It was used to check whether simply increasing the number of reference stations could have a positive impact on the positioning. GPS L1/L2 IF combination was used to eliminate ionospheric effect, and the empirical tropospheric model Saastamoinen was used to reduce the effect of tropospheric delay error. The length of static solution was 2 h. Daily static solutions were determined using software Bernese 5.0 as the true values.

Table 1. Medium-long baseline experimental processing strategies.

Strategy	Reference Stations	Rover Station	Model	Baseline Length/km
S_{XJ12}^{XJ06}	XJ12	XJ06	SBS	16
S_{WJ01}^{XJ06}	WJ01	XJ06	SBS	20
S_{XJ05}^{XJ06}	XJ05	XJ06	SBS	23
$M_{WJ01+XJ12}^{XJ06}$	WJ01 + XJ12	XJ06	MBS	-
$M_{WJ01+XJ05}^{XJ06}$	WJ01 + XJ05	XJ06	MBS	-
$M_{XJ05+XJ12}^{XJ06}$	XJ05 + XJ12	XJ06	MBS	-
$M_{WJ01+XJ05+XJ12}^{XJ06}$	WJ01 + XJ05 + XJ12	XJ06	MBS	-

3.1.2. Analysis of Monitoring Accuracy for Different Reference Stations

Figure 5 shows the baseline positioning errors of S_{XJ12}^{XJ06} and $M_{WJ01+XJ05+XJ12}^{XJ06}$. Figure 6 shows the STD and root-mean-square (RMS) values of the positioning errors for the seven processing strategies in the north (N), east (E), and up (U) components. The corresponding statistics are listed in Table 2, where the improvement rate of the MBS strategy was calculated compared to the SBS strategy S_{XJ12}^{XJ06} with the shortest baseline length. From Figure 5, it can be seen that the positioning error fluctuation of strategy $M_{WJ01+XJ05+XJ12}^{XJ06}$ was significantly smaller than that of SBS strategy S_{XJ12}^{XJ06}, which fluctuated in the N/E/U direction within about 1/1/2 cm, and the rest of the strategies had similar results. The experimental results showed that the positioning accuracy of the MBS model was higher than that of the SBS model. The average improvement of coordinate STD values of the MBS model was approximately 25.7, 19.0, and 21.5% in the N, E, and U components, respectively, with the maximum improvement of about 31.0%. The average improvement of the RMS values was about 22.8, 24.2, and 40.0% in the N, E, and U components, respectively, with a maximum improvement of about 44.8%.

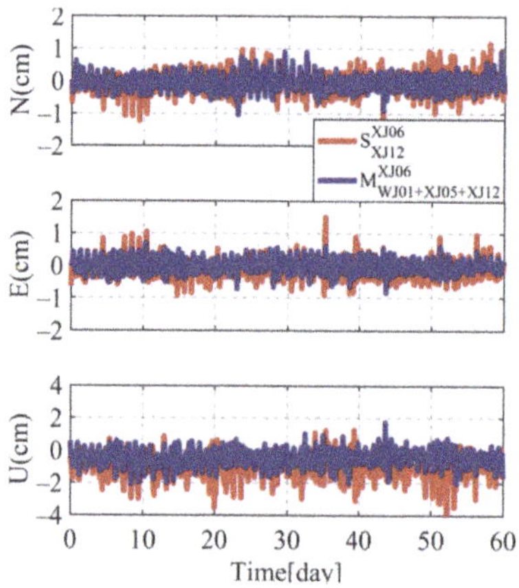

Figure 5. The positioning errors of strategies S_{XJ12}^{XJ06} and $M_{WJ01+XJ05+XJ12}^{XJ06}$.

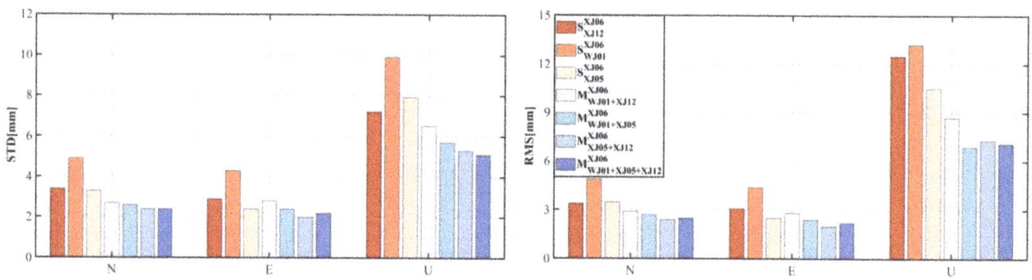

(a) (b)

Figure 6. The STD (**a**) and RMS (**b**) values of the positioning error for SBS and MBS processing strategies.

It was also clear that the improvement rate of the positioning results was related to the quality of baselines involved in the MBS solution. Most notably, the positioning accuracy of strategy $M_{WJ01+XJ05}^{XJ06}$ was better than that of the single-baseline strategy. It indicates that in actual deformation monitoring, if one cannot select a reference station in the near distance, a higher positioning accuracy can still be achieved by using the MBS model to combined several distant reference stations.

The positioning accuracy of the triple-baseline strategy was generally better than that of the double-baseline strategy, indicating that increasing the number of reference stations benefits the positioning due to an increased observation redundancy. However, the difference of positioning accuracy between strategy $M_{WJ01+XJ05+XJ12}^{XJ06}$ and strategy $M_{XJ05+XJ12}^{XJ06}$ was small. This is probably because strategy $M_{WJ01+XJ05+XJ12}^{XJ06}$ included baseline WJ01-XJ06, whose quality was the worst of the three baselines as shown in the single-baseline strategy results (see Table 2).

Table 2. The positioning error statistics for SBS and MBS processing strategies with different reference stations at rover station XJ06.

Strategy	STD/mm			Improvement/%			RMS/mm			Improvement/%		
	N	E	U	N	E	U	N	E	U	N	E	U
S_{XJ12}^{XJ06}	3.4	2.9	7.2	-	-	-	3.4	3.1	12.5	-	-	-
S_{WJ01}^{XJ06}	4.9	4.3	9.9	-	-	-	4.9	4.4	13.2	-	-	-
S_{XJ05}^{XJ06}	3.3	2.4	7.9	-	-	-	3.5	2.5	10.5	-	-	-
$M_{WJ01+XJ12}^{XJ06}$	2.7	2.8	6.5	20.6	3.5	9.7	2.9	2.8	8.7	14.7	9.7	30.4
$M_{WJ01+XJ05}^{XJ06}$	2.6	2.4	5.7	23.5	17.2	20.8	2.7	2.4	6.9	20.6	22.6	44.8
$M_{XJ05+XJ12}^{XJ06}$	2.4	2.0	5.3	29.4	31.0	26.4	2.4	2.0	7.3	29.4	35.5	41.6
$M_{WJ01+XJ05+XJ12}^{XJ06}$	2.4	2.2	5.1	29.4	24.1	29.2	2.5	2.2	7.1	26.5	29.0	43.2
Average improvement rate	-	-	-	25.7	19.0	21.5	-	-		22.8	24.2	40.0

3.2. Large Height Difference Experiment

3.2.1. Experimental Design

Three monitoring stations (YEY2, YEY3, YEY5) equipped with single-frequency receivers in the ground settlement project were selected to analyze the positioning performance of the proposed MBS model in monitoring baselines with large height differences. Six independent baselines with different height differences were formed with two reference stations (MDT1, NS01). From the rover to reference stations, the baseline lengths were about 15 km and the height differences were about 200 m. Figure 7 shows the distribution map of the stations. The datasets were collected continuously for 60 days, from 15 February 2020 to 14 April 2020, with a sampling interval of 30 s and a satellite cut-off elevation angle of 15°. Figure 8 shows the number of visible satellites and PDOPs of station YEY2 on 15 February 2020. In this contribution, G and C stand for GPS and BDS respectively, PDOP.G and PDOP.C are PDOPs for GPS and BDS. It can be seen that the average number of visible satellites for the combined GPS/BDS reached 17, while the number of GPS satellites was only about 7. The PDOP values for GPS/BDS were significantly smaller than those of GPS-only and BDS-only solutions, indicating that combining GPS/BDS observations enhances the satellite visibility.

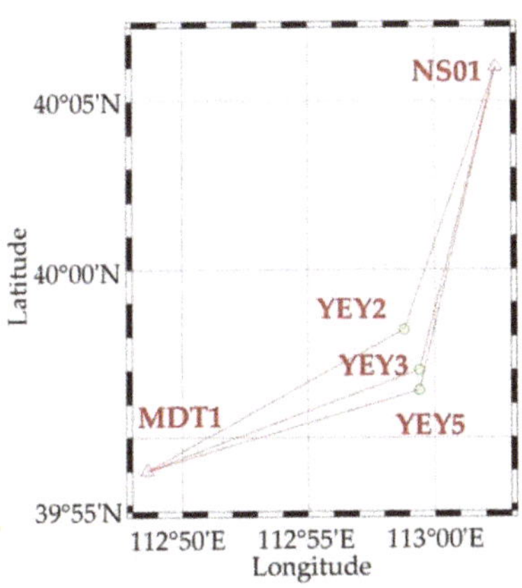

Figure 7. Distribution of large height difference experimental stations.

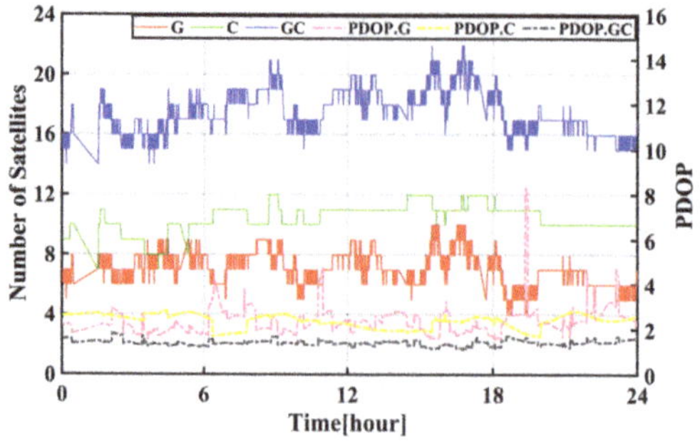

Figure 8. Number of tracked satellites and PDOP values of station YEY2.

Table 3 lists eleven strategies to compare the positioning performance of SBS strategies (e.g., S_{MDT1}^{YEY2}), single-GNSS strategies ($M_{MDT1+NS01}^{YEY2(G)}$ and $M_{MDT1+NS01}^{YEY2(C)}$ that respectively using GPS-only and BDS-only observations), MBS strategies (e.g., $M_{MDT1+NS01}^{YEY2}$) and corresponding CMBS strategies (e.g., $M_{MDT1+NS01}^{YEY2-constraint}$). The session length was 3 h. The reference coordinates of each station were obtained in the same way as introduced in Section 3.1. Table 3 also shows the height difference between the rover station and each reference station. For example, (263 m, 152 m) in strategy $M_{MDT1+NS01}^{YEY2}$ indicates that the height differences between reference stations MDT1 and NS01 and rover station YEY2 were 263 and 152 m, respectively.

Table 3. Large height difference experimental processing strategies.

Strategy	Reference Stations	Rover Station	Height Differences w.r.t. Reference Stations/m	Model	Tropospheric Constraint
S_{MDT1}^{YEY2}	MDT1	YEY2	263	SBS	No
S_{MDT1}^{YEY3}	MDT1	YEY3	200	SBS	No
S_{MDT1}^{YEY5}	MDT1	YEY5	112	SBS	No
$M_{MDT1+NS01}^{YEY2(G)}$	MDT1 + NS01	YEY2	(263, 152)	MBS	No
$M_{MDT1+NS01}^{YEY2(C)}$	MDT1 + NS01	YEY2	(263, 152)	MBS	No
$M_{MDT1+NS01}^{YEY2}$	MDT1 + NS01	YEY2	(263, 152)	MBS	No
$M_{MDT1+NS01}^{YEY2-constraint}$	MDT1 + NS01	YEY2	(263, 152)	CMBS	Yes
$M_{MDT1+NS01}^{YEY3}$	MDT1 + NS01	YEY3	(200, 215)	MBS	No
$M_{MDT1+NS01}^{YEY3-constraint}$	MDT1 + NS01	YEY3	(200, 215)	CMBS	Yes
$M_{MDT1+NS01}^{YEY5}$	MDT1 + NS01	YEY5	(112, 303)	MBS	No
$M_{MDT1+NS01}^{YEY5-constraint}$	MDT1 + NS01	YEY5	(112, 303)	CMBS	Yes

3.2.2. Analysis of Monitoring Accuracy for Different Satellite Systems

Figure 9 presents the positioning errors of YEY2 in N, E, and U components of SBS processing strategies $M_{MDT1+NS01}^{YEY2(G)}$, $M_{MDT1+NS01}^{YEY2(C)}$ and $M_{MDT1+NS01}^{YEY2}$, and the corresponding statistics are listed in Table 4. The experimental results showed that the positioning accuracy of the combined GPS/BDS solutions was superior to those of GPS-only and BDS-only solutions. Compared with the GPS-only solution, the improvement rates of the STD values of the combined GPS/BDS were approximately 23.1, 24.0, and 18.4% in the N, E, and U components, respectively, and the improvement of the RMS values in the U component was more significant than the horizontal components. Compared with the BDS-only solution, the improvement rates of the STD values of the combined GPS/BDS were about 33.3, 13.6, and 41.6% in the N, E, and U components, respectively, and the improvement rates of the RMS values were about 14.6, 7.7, and 43.8% in the N, E, and U components, respectively.

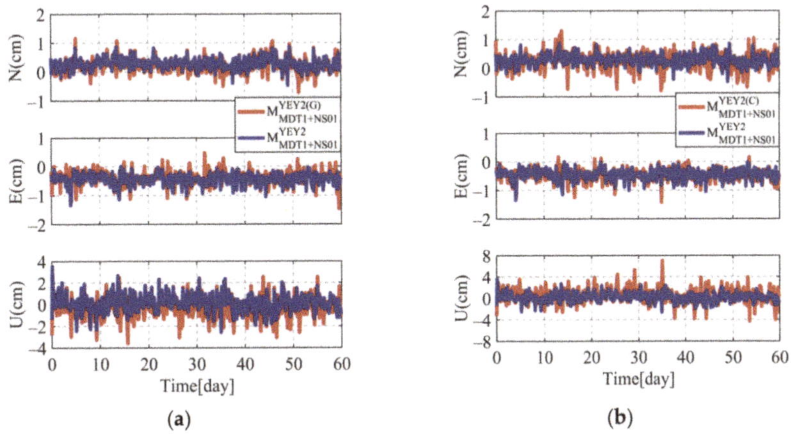

Figure 9. The positioning errors of solutions using different satellite systems including GPS-only (**a**) and BDS-only (**b**) at station YEY2.

Table 4. The positioning error statistics of single-GNSS and dual-GNSS strategies.

Strategy	STD/mm			Improvement/%			RMS/mm			Improvement/%		
	N	E	U	N	E	U	N	E	U	N	E	U
$M_{MDT1+NS01}^{YEY2(G)}$	2.6	2.5	9.8	-	-	-	3.4	4.7	10.4	-	-	-
$M_{MDT1+NS01}^{YEY2(C)}$	3.0	2.2	13.7	-	-	-	4.1	5.2	14.4	-	-	-
$M_{MDT1+NS01}^{YEY2}$	2.0	1.9	8.0	23.1	24.0	18.4	3.5	4.8	8.1	−2.9	−2.1	22.1

3.2.3. Analysis of Monitoring Accuracy for Large Height Difference

Figure 10 shows the positioning errors at YEY2 station when using processing strategies S_{MDT1}^{YEY2} and $M_{MDT1+NS01}^{YEY2}$.Table 5 lists the STD and RMS values of the nine strategies, as well as the improvement rates of six MBS strategies in comparison to single-baseline strategies. The results show that the positioning errors of the MBS and CMBS models were smaller than those of the SBS model. The average performance improvement of STD values of the MBS and CMBS models was about 40.6, 48.3, and 44.7% in the N, E, and U components, respectively, with a maximum improvement rate of 66.2%. The average improvement rate of RMS values was about 36.8, 53.7, and 33.7% in the N, E, and U components, respectively, with a maximum improvement rate of 60.4%. Thus, in the deformation monitoring environment with large height differences, the proposed MBS and CMBS models could have superior positioning performance to the SBS model.

Figure 11 shows the positioning errors of strategies $M_{MDT1+NS01}^{YEY2}$ and $M_{MDT1+NS01}^{YEY2-constraint}$ in the U component. The STD and RMS values in the U component of the six strategies are listed in Table 6. It can be seen that the positioning accuracy of the CMBS model was higher than that of the MBS model, with a maximum improvement of 45.0% for STD values and 29.6% for RMS values. Therefore, the CMBS model could improve the positioning accuracy in the U direction in the deformation monitoring environment with large height differences.

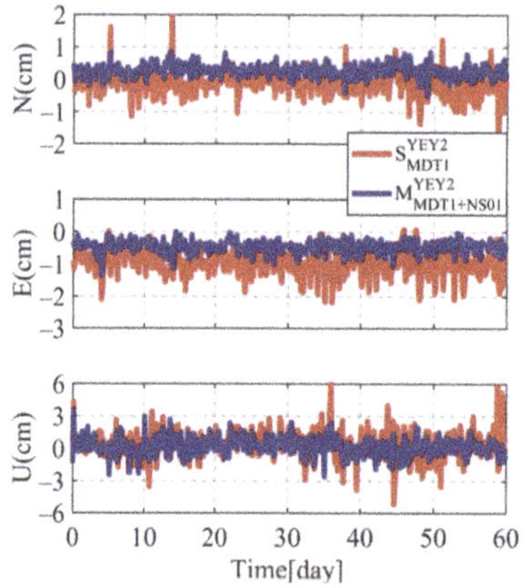

Figure 10. The positioning errors of strategies S_{MDT1}^{YEY2} and $M_{MDT1+NS01}^{YEY2}$.

Table 5. The positioning error statistics of SBS, MBS and CMBS processing strategies with different height differences.

Strategy	STD/mm			Improvement/%			RMS/mm			Improvement/%		
	N	E	U	N	E	U	N	E	U	N	E	U
S_{MDT1}^{YEY2}	3.8	3.9	13.0	-	-	-	4.1	10.4	13.7	-	-	-
S_{MDT1}^{YEY3}	3.4	4.4	13.8	-	-	-	5.1	12.0	14.4	-	-	-
S_{MDT1}^{YEY5}	3.6	4.6	13.2	-	-	-	12.2	9.1	13.3	-	-	-
$M_{MDT1+NS01}^{YEY2}$	2.0	1.9	8.0	47.4	51.3	38.5	3.5	4.8	8.1	14.6	53.9	40.9
$M_{MDT1+NS01}^{YEY2-constraint}$	2.0	1.9	4.4	47.4	51.3	66.2	3.4	4.8	5.7	17.1	53.9	58.4
$M_{MDT1+NS01}^{YEY3}$	2.1	2.2	8.6	38.2	50.0	37.7	2.2	6.4	10.9	56.9	46.7	24.3
$M_{MDT1+NS01}^{YEY3-constraint}$	2.1	2.2	5.6	38.2	50.0	59.4	2.2	6.4	9.8	56.9	46.7	31.9
$M_{MDT1+NS01}^{YEY5}$	2.3	2.6	8.8	36.1	43.5	33.3	7.6	3.6	10.2	37.7	60.4	23.3
$M_{MDT1+NS01}^{YEY5-constraint}$	2.3	2.6	8.8	36.1	43.5	33.3	7.6	3.6	10.2	37.7	60.4	23.3
Average improvement rate	-	-	-	40.6	48.3	44.7	-	-	-	36.8	53.7	33.7

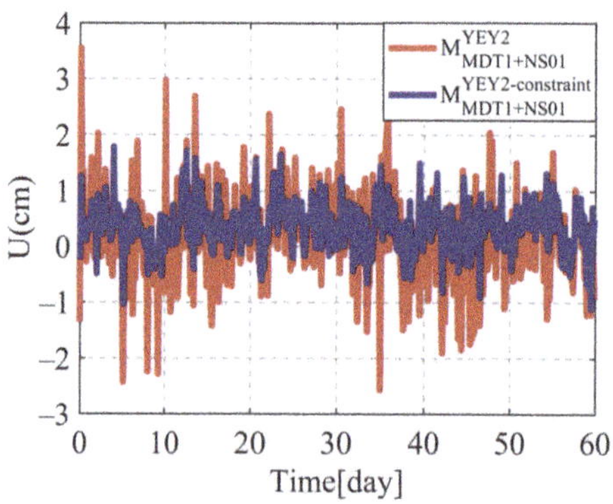

Figure 11. The positioning error of strategy $M_{MDT1+NS01}^{YEY2}$ and $M_{MDT1+NS01}^{YEY2-constraint}$ in the U component.

Table 6. The positioning error of MBS and CMBS processing strategies in the U component.

Strategy	STD/mm	Improvement/%	RMS/mm	Improvement/%
$M_{MDT1+NS01}^{YEY2}$	8.0	-	8.1	-
$M_{MDT1+NS01}^{YEY2-constraint}$	4.4	45.0	5.7	29.6
$M_{MDT1+NS01}^{YEY3}$	8.6	-	10.9	-
$M_{MDT1+NS01}^{YEY3-constraint}$	5.6	34.9	9.8	10.1
$M_{MDT1+NS01}^{YEY5}$	8.8	-	10.2	-
$M_{MDT1+NS01}^{YEY5-constraint}$	8.8	0.0	10.2	0.0

4. Conclusions

The existing GNSS deformation monitoring systems often use the SBS model, which ignores baseline correlation and has limited positioning accuracy when processing baselines with medium-long lengths or large height differences. In this study, we constructed a multi-baseline mathematical model based on DD observation equations and adopted a multi-baseline tropospheric delay estimation method based on a priori constraints on the tropospheric delay parameters.

Datasets from baselines with medium-long lengths and large height differences were used to verify the feasibility of the proposed MBS model, and the following conclusions were obtained from the results:

1. For baselines with medium-long lengths and large height differences, the proposed MBS model can provide better monitoring performance than the SBS model. Compared with the SBS model, the MBS model can improve the positioning accuracy of medium-long baselines with an average improvement of about (25.7/19.0/21.5%) and (22.8/24.2/40.0%) in the N/E/U components, with the highest improvement of about (29.4/31.0/29.2%) and (29.4/35.5/44.8%) in the N/E/U components, respectively. For baselines with large height differences, compared with the SBS model, the MBS model can improve the positioning accuracy with an average improvement of about (40.6/48.3/44.7%) and (36.8/53.7/33.7%) in the N/E/U components, with the highest improvement of about (47.4/51.3/66.2%) and (56.9/60.4/58.4%) in the N/E/U components, respectively. The MBS model uses multiple reference stations thus can improve the positioning model strength and observation redundancy. This is especially beneficial for applying GNSS in complex monitoring environments such as canyons, open pits, slopes, large-area ground settlement, and long-spanned bridges and railroads.
2. The accuracy of the MBS model is related to the number of reference stations and the quality of the baselines. With comparable baseline quality, the accuracy of the MBS model improves as the number of reference stations increases. Medium-long baseline experimental results show that compared with the SBS model, the MBS model using double reference stations can achieve an average improvement rate of about 24.0%, while the MBS model using triple reference stations can achieve an average improvement rate of about 30.2%.
3. Compared with GPS-only and BDS-only positioning, the combined GPS/BDS positioning has an accuracy improvement of an average of 13.8 and 25.8% in the baseline components. Meanwhile, the proposed CMBS model can improve accuracy in the U direction and reach up to 45.0%.

Future research will be focused on the application of the MBS approach to near-real-time dynamic monitoring. Baselines with larger tropospheric residual errors will be further investigated.

Author Contributions: Conceptualization, H.W. and W.Y.; methodology, H.W.; software, H.W.; validation, W.Y. and W.D.; formal analysis, H.W.; investigation, W.Y. and W.D.; data curation, H.W.; writing—original draft preparation, H.W.; writing—review and editing, W.Y. and W.D.; supervision, W.D.; funding acquisition, H.W. and W.Y. All authors have read and agreed to the published version of the manuscript.

Funding: The work is supported by National Natural Science Foundation of China (Grant No. 42104022), Natural Science Foundation of Hunan Province (Grant No. 2021JJ40728), the Fundamental Research Funds for the Central Universities of Central South University (Grant No. 2021zzts0800) and the Hunan Provincial Innovation Foundation for Postgraduate (Grant No. CX20210287).

Conflicts of Interest: The authors declare no conflict of interest.

References

1. Yu, W.; Peng, H.; Pan, L.; Dai, W.; Qu, X.; Ren, Z. Performance Assessment of High-Rate GPS/BDS Precise Point Positioning for Vibration Monitoring Based on Shaking Table Tests. *Adv. Space Res.* **2022**, *69*, 2362–2375. [CrossRef]
2. Yan, H.; Dai, W.; Xie, L.; Xu, W. Fusion of GNSS and InSAR Time Series Using the Improved STRE Model: Applications to the San Francisco Bay Area and Southern California. *J. Geod.* **2022**, *96*, 47. [CrossRef]
3. Shen, N.; Chen, L.; Liu, J.; Wang, L.; Tao, T.; Wu, D.; Chen, R. A Review of Global Navigation Satellite System (GNSS)-Based Dynamic Monitoring Technologies for Structural Health Monitoring. *Remote Sens.* **2019**, *11*, 1001. [CrossRef]
4. Zhong, P.; Ding, X.; Yuan, L.; Xu, Y.; Kwok, K.; Chen, Y. Sidereal Filtering Based on Single Differences for Mitigating GPS Multipath Effects on Short Baselines. *J. Geod.* **2010**, *84*, 145–158. [CrossRef]
5. Dai, W.; Huang, D.; Cai, C. Multipath Mitigation via Component Analysis Methods for GPS Dynamic Deformation Monitoring. *GPS Solut.* **2014**, *18*, 417–428. [CrossRef]
6. Chen, J.; Yue, D.; Liu, Z.; Xu, C.; Zhu, S.; Chen, H. Experimental Research on Daily Deformation Monitoring of Bridge Using BDS/GPS. *Surv. Rev.* **2019**, *51*, 472–482. [CrossRef]
7. Msaewe, H.A.; Hancock, C.M.; Psimoulis, P.A.; Roberts, G.W.; Bonenberg, L.; de Ligt, H. Investigating Multi-GNSS Performance in the UK and China Based on a Zero-Baseline Measurement Approach. *Measurement* **2017**, *102*, 186–199. [CrossRef]
8. Quan, Y.; Lau, L.; Roberts, G.W.; Meng, X. Measurement Signal Quality Assessment on All Available and New Signals of Multi-GNSS (GPS, GLONASS, Galileo, BDS, and QZSS) with Real Data. *J. Navig.* **2016**, *69*, 313–334. [CrossRef]
9. Roberts, G.W.; Tang, X. The Use of PSD Analysis on BeiDou and GPS 10 Hz Dynamic Data for Change Detection. *Adv. Space Res.* **2017**, *59*, 2794–2808. [CrossRef]
10. Huang, G.; Du, Y.; Meng, L.; Huang, G.; Wang, J.; Han, J. Application Performance Analysis of Three GNSS Precise Positioning Technology in Landslide Monitoring. In *China Satellite Navigation Conference (CSNC) 2017 Proceedings: Volume I*; Sun, J., Liu, J., Yang, Y., Fan, S., Yu, W., Eds.; Lecture Notes in Electrical Engineering; Springer: Singapore, 2017; Volume 437, pp. 139–150. ISBN 978-981-10-4587-5.
11. Xi, R.; Chen, H.; Meng, X.; Jiang, W.; Chen, Q. Reliable Dynamic Monitoring of Bridges with Integrated GPS and BeiDou. *J. Surv. Eng.* **2018**, *144*, 04018008. [CrossRef]
12. Li, B.; Shen, Y.; Feng, Y.; Gao, W.; Yang, L. GNSS Ambiguity Resolution with Controllable Failure Rate for Long Baseline Network RTK. *J. Geod.* **2014**, *88*, 99–112. [CrossRef]
13. Wilgan, K.; Hurter, F.; Geiger, A.; Rohm, W.; Bosy, J. Tropospheric Refractivity and Zenith Path Delays from Least-Squares Collocation of Meteorological and GNSS Data. *J. Geod.* **2017**, *91*, 117–134. [CrossRef]
14. Wielgosz, P.; Paziewski, J.; Krankowski, A.; Kroszczyński, K.; Figurski, M. Results of the Application of Tropospheric Corrections from Different Troposphere Models for Precise GPS Rapid Static Positioning. *Acta Geophys.* **2012**, *60*, 1236–1257. [CrossRef]
15. Schüler, T. Impact of Systematic Errors on Precise Long-Baseline Kinematic GPS Positioning. *GPS Solut.* **2006**, *10*, 108–125. [CrossRef]
16. Luo, X.; Ou, J.; Yuan, Y.; Gao, J.; Jin, X.; Zhang, K.; Xu, H. An Improved Regularization Method for Estimating near Real-Time Systematic Errors Suitable for Medium-Long GPS Baseline Solutions. *Earth Planets Space* **2008**, *60*, 793–800. [CrossRef]
17. Tang, W.; Shen, M.; Deng, C.; Cui, J.; Yang, J. Network-Based Triple-Frequency Carrier Phase Ambiguity Resolution between Reference Stations Using BDS Data for Long Baselines. *GPS Solut.* **2018**, *22*, 73. [CrossRef]
18. Assiadi, M.; Edwards, S.J.; Clarke, P.J. Enhancement of the Accuracy of Single-Epoch GPS Positioning for Long Baselines by Local Ionospheric Modelling. *GPS Solut.* **2014**, *18*, 453–460. [CrossRef]
19. Chu, F.-Y.; Yang, M. GPS/Galileo Long Baseline Computation: Method and Performance Analyses. *GPS Solut.* **2014**, *18*, 263–272. [CrossRef]
20. Zhang, Y.; Kubo, N.; Chen, J.; Chu, F.-Y.; Wang, H.; Wang, J. Contribution of QZSS with Four Satellites to Multi-GNSS Long Baseline RTK. *J. Spat. Sci.* **2020**, *65*, 41–60. [CrossRef]
21. Li, X.; Huang, G.; Zhang, Q.; Zhao, Q. A New GPS/BDS Tropospheric Delay Resolution Approach for Monitoring Deformation in Super High-Rise Buildings. *GPS Solut.* **2018**, *22*, 90. [CrossRef]
22. Paziewski, J. Precise GNSS Single Epoch Positioning with Multiple Receiver Configuration for Medium-Length Baselines: Methodology and Performance Analysis. *Meas. Sci. Technol.* **2015**, *26*, 035002. [CrossRef]

23. Chen, H.; Jiang, W.; Ge, M.; Wickert, J.; Schuh, H. An Enhanced Strategy for GNSS Data Processing of Massive Networks. *J. Geod.* **2014**, *88*, 857–867. [CrossRef]
24. Borko, A.; Even-Tzur, G. Stochastic Model Reliability in GNSS Baseline Solution. *J. Geod.* **2021**, *95*, 20. [CrossRef]
25. Alizadeh-Khameneh, M.A.; Sjöberg, L.E.; Jensen, A.B.O. Optimisation of GNSS Networks—Considering Baseline Correlations. *Surv. Rev.* **2019**, *51*, 35–42. [CrossRef]
26. Wang, J.; Xu, T.; Nie, W.; Xu, G. GPS/BDS RTK Positioning Based on Equivalence Principle Using Multiple Reference Stations. *Remote Sens.* **2020**, *12*, 3178. [CrossRef]
27. Fan, L.; Cui, X.; Lu, M. Precise and Robust RTK-GNSS Positioning in Urban Environments with Dual-Antenna Configuration. *Sensors* **2019**, *19*, 3586. [CrossRef]
28. Wang, J.; Xu, T.; Nie, W.; Xu, G. A Simplified Processing Algorithm for Multi-Baseline RTK Positioning in Urban Environments. *Measurement* **2021**, *179*, 109446. [CrossRef]
29. Dardanelli, G.; Maltese, A.; Pipitone, C.; Pisciotta, A.; Lo Brutto, M. NRTK, PPP or Static, That Is the Question. Testing Different Positioning Solutions for GNSS Survey. *Remote Sens.* **2021**, *13*, 1406. [CrossRef]
30. Mageed, K.M.A. Comparison of GPS Commercial Software Packages to Processing Static Baselines up to 30 KM. *ARPN J. Eng. Appl. Sci.* **2015**, *10*, 11.
31. Andritsanos, V.D.; Arabatzi, O.; Gianniou, M.; Pagounis, V.; Tziavos, I.N.; Vergos, G.S.; Zacharis, E. Comparison of Various GPS Processing Solutions toward an Efficient Validation of the Hellenic Vertical Network: The ELEVATION Project. *J. Surv. Eng.* **2016**, *142*, 04015007. [CrossRef]
32. Xu, G. GPS Data Processing with Equivalent Observation Equations. *GPS Solut.* **2002**, *6*, 28–33. [CrossRef]
33. Shen, Y.; Xu, G. Simplified Equivalent Representation of GPS Observation Equations. *GPS Solut.* **2008**, *12*, 99–108. [CrossRef]
34. Shen, Y.; Li, B.; Xu, G. Simplified Equivalent Multiple Baseline Solutions with Elevation-Dependent Weights. *GPS Solut.* **2009**, *13*, 165–171. [CrossRef]
35. King, R.W.; Herring, T.A.; McCluscy, S.C. (Eds.) *Documentation for the GAMIT GPS Analysis Software 10.6*; Technology Report; Massachusetts Institute of Technology: Cambridge, MA, USA, 2015.
36. Tralli, D.M.; Lichten, S.M. Stochastic Estimation of Tropospheric Path Delays in Global Positioning System Geodetic Measurements. *Bull. Géod.* **1990**, *64*, 127–159. [CrossRef]
37. Böhm, J.; Möller, G.; Schindelegger, M.; Pain, G.; Weber, R. Development of an Improved Empirical Model for Slant Delays in the Troposphere (GPT2w). *GPS Solut.* **2015**, *19*, 433–441. [CrossRef]
38. Boehm, J.; Werl, B.; Schuh, H. Troposphere Mapping Functions for GPS and Very Long Baseline Interferometry from European Centre for Medium-Range Weather Forecasts Operational Analysis Data: TROPOSPHERE MAPPING FUNCTIONS FROM ECMWF. *J. Geophys. Res. Solid Earth* **2006**, *111*. [CrossRef]
39. Song, J.; Park, B.; Kee, C. Comparative Analysis of Height-Related Multiple Correction Interpolation Methods with Constraints for Network RTK in Mountainous Areas. *J. Navig.* **2016**, *69*, 991–1010. [CrossRef]
40. Lou, Y.; Huang, J.; Zhang, W.; Liang, H.; Zheng, F.; Liu, J. A New Zenith Tropospheric Delay Grid Product for Real-Time PPP Applications over China. *Sensors* **2017**, *18*, 65. [CrossRef]
41. Song, J.; Kee, C.; Park, B.; Park, H.; Seo, S. Correction Combination of Compact Network RTK Considering Tropospheric Delay Variation over Height. In Proceedings of the 2014 IEEE/ION Position, Location and Navigation Symposium—PLANS 2014, Monterey, CA, USA, 5–8 May 2014; IEEE: Monterey, CA, USA, May, 2014; pp. 95–101.
42. Tang, W.; Liu, J.; Shi, C.; Lou, Y. Three Steps Method to Determine Double Difference Ambiguities Resolution of Network RTK Reference Station. *Geomat. Inf. Sci. Wuhan Univ.* **2007**, *32*, 305–308.
43. Hatch, R. The Synergism of GPS Code and Carrier Measurements. In Proceedings of the Third International Geodetic Symposium on Satellite Doppler Positioning, Las Cruces, NM, USA, 8–12 February 1982; Volume 2, pp. 1213–1231.
44. Lu, L.; Liu, W.; Zhang, X. An Effective QR-Based Reduction Algorithm for the Fast Estimation of GNSS High-Dimensional Ambiguity Resolution. *Surv. Rev.* **2018**, *50*, 57–68. [CrossRef]

remote sensing MDPI

Communication

A Beacons Selection Method under Random Interference for Indoor Positioning

Yu Guo [1], Jiazhu Zheng [1,*], Shaoning Di [2], Guiqiu Xiang [1,3] and Fei Guo [4]

1 Civil Engineering College, Nanjing Forestry University, Nanjing 210037, China
2 School of Geosciences and Info-Physics, Central South University, Changsha 410083, China
3 Zhejiang Institute of Surveying and Mapping Science and Technology, Hangzhou 310000, China
4 School of Geodesy and Geomatics, Wuhan University, Wuhan 430079, China
* Correspondence: zjz90139@njfu.edu.cn

Abstract: GNSS is still not well applied in indoor environments. This is an important challenge for seamless positioning and navigation. Using other sensors to replace and connect is the mainstream practice at present. No matter what technology is used, the problem of real-time optimal station selection is faced in complex indoor environments. In this paper, we first verified the impact of random interference from walkers on positioning signals in an indoor environment. Based on this phenomenon, we proposed a novel real-time dynamic Beacons selection method (RD) in the field of indoor positioning. First, we introduced a machine learning algorithm for real-time anomaly detection of received signals from different Beacons. Then the Beacon selection is completed based on the real-time anomaly detection results and RSSI. In an indoor scene, we verified the positioning accuracy of three other methods when selecting various numbers of Beacons. Then we used the best selection strategies to compare with the RD method. Experiments showed that the RD method can use the least Beacons to obtain higher accuracy and stable positioning results. This paper provides a new idea for real-time optimal selection of signal sources in a complex indoor environment.

Keywords: positioning and navigation; indoor positioning; Beacons selection; signal anomaly detection

Citation: Guo, Y.; Zheng, J.; Di, S.; Xiang, G.; Guo, F. A Beacons Selection Method under Random Interference for Indoor Positioning. *Remote Sens.* **2022**, *14*, 4323. https://doi.org/10.3390/rs14174323

Academic Editors: Giuseppe Casula, Zhetao Zhang and Wenkun Yu

Received: 17 August 2022
Accepted: 25 August 2022
Published: 1 September 2022

1. Introduction

In recent years, location-based services (LBS) have gradually become a research hotspot [1,2]. In outdoor scenarios, the GNSS has been able to provide accurate and stable real-time positioning and navigation services [3,4]. Delivering effective location services in indoor scenarios is challenging because satellite signals are difficult to be obtained in indoor environments [5,6]. The main challenges faced by GNSS are shown in Figure 1. Therefore, it is urgent to find a reliable and simple alternative technology in the indoor positioning and navigation stage. Now, the realization of efficient indoor positioning and navigation is inseparable from the use of sensors such as Wireless Local Area Network (WLAN) [7,8], Bluetooth [9,10], Ultra Wide Band (UWB) [11,12], etc. The basic technologies these rely on Time of Arrival (TOA) [13], Time Difference of Arrival (TDOA) [14], Angle of Arrival (AOA) [15], Received Signal Strength Indication (RSSI) [16], etc. Of these, the RSSI technology based on Bluetooth Low Energy (BLE) [9] has many advantages, such as low cost and easy implementation, and has thus attracted the attention of many scholars and enterprises.

There are two mainstream technologies for RSSI-based indoor positioning: trilateration and fingerprinting [17,18]. Regardless of the basic technology used, however, choosing an optimal station may be challenging in an indoor environment. Some research results showed that it is not that more stations used can achieve higher positioning accuracy and stability [19,20]. The current choices for real-time Beacons mainly include Max Mean RSSI (MM) [21], Loss Rate (LR) [22], Minimum Variance (MV) [23], etc.

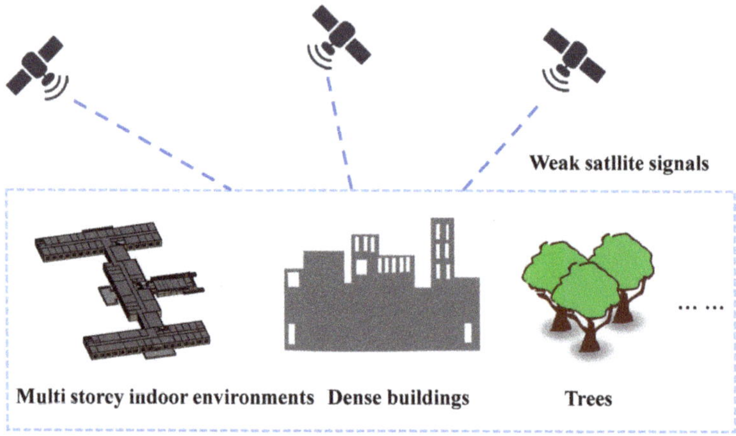

Figure 1. Main challenges faced by GNSS.

In the fingerprinting technology for indoor positioning, the selection methods of Access Points (AP) or Beacons are often static [24–26]. This is because the RSSI map needs to be generated in the offline stage in advance for positioning based on the database. And then perform positioning based on KNN and other methods in the online stage [27,28]. This means that some fixed signal base stations must be selected before positioning for RSSI map generation. In recent years, in addition to the above-mentioned methods (MM, MV, LR), base station selection methods based on information theory (InfoGain) and machine learning (PCA) have also been proposed by many scholars for fingerprinting [24,29]. However, these methods can't achieve a real-time selection of APs or Beacons. On the other hand, fingerprinting needs to make RSSI maps in advance, which increases the cost and makes it difficult to deploy applications quickly. In contrast, trilateration can achieve rapid and low-cost deployment and application. Now, there are few published works on real-time APs/Beacons selection for trilateration.

At present, the methods for selecting the best Beacons are mostly based on the absolute value of RSSI or the distribution of Beacons' positions (whether it is fingerprinting or trilateration). The RSSI value transmitted by BLE is extremely susceptible to the influence of pedestrians and other environmental factors, and these influences often are volatile and unpredictable [30]. Simply using the RSSI value or spatial distribution to select Beacons, therefore, means that it is nearly impossible to avoid these random influences. In the literature [30], a method of weighting based on real-time abnormal ratios was proposed. However, it did not discuss Beacons selection in detail.

In this paper, we focused on the real-time Beacons selection problem when the number of base stations in the scene is large. To address the phenomenon that the RSSI of the signal is easily affected by environmental factors, an automatic anomaly detection algorithm (isolation forest [31]) is used to detect anomalies in each Beacon's signals in the scene in real-time. The anomaly detection results and RSSI values are used to select the optimal Beacons in real-time. Finally, in order to verify the positioning accuracy of the RD method, these optimal Beacons selected are used to calculate indoor coordinates based on nonlinear least squares. The main differences between the proposed work are as follows:

(1) the effect of pedestrian random walking on real-time indoor positioning was verified;
(2) a fast and real-time Beacons selection method was proposed and verified.

2. Materials and Methods

2.1. Indoor Positioning Based on BLE Technology and RSSI

Since this paper uses RSSI for indoor positioning, a proper signal path attenuation model is essential. The widely used logarithmic decay factor model is used for this, as shown in Equation (1) [32]:

$$d = 10^{\frac{A - RSSI}{10n}} \tag{1}$$

where, the parameters A and n are variables that need to be fitted and calculated according to the current indoor environment, and A is the RSSI at 1 m from the base station, n is the signal attenuation parameter in the current indoor scene, d is the distance from the base station.

The distance between an unknown point and the BLE base station based on RSSI can be used to calculate indoor coordinates. In theory, only three BLE stations ((x_1, y_1), (x_2, y_2), (x_3, y_3)) are required to solve the spatial plane coordinates of unknown points, but, in practice, this makes it difficult to obtain the most accurate coordinates. In actual situations, RSSI signals from several base stations are often required to participate in the final coordinate calculation, so as to improve the accuracy and stability of indoor positioning. Equation (2) is the calculation to solve the unknown point, and this is further represented in the schematic diagram shown in Figure 2:

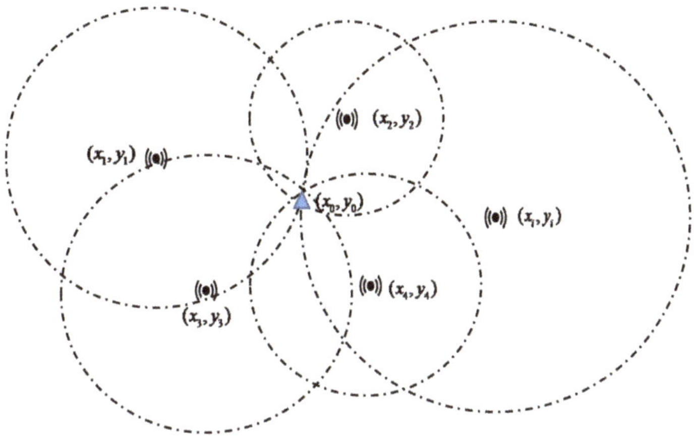

Figure 2. Indoor positioning with multiple BLE base stations.

$$\begin{cases} (x_0 - x_1)^2 + (y_0 - y_1)^2 = d_1^2 \\ (x_0 - x_2)^2 + (y_0 - y_2)^2 = d_2^2 \\ \vdots \\ (x_0 - x_i)^2 + (y_0 - y_i)^2 = d_i^2 \end{cases} \tag{2}$$

This paper used the nonlinear least squares Gauss-Newton (GN) [33] algorithm for the coordinate calculation (Equation (2)), the iterative solution of which is shown in Equation (3):

$$F^{s+1} = F^s + \Delta \tag{3}$$

where, F is the coordinate $[x, y]$ to be solved, s is the number of iterations, and Δ is as in Equation (4):

$$\Delta = -\left(J^T J\right)^{-1} J^T \varepsilon \tag{4}$$

where, J is as in Equation (5), ε is the residual vector calculated previously, as in Equation (6):

$$J = \begin{bmatrix} \frac{\partial f_1}{\partial x}, \frac{\partial f_1}{\partial y} \\ \frac{\partial f_2}{\partial x}, \frac{\partial f_2}{\partial y} \\ \vdots \ddots \vdots \\ \frac{\partial f_i}{\partial x}, \frac{\partial f_i}{\partial y} \end{bmatrix} \quad (5)$$

$$\varepsilon_i = d_i - f_i(x_0, y_0) \quad (6)$$

The f in Equations (5) and (6) is as Equation (7):

$$f_i(x_0, y_0) = \sqrt{(x_0 - x_i)^2 + (y_0 - y_i)^2} \quad (7)$$

We then carried out iterative calculations according to Equation (3), and the final optimal coordinates are calculated as in Equation (8):

$$(x_0, y_0) = \text{argmin}(||\varepsilon^T \varepsilon||) \quad (8)$$

As shown in Equation (2), when using the GN algorithm, a selection of base stations must be used in the iterative solution. The choice of base stations affects the exact coordinates that are calculated as a result. Furthermore, as Xue et al. [19] and Zhang et al. [20] have shown, when undertaking indoor positioning based on RSSI, using more BLE base stations does not deliver greater positioning accuracy. This means that it is very important to choose suitable Beacons to participate in the coordinate calculation.

2.2. Signal Real-Time Anomaly Detection and Beacons Selection

As discussed in the Introduction section, however, most of these Beacons selection strategies do not take into account the real-time fluctuations of the base station for example, due to the influence of pedestrians. In the experimental scenario in this paper, the influence of a pedestrian on the RSSI is shown in Figure 3 (a real experiment in the scene). Thus, it is unreasonable to select a Beacon based solely on the absolute value of the RSSI or the position of the BLE base station.

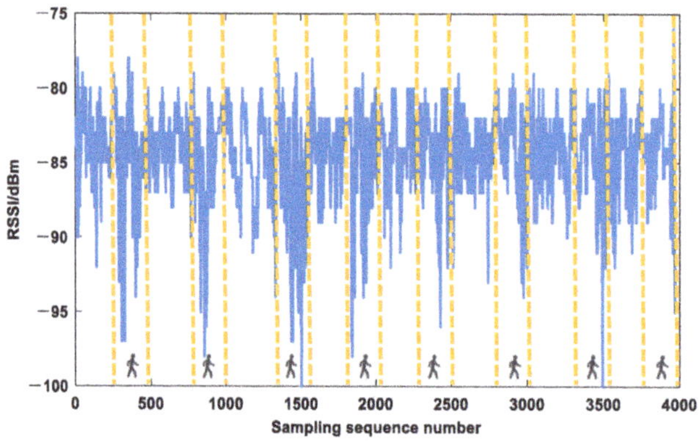

Figure 3. Pedestrian walking affects RSSI.

As shown in Figure 3, this paper showed through experiments that the influence of pedestrians will have a great impact on the RSSI transmitted by the BLE base station. Specifically, we collected RSSI signals at a distance of 3m from a Beacon station. Taking 60 s as the time unit, there is no interference between the mobile phone and the Beacon in about 40 s, and pedestrians are arranged to walk randomly between the mobile phone and the Beacon in about 20 s. Experiments showed that the random interference of pedestrians can have a strong impact on RSSI signals. Compared with the case without interference, the attenuation of the RSSI signals can exceed 20%. In turn, this will affect the subsequent calculation of indoor coordinates, causing fluctuations in positioning accuracy. This paper addressed the shortcomings of relying only on the value of RSSI for Beacons selection by detecting anomalies in the RSSI of Beacons and then using these anomaly results to inform the selection of the optimum Beacons. We proposed the selection method that considered the real-time status of the RSSI for the situation where there are many Beacons in the scene. The RD method proposed in this paper mainly includes the following steps: (a) calculate the number of all Beacons in the scene by the Media Access Control Address (MAC); (b) implement real-time anomaly detection for each Beacon based on a machine learning method; (c) select some of the Beacons with lower real-time abnormal ratios; (d) on the basis of the third step, selecting some of the Beacons with the larger RSSI as the best optimal strategy. The detailed flow chart is shown in Figure 4.

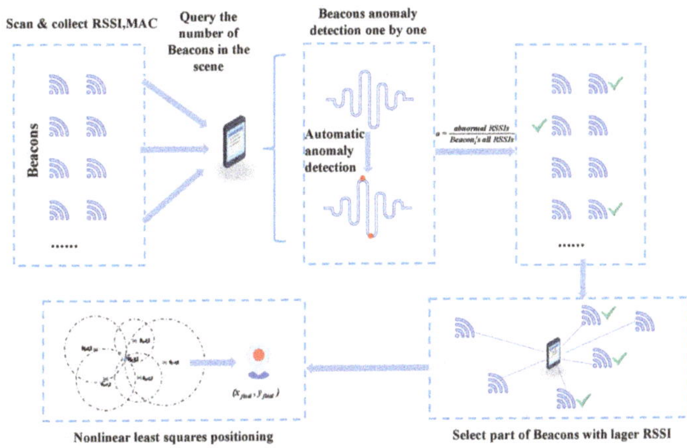

Figure 4. Flow chart of Beacons selection and positioning.

As shown in Figure 4, we need to choose a suitable and convenient real-time RSSI anomaly detection method. Thus, an unsupervised machine learning algorithm is used to detect abnormality in RSSI fluctuations over a period of time. The commonly used and effective methods for signal fluctuation anomaly detection include isolation forest, Support Vector Machine (SVM), etc. In this paper, we chose the one-dimensional isolation forest algorithm for RSSI real-time fluctuation detection. The main steps of the isolation forest used in this paper include: (a) generating 100 isolated binary trees to form an isolation forest; (b) then, the RSSIs are segmented using the isolation trees until each RSSI is in a separate space; (c) calculate the anomaly score for each RSSI using Equation (9) ($A(h(x))$ is the average depth of an RSSI signal in 100 isolated trees, and $C(\psi)$ is the average path length of isolated trees generated); (d) obtain abnormal signals (the closer Q is to 1, the greater the probability the signal is an abnormal value, the threshold set in this paper is 0.7). As Liu et al. [31] have shown in their detailed account of the specific principles and implementation process of isolation forests, this algorithm can ensure a good anomaly detection result when the data is limited.

$$Q = 2^{-\frac{A(h(x))}{C(\psi)}} \tag{9}$$

Given the scenario of RSSIs received from different Beacons stations at a certain point, this paper implemented the isolation forest algorithm to conduct real-time outlier detection in turn. Firstly, generate a hyper-plane segmented isolation forest of the one-dimensional space where each group of RSSI is located. Secondly, to get the real-time abnormal RSSI of each Beacon, we used the generated isolation forest hyper-plane to perform anomaly detection on each Beacon and then judged abnormal points based on ensemble theory. Thirdly, the signal anomaly rate a is calculated in the current time period. The calculation method of a can be expressed as Equation (10)

$$a = \frac{abnormal\ RSSIs}{Beacon_i^{\prime}s\ all\ RSSIs} \tag{10}$$

The non-abnormal rate is calculated as $b = 1 - a$ (that is, the non-volatile RSSI ratio). Finally, the Beacons are sorted for selection based on b. The pseudo-code for Beacons selection is given in Algorithm 1.

Algorithm 1 Beacons selection based on isolation forest

Require: RSSI signals and MACs by Beacons.
Ensure: Beacons are real coordinates.

1: Match and group all RSSIs according to MACs:
$R = \{RSSI_{Beacon1}, RSSI_{Beacon2}, \ldots, RSSI_{Beaconi}\}$;
2: Set to generate isolation forest T_i
3: Use T_i to perform anomaly detection on each grouped R in turn;
4: Calculate Beacon's non-abnormal rate b;
5: Beacons are sorted by b and push Beacons' MAC into F;
6: **Return:** F and R.

We first selected part of the Beacons with smaller abnormal fluctuations (60%, 50%, 40%, etc.). Then, we selected part of the Beacons with a larger average RSSI in the first selection result (60%, 50%, 40%, etc.). Each spatial distance is calculated according to Equation (1). Then we take $1/RSSI^2$ as the weighted matrix and the coordinates are calculated in the experimental scene by using Equation (8). The experiment proves the feasibility and effectiveness of Beacons selection based on the RD method. The results will be explained in detail in the experimental section.

In order to prove the phenomenon (that indoor positioning accuracy is affected by real-time pedestrian movement) and the effectiveness, rationality and superiority of the RD method, an experiment was devised to compare the results from several other methods. We set up three other comparison methods as follows:

(A) Max Mean RSSI (MM): Beacons selection based on the mean maximum RSSI over a period of time.

(B) Loss Rate (LR): The basic idea of the LR method is that we should choose the Beacons with the lowest RSSI loss rate within a period of time. It can be calculated by Beacon transmit frequency and acquisition time. It can be expressed as Equation (11):

$$LR_{\text{Beacons}_i} = (s - s_i)/s \tag{11}$$

where, $Beacon_i$ represents the ith Beacon, s represents the maximum number of RSSIs that can be received during this period, and s_i represents the actual number of RSSIs received by this Beacon.

(C) Minimum Variance (MV): Use the collected RSSI variance as the basis for selecting Beacons.

In the experimental stage of this paper, a comparison was set up between the use of MM, LR, MV and our RD method. Because the three comparison methods are difficult to directly determine the optimal Beacons number selection strategy. Therefore, we first calculated the positioning accuracy of 42 experimental points when selecting 3~18 Beacons (MM, LR, MV). Then choose the optimal number of Beacons among them to compare with the RD method in this paper.

3. Experiment and Discussion

The experiment was conducted in an indoor scene with an area of about 8 × 20 m, and in which we deployed 18 BLE base stations. We then used a mobile phone (HUAWEI P20, Shenzhen, China) along the routes shown in Figure 4 to collect the RSSIs data at 42 fixed points (their coordinates gotten by the traditional measurement method (the errors are in millimeters)). As shown in Figures 5 and 6, we engaged researchers to walk completely randomly and irregularly across the scene, allowing us to simulate the presence of pedestrians and their influence on the collected RSSI.

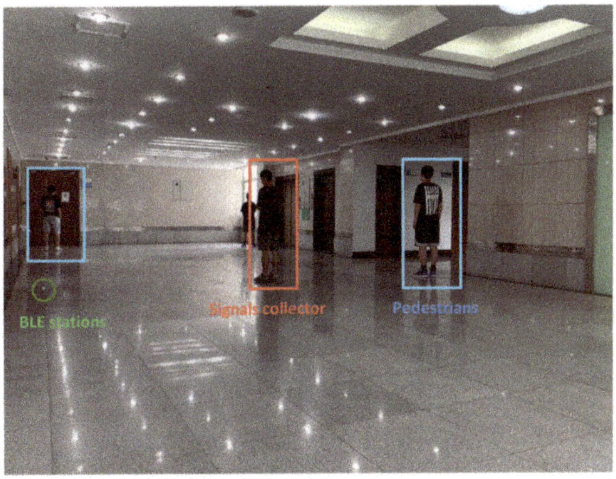

Figure 5. Experimental process.

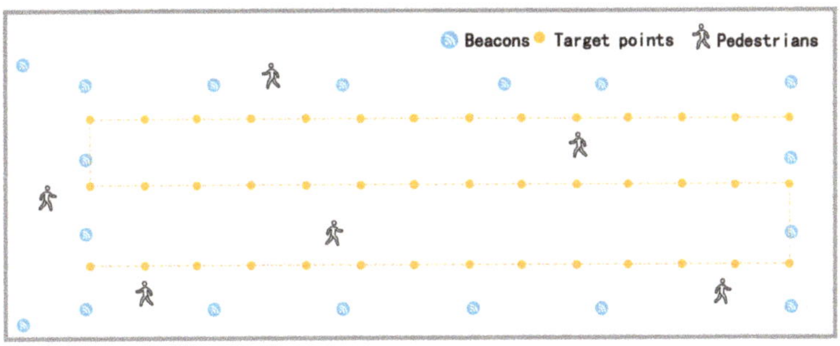

Figure 6. Schematic diagram of the experimental scene.

As shown in Equation (1), in order to calculate the collected RSSI value as a spatial distance, it is necessary to obtain the parameters A and n. Accordingly, the RSSI values were collected between 0.6 m and 9.6 m from the base stations at 0.6 m intervals. From this, we found that $A = -66.25\ dBm$ and $n = 2.877$. The RSSI attenuation curve is shown in Figure 7.

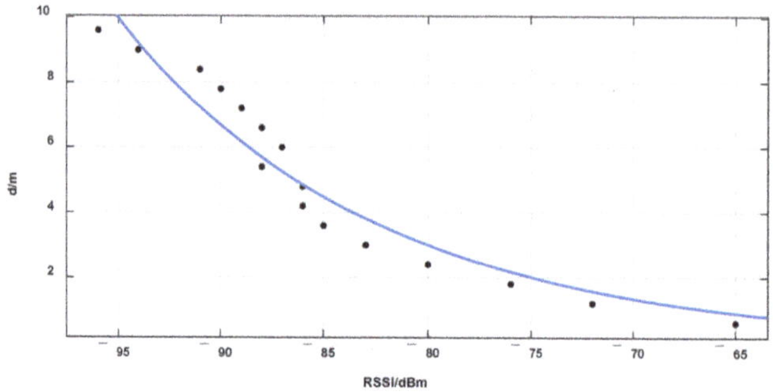

Figure 7. Attenuation model fitting.

After completing the above preparatory work, we collected the RSSIs of 42 fixed positions in the scene and calculated the coordinates of each position using Equation (8).

4. Results

Equation (12) (RMSE) and Equation (13) (ME) are used to calculate the overall accuracy of the 42 positioning points.

$$RMSE = \sqrt{\frac{\sum_{i=1}^{M} [(x_i - x_i^*)^2 + (y_i - y_i^*)^2]}{M}} \tag{12}$$

$$ME = \frac{\sum_{i=1}^{M} \sqrt{(x_i - x_i^*)^2 + (y_i - y_i^*)^2}}{M} \tag{13}$$

where, (x_i, y_i) is the true value of the coordinate, (x_i^*, y_i^*) is the positioning results, and M is the number of measurement points (42). We first calculated the positioning accuracy when selecting 3~18 Beacons by three comparison methods (Figure 8, Table 1).

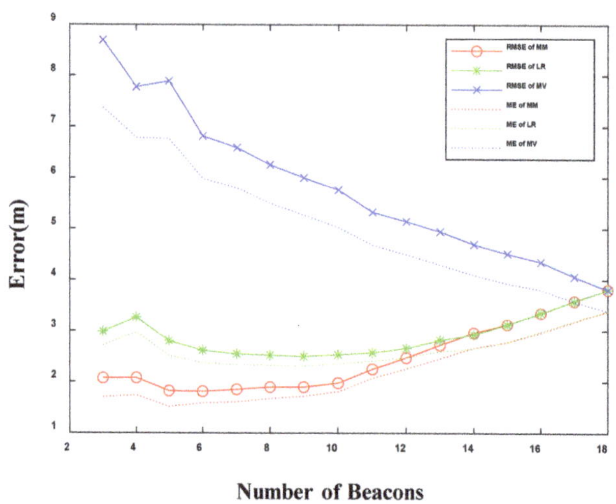

Figure 8. Errors when selecting different amounts of Beacons.

Table 1. Comparison of method positioning accuracy.

Errors (m)	3Beacons	4Beacons	5Beacons	6Beacons	7Beacons	8Beacons	9Beacons	10Beacons
RMSE (MM)	2.07	2.08	1.82	1.81	1.85	1.90	1.90	1.98
RMSE (LR)	2.99	3.27	2.80	2.61	2.55	2.52	2.50	2.53
RMSE (MV)	8.70	7.78	7.89	6.82	6.59	6.26	6.00	5.76
ME (MM)	1.70	1.74	1.51	1.59	1.61	1.67	1.72	1.81
ME (LR)	2.71	2.97	2.51	2.37	2.34	2.32	2.31	2.36
ME (MV)	7.37	6.79	6.77	5.99	5.80	5.50	5.27	5.03
Errors (m)	**11Beacons**	**12Beacons**	**13Beacons**	**14Beacons**	**15Beacons**	**16Beacons**	**17Beacons**	**18Beacons**
RMSE (MM)	2.25	2.47	2.72	2.96	3.13	3.35	3.59	3.81
RMSE (LR)	2.57	2.66	2.82	2.92	3.12	3.35	3.59	3.81
RMSE (MV)	5.32	5.14	4.95	4.70	4.51	4.35	4.07	3.81
ME (MM)	2.07	2.26	2.46	2.67	2.78	2.96	3.18	3.39
ME (LR)	2.41	2.47	2.60	2.65	2.77	2.99	3.19	3.39
ME (MV)	4.68	4.50	4.29	4.10	3.93	3.81	3.58	3.39

For the RD method, we set up three different selection ratios for verification. The optimal selection strategies of the above three comparison methods and the results of the RD method are shown in Table 2.

Table 2. Comparison methods positioning accuracy.

Errors (m)	RMSE	ME
MM (Beacons = 6)	1.81	1.59
LR (Beacons = 9)	2.50	2.31
MV (Beacons = 18)	3.81	3.39
RD (Beacons = 7 (60%, 60%))	1.88	1.73
RD (Beacons = 5 (50%, 50%))	**1.66**	**1.51**
RD (Beacons = 4 (40%, 40%))	1.82	1.68

From Tables 1 and 2, it can be seen that the RD method delivers the most accurate positioning when 5 Beacons are selected (an RMSE of 1.66 m and a ME of 1.51 m). Meanwhile, the MM, LR and MV methods deliver their most accurate positioning when 6, 9 and 18 Beacons are selected. The other two ratio choices (Beacons = 4, 7) verified in this paper are similar to the optimal choice of MM. Thus, using the method proposed in this paper for Beacons selection and positioning can reduce RMSE by 8.3%, 33.6% and 56.4% compared with MM, LR and MV. We also calculated the cumulative distribution of errors (Figure 9) and error frequency (Figure 10) for the RD method, the best case with the MM method (Beacons = 6), the LR method (Beacons = 9) and the MV method (Beacons = 18). The results showed that the RD method achieved a better positioning accuracy with a smaller number of Beacons. Using the RD method showed a more significant improvement compared with the other three methods, especially in reducing the occurrence of fluctuation errors (Figure 9). Specifically, 95.24% of the 42 points were located with an error of less than 3 m (Beacons = 5).

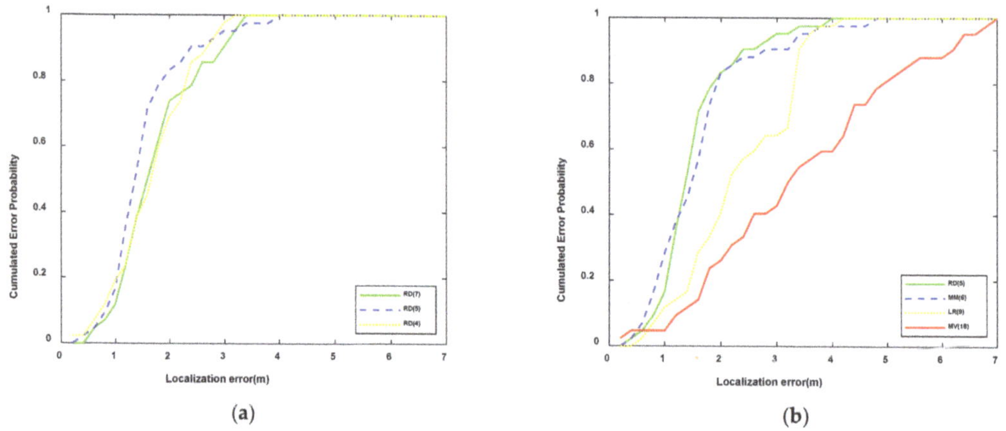

Figure 9. Position error CDF. (**a**) CDF of proposed. (**b**) Optimal choices CDF comparison.

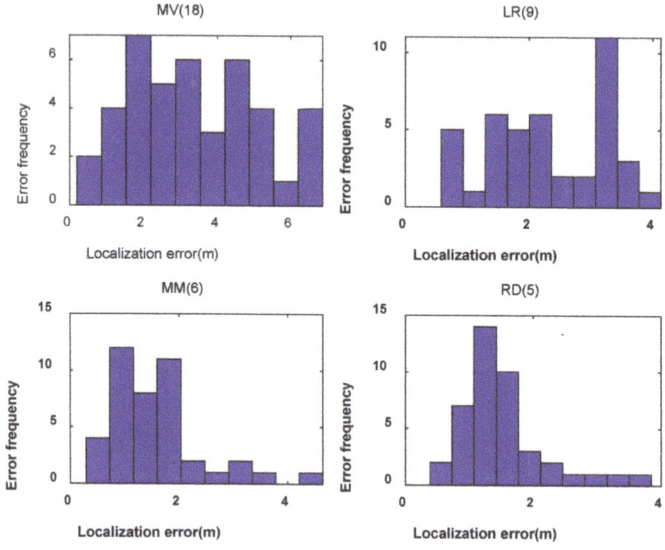

Figure 10. Error distribution frequency.

In order to show the error of each target point more clearly, we draw the plane distance graph between the results of the four methods and the target points, as well as the RMSE of each point (Figure 11). It is not difficult to see that the method RD in this paper shows more accurate results in both in-plane distribution and RMSE. Especially when the Beacon's performance selected by other methods fluctuates violently, it still maintains high robustness.

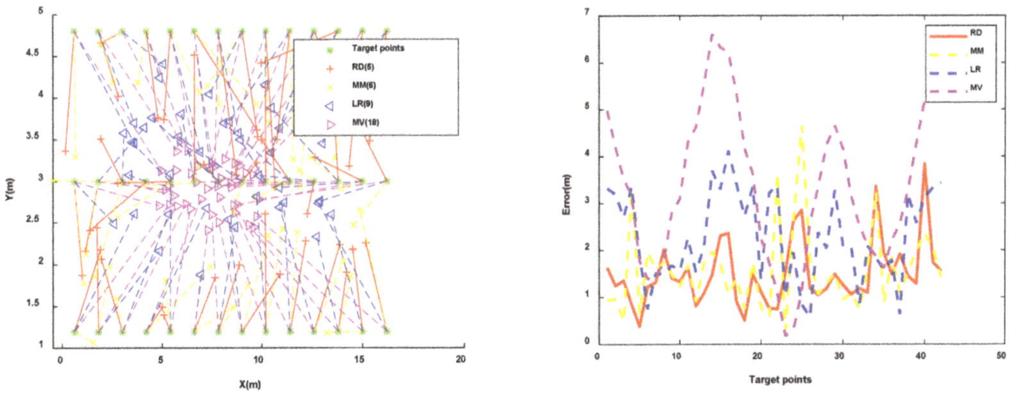

(**a**) (**b**)

Figure 11. Error of each point. (**a**) Plane distribution of errors. (**b**) RMSE of each point.

The main contributions of this paper are as follows: (1) We verified that the choice of Beacons in a complex indoor environment will affect the positioning accuracy; (2) using more Beacons does not straightforwardly deliver more accurate indoor positioning; (3) relying solely on the absolute RSSI cannot avoid the influence of pedestrians on the signals; (4) we proposed a Beacons selection strategy that takes into account the real-time abnormal states of the stations, and verified its superiority and reliability through experiments.

5. Conclusions

In order to solve the problem of fast selection of positioning signals in a complex indoor environment, we proposed a novel method to dynamically select Beacons based on real-time signal anomaly detection results and RSSI (RD). The GN nonlinear least squares algorithm was then used for indoor positioning. Experiments showed that the method proposed in this paper can effectively overcome the interference of pedestrians on signals, and select the best Beacons for indoor positioning in real-time. Compared with MM, LR, and MV methods, the RD method can greatly improve positioning accuracy and stability. Further, the number of Beacons required can be greatly decreased.

Author Contributions: Conceptualization, Y.G.; methodology, Y.G. and J.Z.; formal analysis, S.D. and G.X.; data curation, Y.G. and S.D.; writing—original draft preparation, Y.G.; writing—review and editing, Y.G., J.Z. and F.G. All authors have read and agreed to the published version of the manuscript.

Funding: This research was supported by the Key Research and Development Program of Hubei Province (Grant No. 2022BAA054), and the Natural Science Foundation of Hubei Province for Distinguished Young Scholars (Grant No. 2021CFA039).

Data Availability Statement: The data presented in this study are available on request from the corresponding author. The data are not publicly available due to privacy.

Acknowledgments: The author thanks the Infrastructure Department of Nanjing Forestry University for providing some plane drawings of the experimental area.

Conflicts of Interest: The authors declare no conflict of interest.

References

1. Yassin, A.; Nasser, Y.; Awad, M.; Al-Dubai, A.; Liu, R.; Yuen, C.; Raulefs, R.; Aboutanios, E. Recent Advances in Indoor Localization: A Survey on Theoretical Approaches and Applications. *IEEE Commun. Surv. Tutor.* **2017**, *19*, 1327–1346. [CrossRef]
2. Farahsari, P.S.; Farahzadi, A.; Rezazadeh, J.; Bagheri, A. A Survey on Indoor Positioning Systems for IoT-Based Applications. *IEEE Internet Things J.* **2022**, *9*, 7680–7699. [CrossRef]
3. Hegarty, C.J.; Chatre, E. Evolution of the Global Navigation Satellite System (GNSS). *Proc. IEEE* **2008**, *96*, 1902–1917. [CrossRef]

4. Asaad, S.M.; Maghdid, H.S. A Comprehensive Review of Indoor/Outdoor Localization Solutions in IoT era: Research Challenges and Future Perspectives. *Comput. Netw.* **2022**, *212*, 109041. [CrossRef]
5. Mendoza-Silva, G.M.; Torres-Sospedra, J.; Huerta, J. A Meta-Review of Indoor Positioning Systems. *Sensors* **2019**, *19*, 4507. [CrossRef] [PubMed]
6. Vanclooster, A.; Van de Weghe, N.; De Maeyer, P. Integrating Indoor and Outdoor Spaces for Pedestrian Navigation Guidance: A Review. *Trans. GIS* **2016**, *20*, 491–525. [CrossRef]
7. Yang, C.C.; Shao, H.R. WiFi-Based Indoor Positioning. *IEEE Commun. Mag.* **2015**, *53*, 150–157. [CrossRef]
8. Chen, X.; Li, H.; Zhou, C.Y.; Liu, X.; Wu, D.; Dudek, G. Fidora: Robust WiFi-Based Indoor Localization via Unsupervised Domain Adaptation. *IEEE Internet Things J.* **2022**, *9*, 9872–9888. [CrossRef]
9. Faragher, R.; Harle, R. Location Fingerprinting with Bluetooth Low Energy Beacons. *IEEE J. Sel. Areas Commun.* **2015**, *33*, 2418–2428. [CrossRef]
10. Aranda, F.J.; Parralejo, F.; Alvarez, F.J.; Paredes, J.A. Performance analysis of fingerprinting indoor positioning methods with BLE. *Expert Syst. Appl.* **2022**, *202*, 117095. [CrossRef]
11. Alarifi, A.; Al-Salman, A.; Alsaleh, M.; Alnafessah, A.; Al-Hadhrami, S.; Al-Ammar, M.A.; Al-Khalifa, H.S. Ultra Wideband Indoor Positioning Technologies: Analysis and Recent Advances. *Sensors* **2016**, *16*, 707. [CrossRef] [PubMed]
12. Yang, B.; Li, J.; Shao, Z.P.; Zhang, H. Robust UWB Indoor Localization for NLOS Scenes via Learning Spatial-Temporal Features. *IEEE Sens. J.* **2022**, *22*, 7990–8000. [CrossRef]
13. Rogel, N.; Raphaeli, D.; Bialer, O. Time of Arrival and Angle of Arrival Estimation Algorithm in Dense Multipath. *IEEE Trans. Signal Process.* **2021**, *69*, 5907–5919. [CrossRef]
14. He, J.J.; So, H.C. A Hybrid TDOA-Fingerprinting-Based Localization System for LTE Network. *IEEE Sens. J.* **2020**, *20*, 13653–13665. [CrossRef]
15. Ye, H.Y.; Yang, B.; Long, Z.Q.; Dai, C.H. A Method of Indoor Positioning by Signal Fitting and PDDA Algorithm Using BLE AOA Device. *IEEE Sens. J.* **2022**, *22*, 7877–7887. [CrossRef]
16. Wu, S.X.; Huang, W.X.; Li, M.; Xu, K. A Novel RSSI Fingerprint Positioning Method Based on Virtual AP and Convolutional Neural Network. *IEEE Sens. J.* **2022**, *22*, 6898–6909. [CrossRef]
17. Yang, B.; Guo, L.Y.; Guo, R.J.; Zhao, M.M.; Zhao, T.T. A Novel Trilateration Algorithm for RSSI-Based Indoor Localization. *IEEE Sens. J.* **2020**, *20*, 8164–8172. [CrossRef]
18. Shang, S.; Wang, L.X. Overview of WIFI fingerprinting-based indoor positioning. *IET Commun.* **2022**, *16*, 725–733. [CrossRef]
19. Xue, W.X.; Yu, K.G.; Li, Q.Q.; Zhou, B.D.; Zhu, J.S.; Chen, Y.W.; Qiu, W.N.; Hua, X.H.; Ma, W.; Chen, Z.P. Eight-Diagram Based Access Point Selection Algorithm for Indoor Localization. *IEEE Trans. Veh. Technol.* **2020**, *69*, 13196–13205. [CrossRef]
20. Zhang, W.; Yu, K.G.; Wang, W.X.; Li, X.M. A Self-Adaptive AP Selection Algorithm Based on Multiobjective Optimization for Indoor WiFi Positioning. *IEEE Internet Things J.* **2021**, *8*, 1406–1416. [CrossRef]
21. Youssef, M.A.; Agrawala, A.; Shankar, A.U. WLAN location determination via clustering and probability distributions. In Proceedings of the First IEEE International Conference on Pervasive Computing and Communications (Percom 2003), Washington, DC, USA, 23–26 March 2003; pp. 143–150. [CrossRef]
22. Li, J.H.; Gao, X.X.; Hu, Z.Y.; Wang, H.J.; Cao, T.; Yu, L. Indoor Localization Method Based on Regional Division with IFCM. *Electronics* **2019**, *8*, 559. [CrossRef]
23. Li, C.G.; Huang, H.; Liao, B.W. An Improved Fingerprint Algorithm with Access Point Selection and Reference Point Selection Strategies for Indoor Positioning. *J. Navig.* **2020**, *73*, 1182–1201. [CrossRef]
24. Chen, Y.Q.; Yang, Q.; Yin, J.; Chai, X.Y. Power-efficient access-point selection for indoor location estimation. *IEEE Trans. Knowl. Data Eng.* **2006**, *18*, 877–888. [CrossRef]
25. Laitinen, E.; Lohan, E.S. On the Choice of Access Point Selection Criterion and Other Position Estimation Characteristics for WLAN-Based Indoor Positioning. *Sensors* **2016**, *16*, 737. [CrossRef]
26. Deng, Z.A.; Ma, L.; Xu, Y.B. Intelligent AP Selection for Indoor Positioning in Wireless Local Area Network. In Proceedings of the 2011 6th International Icst Conference on Communications and Networking in China (Chinacom), Harbin, China, 17–19 August 2011; pp. 257–261.
27. Cui, X.R.; Wang, M.Y.; Li, J.; Ji, M.Q.; Yang, J.; Liu, J.H.; Huang, T.P.; Chen, H.H. Indoor Wi-Fi Positioning Algorithm Based on Location Fingerprint. *Mobile Netw. Appl.* **2021**, *26*, 146–155. [CrossRef]
28. Zhang, H.; Wang, Z.K.; Xia, W.C.; Ni, Y.Y.; Zhao, H.T. Weighted Adaptive KNN Algorithm with Historical Information Fusion for Fingerprint Positioning. *IEEE Wirel. Commun. Lett.* **2022**, *11*, 1002–1006. [CrossRef]
29. Fang, S.H.; Lin, T.N. Principal Component Localization in Indoor WLAN Environments. *IEEE. Trans. Mob. Comput.* **2012**, *11*, 100–110. [CrossRef]
30. Guo, Y.; Zheng, J.Z.; Zhu, W.Z.; Xiang, G.Q.; Di, S.N. iBeacon Indoor Positioning Method Combined with Real-Time Anomaly Rate to Determine Weight Matrix. *Sensors* **2021**, *21*, 120. [CrossRef]
31. Liu, F.T.; Ting, K.M.; Zhou, Z.H. Isolation-Based Anomaly Detection. *Acm Trans. Knowl. Discov. Data* **2012**, *6*, 1–39. [CrossRef]
32. Li, X.R. RSS-based location estimation with unknown pathloss model. *IEEE Trans. Wirel. Commun.* **2006**, *5*, 3626–3633. [CrossRef]
33. Gratton, S.; Lawless, A.S.; Nichols, N.K. Approximate Gauss-Newton methods for nonlinear least squares problems. *SIAM J. Optim.* **2007**, *18*, 106–132. [CrossRef]

remote sensing

Technical Note

The Bias in PPP-B2b Real-Time Clock Offset and the Strategy to Reduce It

Jinhuo Liu [1,2], Chengpan Tang [1,*], Shanshi Zhou [1], Xiaogong Hu [1], Yufei Yang [3], Jianhua Yang [1,2] and Yuchen Liu [1,2]

1 Shanghai Astronomical Observatory, Chinese Academy of Sciences, Shanghai 200030, China
2 University of Chinese Academy of Sciences, Beijing 100049, China
3 Beijing Satellite Navigation Center, Beijing 100094, China
* Correspondence: cptang@shao.ac.cn

Abstract: Precise point positioning can provide accurate coordinates to users without reference stations, and the high-precision real-time clock offset product is necessary for real-time precise point positioning application. As an integral part of the third generation BeiDou Navigation Satellite System, Precision Point Positioning Service provides dual systems (BDS-3 and GPS) real-time PPP services with centimeter- and decimeter-level accuracy for static and kinematic positioning users around China, respectively. However, there exist inconsistent biases in the clock offset of Precision Point Positioning Service, which will negatively affect the positioning and timing performance of the service. By comparing with the post-processing clock offset, this paper verifies that the broadcast clock offset has smaller and more stable biases in the long term and proposes a regional clock offset estimation strategy using broadcast clock offset for a priori constraint. The results show that the new algorithm can effectively reduce the bias in PPP-B2b clock offset. The new clock offset product could improve convergence speed by 25% and 10% in the horizontal and vertical directions. For positioning accuracy, the improvement is 22% and 17%. The absolute error of timing can also be reduced by 60%.

Keywords: BeiDou-3 Navigation Satellite System; real-time clock offset; PPP-B2b service

Citation: Liu, J.; Tang, C.; Zhou, S.; Hu, X.; Yang, Y.; Yang, J.; Liu, Y. The Bias in PPP-B2b Real-Time Clock Offset and the Strategy to Reduce It. *Remote Sens.* **2022**, *14*, 4569. https://doi.org/10.3390/rs14184569

Academic Editors: Giuseppe Casula, Zhetao Zhang and Wenkun Yu

Received: 13 August 2022
Accepted: 12 September 2022
Published: 13 September 2022

1. Introduction

Precise point positioning (PPP) can provide accurate coordinates to users without reference stations [1] and is widely used in scientific studies, such as oceanography [2] and earthquake early warning [3]. The precision orbit and clock offset for PPP are usually provided by IGS analysis centers. The consistency of GPS, GLONASS, Galileo, and Beidou-2 satellite post-processed clock offset are 2 cm, 5 cm, 5 cm, and 10 cm, respectively [4]. However, the release time is generally delayed by several days. In recent years, with the exploration of GNSS applications, precise products are needed more timely for users of some emerging fields, such as LEO satellite orbit determination [5], unmanned aerial vehicle photogrammetry [6], GNSS meteorology [7], etc.

As an integral part of the third generation BeiDou Navigation Satellite System (BDS-3), Precision Point Positioning Service (PPP-B2b) provides dual system (BDS-3 and GPS) real-time PPP services with centimeter- and decimeter-level accuracy for static and kinematic positioning users, respectively, around China [8,9]. Through the GEO satellite in BDS-3, users of PPP-B2b can obtain PPP-B2b precise clock offset (B2bCO) products with a precision better than 0.3 ns.

Previously, some scholars have evaluated the B2bCO by comparing it with the post-processed product of the analysis center [10,11]. Although there are some numerical differences between these assessments, qualitative conclusions can be drawn: Compared with external products, the B2bCO of each satellite has a different bias, ranging from decimeter to meter, though it is precise enough to support the official positioning accuracy standard. This bias has not been studied in depth before. It is generally believed that the bias in clock offset may be related to the receiver type [12], algorithm, and the calibration of equipment delay.

The satellite clock offset parameters in the broadcast message of BDS-3 Positioning, Navigation and Timing Service (PNTS) is estimated differently from PPP-B2b. There are two time synchronization methods used in the calculation. The L-band two-way satellite time and frequency transfer (TWSTFT) establishes the satellite ground connection, and the time division multiple access inter-satellite link (ISL) clock offset estimation establishes the inter-satellite connection [13,14]. The former has been successfully applied in the construction of the Beidou-2 regional navigation system [15], but it can only cover 40% arc of the MEO satellite with regional observation [16]. The latter effectively improves the clock offset estimation accuracy of outland BDS-3 satellites without satellite-ground observation. According to the long-term analysis, there is no apparent bias between the clock offset of PNTS and the post-processing.

Based on the characteristics of PNTS clock offset and B2bCO, a regional network clock offset estimation strategy with broadcast clock offset for priori constraint is proposed. According to the strategy, we reconstructed the computing process of PPP-B2b, and obtained then new clock offset product. The results show that the new algorithm effectively reduces the bias in B2bCO. The new clock offset product can improve the convergence speed by about 25% and 10%, and the positioning accuracy after convergence by 22% and 17% in the horizontal and vertical directions, respectively. It can also reduce the absolute error of timing by 60%.

This paper is organized as follows. First of all, in Section 2, the estimation process and the bias of PPP-B2b clock offset is briefly described. Section 3 introduces the new clock offset estimation process and evaluates the new clock offset. In Section 4, the positioning performance and timing performance of the new clock offset and B2bCO are compared and analyzed. The summary and discussion are in the final part.

2. The Estimation Process and Bias of PPP-B2b Clock Offset Product

2.1. Estimation Process of PPP-B2b Clock Offset Product

The estimation process of the PPP-B2b clock offset product is given in Figure 1. The process is roughly divided into four modules, multi-satellite orbit determination module, real-time OC (Observation minus Calculation) calculation module, clock offset estimation module, and correction calculation module.

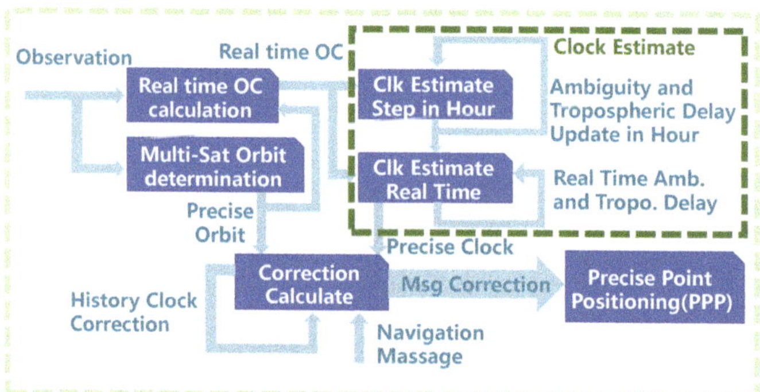

Figure 1. Estimation Process of PPP-B2b Clock Offset Product.

In a whole process, the orbit determination was carried out at first using the observation data of the previous three days. The obtained satellite's precise orbit and prediction orbit for several days are saved in memory. Driven by the real-time observation data stream, the real-time OC calculation module uses the orbit in memory to deduct various errors in the observation:

$$\begin{cases} l^s_{PC,r} = PC^s_r - \rho^s_r - T^s_r - (b^s_{IF} + b_{r,IF}) - \epsilon_{PC} \\ l^s_{LC,r} = LC^s_r - \rho^s_r - T^s_r - (\phi^s_{IF} + \phi_{r,IF}) - \lambda N^s_{IF,r} - \epsilon_{LC} \end{cases} \quad (1)$$

The r and s denote the specific receiver and satellite. The ρ, T, and N represent the geometric distance, tropospheric delay, and ambiguity, respectively. The l_{PC} and l_{LC} is the OC (including satellite and receiver clock offsets) calculated in this module. The real-time OC stream is injected into the clock offset estimation module. The clock offset estimation module includes two sub-modules: hourly estimation and real-time estimation. The hourly estimation sub-module caches the residual stream, filters it in hours, and sends the troposphere and ambiguity information to the real-time estimation module. With the constraint of such corresponding information, the real-time clock offset estimation module, which is driven by the residual stream, estimates the real-time clock offset and saves this unoptimized clock offset into the memory. Finally, using the clock offset, precise orbit, and navigation messages, the correction products are generated after judgment and elimination with historical data in the correction product calculation module.

Among these modules, the multi-satellite orbit determination supported by ISL has been thoroughly studied, and its stability and the accuracy of prediction products can be guaranteed [17–20]. The crux of PPP-B2b is to generate continuous real-time clock offset products with high precision stably. The combination of hourly and real-time filters ensures the stable convergence of the clock offset estimation based on regional observation. Moreover, the estimation process of the PPP-B2b clock offset product is distributed in three independent modules: real-time OC calculation, clock offset estimation, and correction calculation. Even if catastrophic failures occur in the clock offset estimation module (the most probable), the upstream module (real-time OC calculation) and the downstream module (correction calculation) will not be involved. The correction products can be constantly predicted in the correction calculation module within 2 min, and the uninterrupted real-time residual stream facilitates the rapid resumption of the clock offset estimation module.

2.2. PPP-B2b Clock Correction and Bias of Clock Offset

The GEO satellite of BDS-3 broadcasts four types of PPP-B2b correction products to users at a rate of 500 bps [8]. The clock correction of all satellites in the mask is broadcast to the user through Message Type 4 in a period of 6 seconds. If the clock offset of a specific satellite is invalid, it is assigned the value of −26.2128. Through the PPP-B2b clock correction sent by GEO and the broadcast clock offset (BRCO) of the corresponding satellite, users can obtain B2bCO with:

$$C^i_{\mathrm{B2bCO}} = C^i_{\mathrm{BRCO}} - \frac{C^i_0}{c} \quad (2)$$

C^i_0 is the clock correction of PPP-B2b at epoch i, while C^i_{B2bCO} and C^i_{BRCO} are the B2bCO and BRCO at this time; c is the speed of light.

From August 2020 (PPP-B2b on service) to July 2021, a total of 320 days of long-term B2bCO were used for evaluation in order to analyze the value and change of the bias. The post-processed clock offset (PPCO), calculated with the same orbit, PCO, DCB, and the observation data of the same type receiver, was set as the benchmark for evaluation to ensure other effects were eliminated. The bias of B2bCO on a certain day can be extracted as:

$$\begin{cases} \Delta C^i_{\mathrm{B2bCO}} = C^i_{\mathrm{B2bCO}} - C^i_{\mathrm{PPCO}} \\ \Delta C_{\mathrm{B2bCO}} = \frac{1}{n}\sum^n_{i=1} \Delta C^i_{\mathrm{B2bCO}} \end{cases} \quad (3)$$

i is the i_{th} valid epoch of the day (B2bCO and PPCO is both valid); n is the total number of valid epochs in this day; C^i_{B2bCO} and C^i_{PPCO} is the B2bCO and PPCO of the corresponding epoch; $\Delta C_{\mathrm{B2bCO}}$ is the bias between B2bCO and PPCO. For a satellite, if its clock offset has two or more effective arcs in a day, the longest one is selected for statistics.

In Figure 2, the X-axis is the number of days from 1 August 2020, the blue dot is the bias of all satellites, and the red line is the 3σ confidence interval counted by day. The bias still existed after other effects were removed, which indicated that the estimation algorithm or receiver characteristic is a significant source of this bias. The bias can be considered in two parts. The biases of all satellites have the same long-term variation trend, which ranges between ±3 m. After removing the long-term trend, the biases of satellites are not consistent, with the standard deviation (STD) σ_{Bia} of 0.315 m (1.05 ns).

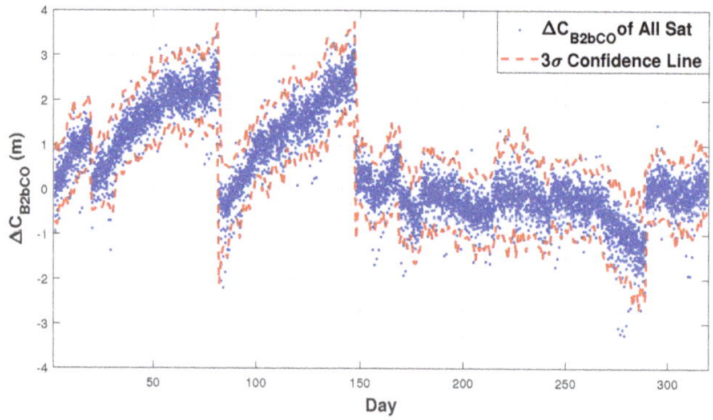

Figure 2. The bias of B2bCO changes in the long term.

Although most of the inconsistent biases in B2bCO will be eventually absorbed by ambiguity coefficients in PPP, it will still affect the convergence time and final accuracy of PPP. For static PPP, the impact can be approximately estimated as:

$$\Delta_P = \frac{SigLC}{SigPC} * PDOP * \sigma_{Bia} \tag{4}$$

$SigPC$ and $SigLC$ are the a priori observation noise of pseudo-range and carrier phase, respectively; $PDOP$ is the average position dilution of precision; Δ_P is the positioning error caused by the inconsistent part of the bias, which is generally around 1 cm to 3 cm. For kinematic PPP, these biases will lead to the divergence of the positioning with the entry and exit of satellites. The order of the divergence is centimeters to decimeters.

The users' receiver clock offset will absorb the consistent part of the bias. It can be inferred from the above that the timing error of PPP-B2b is around ±10 ns, although the timing performance of PPP-B2b is not given officially.

3. Real-Time Algorithm of B2bCO with BRCO Constraint

BRCO and B2bCO are calculated with two different time synchronization systems. BRCO uses L-band TWSTFT and Ka-band Full-ISL to estimate satellite clock offset, while B2bCO is solved with pseudo-range and carrier phase of navigation signal. We are concerned about whether the BRCO is also biased compared with PPCO to confirm the bias is from B2bCO. The comparison is carried out in the same way between BRCO and PPCO:

$$\begin{cases} \Delta C^i_{\mathrm{BRCO}} = C^i_{\mathrm{BRCO}} - C^i_{\mathrm{PPCO}} \\ \Delta C_{\mathrm{BRCO}} = \frac{1}{n}\sum_{i=1}^{n} \Delta C^i_{\mathrm{BRCO}} \end{cases} \tag{5}$$

i is the i_{th} valid epoch of the day; C^i_{BRCO} and C^i_{PPCO} is the BRCO and PPCO of the corresponding epoch; $\Delta C^i_{\mathrm{BRCO}}$ is the difference between BRCO and PPCO on epoch i, while ΔC_{BRCO} is its daily statistic result.

In Figure 3, the X-Axis is the number of days from 1 August 2020, the blue dot is the difference between BRCO and PPCO of all satellites, and the red line is the 3σ confidence interval counted by day. Compared with B2bCO, there is almost no long-term trend in the difference between BRCO and PPCO. After removing the long-term trend, its STD is around 0.32 m (1.06 ns), slightly larger than B2bCO. It should be noted that this is a statistical result that aims to illustrate that BRCO has a more negligible bias. Actually, B2bCO is much more precise than BRCO and has much better positioning performance. Because of its smaller bias, the BRCO could be used as a priori constraint for the estimation filter of B2bCO and using more appropriate constraints on satellite clock coefficient, so as to reduce the bias in B2bCO.

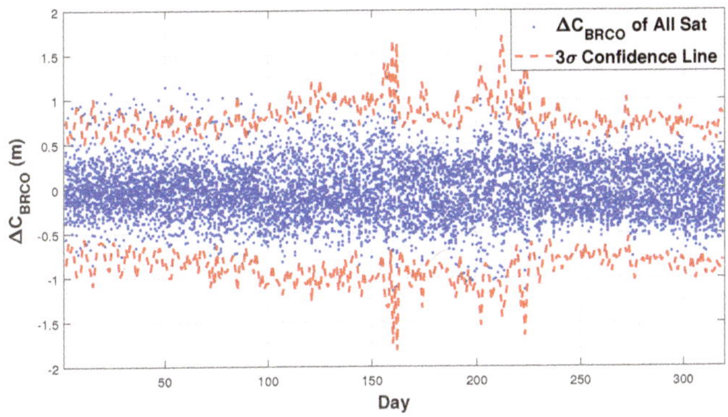

Figure 3. The difference between BRCO and PPCO.

3.1. The Design of Estimation Process

We reconstructed the process of PPP-B2b as shown in Figure 4. The correction calculation module is divided into long-term and short-term correction calculation modules according to the update period of the output message. Message Type 1 to 3 (mask, orbit correction, and DCB) is calculated in the long-term correction calculation module, while Message Type 4 (clock correction) is calculated in the short-term module. The update period is 48 s and 6 s, respectively. The short-term correction calculation module is combined with the clock offset estimation module so that the historical information and statistical results can be used more flexibly and effectively. Since the stability of the estimation can be guaranteed after the prior information is used more appropriately, the hourly clock offset estimation module is removed to eliminate the hourly jump in B2bCO (around 1 cm to 3 cm) caused by it. Clock correction is now directly generated in the clock offset estimation module.

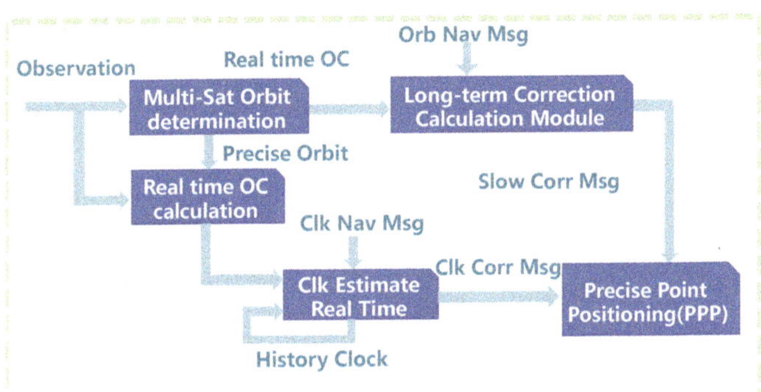

Figure 4. Reconstructed PPP-B2b clock offset estimation process.

In the clock offset estimation module, we use two methods to constrain the clock offset coefficients of satellites. When the clock offset sequence is less than 30 min, the BRCO is used to constrain the coefficients. When the clock offset sequence is longer than 30 min, the historical clock offset sequence is used to predict the clock offset on the current epoch and constrain the coefficients. Different priori noise is needed in different constraint methods.

3.2. Priori Noise of BRCO Constraint Method

The BDS-3 navigation broadcast is updated hourly. In order to restrict the clock coefficient appropriately, the RMS of $\Delta C^{i}_{\mathrm{BRCO}}$ (in Formula (5)) is counted according to the time to TOC (Time of Clock). As illustrated in Figure 5, the X-axis is the time to TOC, while the Y-axis is the RMS of $\Delta C^{i}_{\mathrm{BRCO}}$. The dots, which increase linearly, are the RMS of $\Delta C^{i}_{\mathrm{BRCO}}$ counted in minutes between BRCO and PPCO of all satellites. The solid line is the fitting result, and its intercept (P_0) and slope (P_1) are shown in Table 1.

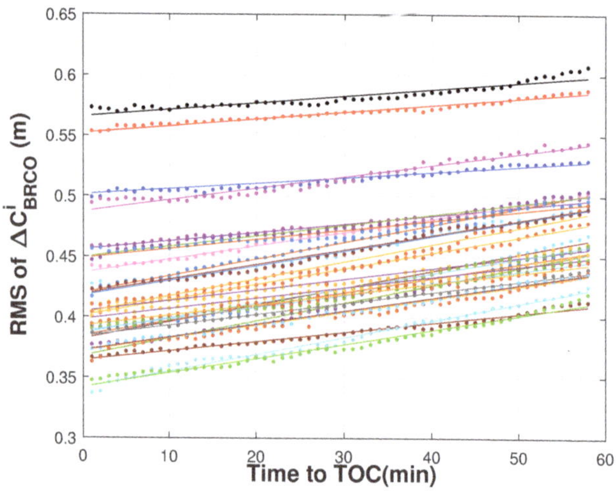

Figure 5. The BRCO errors increase with time.

Table 1. The fitting result of the BRCO error.

PRN	P_1	P_0	PRN	P_1	P_0
C19	0.0012	0.387	C34	0.0011	0.437
C20	0.0009	0.384	C35	0.0014	0.403
C21	0.0011	0.372	C36	0.0009	0.450
C22	0.0005	0.502	C37	0.0011	0.373
C23	0.0012	0.419	C38	0.0006	0.552
C24	0.0014	0.420	C39	0.0007	0.456
C25	0.0014	0.342	C40	0.0008	0.364
C26	0.0013	0.401	C41	0.0008	0.449
C27	0.0015	0.368	C42	0.0010	0.487
C28	0.0014	0.382	C43	0.0010	0.388
C29	0.0012	0.385	C44	0.0012	0.418
C30	0.0009	0.405	C45	0.0009	0.393
C32	0.0009	0.398	C46	0.0012	0.342
C33	0.0006	0.566	Average	0.0010	0.416

According to the fitting results, the average value of P_1 is 0.001. It can be inferred that the error of BRCO increases around 6cm within one hour, which is only about 13% of the average value of the initial error (P_0). We take the 3σ upper limit of P_0 and P_1 to estimate the BRCO error:

$$\delta_{\mathrm{BRCO}} = 0.583 + 0.0019dt; \tag{6}$$

dt is the time to TOC. δ_{BRCO} is the estimated error of BRCO at this time, which can be used as a priori noise of clock coefficients in the filter.

3.3. Priori Noise of Predict Clock Offset Constraint Method

When the clock offset sequence of a satellite is longer than 30 min, the current clock offset value can be predicted with historical information, and the predicted clock offset can be used as a priori constraint of the filter. We assume that the error of the clock offset is composed of the clock speed error and frequency white noise because this prediction is usually shorter than a few minutes. By comparing with the previous epoch, we can obtain the clock speed value $a_1(i)$ of the current epoch and store it. The clock offset in the new epoch can be predicted as:

$$\widetilde{T}_{n+1} = T_n + \frac{dt}{m}\sum_{i=n-m+1}^{n} a_1(i) \tag{7}$$

n is the current epoch; m is the number of epochs used in the prediction; dt is the time interval between epochs; T_n is the clock offset of current epoch, and $\widetilde{T}_{n+1}$ is the predicted clock offset of epoch $n+1$. Considering that the prediction error will be affected by clock drift and white noise, the relations of the errors are:

$$\begin{cases} \sigma_0 = \sqrt{\sigma_{a_2}^2 + \sigma_{\varepsilon}^2} = std(a_1(n-m+1:n)) \\ \sigma_{\widetilde{T}_{n+1},a_2} = \sqrt{\frac{3(m+1)}{m-1}}\sigma_{a_2} \\ \sigma_{\widetilde{T}_{n+1},\varepsilon} = \sigma_{\varepsilon} \\ \sigma_{\widetilde{T}_{n+1}} = \sqrt{\sigma_{\widetilde{T}_{n+1},a_2}^2 + \sigma_{\widetilde{T}_{n+1},\varepsilon}^2} \end{cases} \tag{8}$$

σ_0 is the STD of the clock speed of historical epochs, which is obtained by the saved sequence of a_1, while the $\sigma_{\widetilde{T}_{n+1}}$ is the prediction error of $\widetilde{T}_{n+1}$. The errors caused by clock drift and white noise are represented by subscripts a_2 and ε, respectively. The error caused by these two parts cannot be accurately estimated, but the following relationship can be obtained:

$$\sigma_0 < \sigma_{\widetilde{T}_{n+1}} < \sqrt{\frac{3(m+1)}{m-1}}\sigma_0 \tag{9}$$

Considering that the prediction is a positive feedback process, that is, too tight constraints will cause the clock offset sequence to be too linear, which leads to tighter constraints. In order to avoid triggering this process, the constraint needs to be loose enough (greater than three times of $\sqrt{\frac{3(m+1)}{m-1}}\sigma_0$), while it should also effectively constrain the clock coefficient. We set the constraint as $8\sigma_0$ after a trail.

3.4. Others Coefficients and Strategies of Filter

Except for clock offset, other coefficients of the filter are shown in Table 2. Consistent with the clock coefficients, all types of prior noise and process noise followed the three-sigma rule, such as the observation noise of pseudo-range and carrier phase. Since only observations within China are used, in order to improve the western coverage capability as much as possible, we set the conditions to start estimation as loose as possible. When a satellite is observed by the three receivers above the elevation of 3°, its clock offset starts to be estimated. The clock offset starts to be output after the estimated time is up to 30 min. We name the clock offset product as Broadcast Constraint Filtering Clock Offset (BCFCO).

Table 2. Others coefficients and strategies of the filter.

Type	Strategy/Coefficient
Observation	9 receivers in China
Date	22 March to 5 April 2022
System	BDS-3
Frequency Combine	B1I/B3I PC + LC
Elevation Limit	>3°
Interval	30 s
Orbit Product	PPP-B2b Orbit Product
Observation Noise	PC: 1 m; LC: 0.01 m
Meteorological Model	GPT2.1W + SAAS + VMF1
Trop. Esti. Model	Random Walk,Process Noise 1 cm/h
Rcv. Clock Esti. Model	WN, Priori Noise 2 m
Amb. Priori Noise	1.5 m
Esti. Method	Kalman Filter

Trop.: Troposphere; Esti.: Estimation; Sat.: Satellite; Rcv.: Receiver; WN: White Noise; Amb.: Ambiguity.

3.5. Analysis of BCFCO

With the PPCO as the standard, the bias and precision of BCFCO are discussed in this section. Compared with B2bCO, the accuracy improvement of BCFCO can be verified and analyzed.

$$\begin{cases} \Delta C^i_{\text{BCFCO}} = C^i_{\text{BCFCO}} - C^i_{\text{PPCO}} \\ \Delta C_{\text{BCFCO}} = \frac{1}{n}\sum_{i=1}^{n} \Delta C^i_{\text{BCFCO}} \end{cases} \tag{10}$$

The comparison between the bias of BCFCO and B2bCO is given in Figure 6. The X-axis is the number of hours from 00:00 on 22 March 2022; the blue dot is the bias of clock offset, and the red line is the 3σ confidence interval. The mean value of the bias of BCFCO is stable around 0, while the bias of B2bCO changes around -0.7 m. After removing the trend, the average STD of BCFCO is 0.231 m, while the average STD of B2bCO is 0.417 m. It can be confirmed that BCFCO is more accurate than B2bCO.

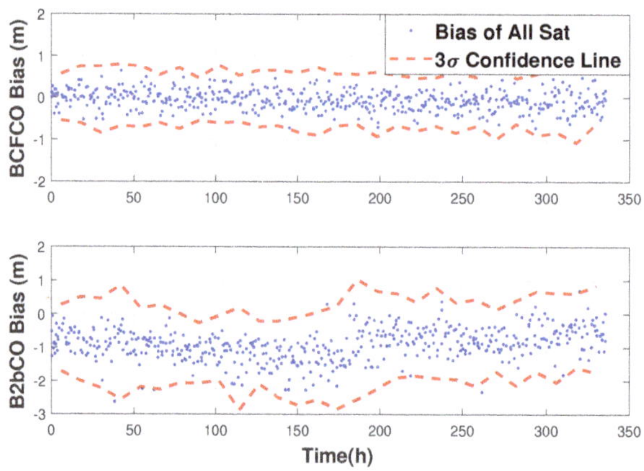

Figure 6. The bias of BCFCO and B2bCO.

The precision of BCFCO and B2bCO can be reflected by STD of $\Delta C^i_{\text{BCFCO}}$ and $\Delta C^i_{\text{B2bCO}}$ after the bias is eliminated. According to the results given in Table 3, the precision of the two kinds of clock offset is basically the same, which illustrates that the new clock offset estimation module can still work stably with appropriate constraints, even if the hourly clock estimation and post-processing module are removed.

Table 3. The precision (STD) of clock offset.

PRN	BCFCO (m)	B2bCO (m)	PRN	BCFCO (m)	B2bCO (m)
19	0.026	0.027	34	0.016	0.014
20	0.034	0.033	35	0.017	0.015
21	0.019	0.018	36	0.019	0.023
22	0.013	0.014	37	0.022	0.022
23	0.015	0.017	38	0.017	0.017
24	0.034	0.034	39	0.017	0.021
25	0.015	0.016	40	0.017	0.017
26	0.037	0.038	41	0.014	0.014
27	0.014	0.019	42	0.016	0.013
28	0.016	0.013	43	0.017	0.017
29	0.017	0.015	44	0.016	0.015
30	0.014	0.017	45	0.025	0.023
32	0.014	0.015	46	0.017	0.017
33	0.015	0.017	mean	0.020	0.021

4. Timing and Positioning Performance

In PPP, the error on clock offset will have different effects on the positioning results under the influence of various factors. In order to verify the positioning performance of BCFCO products, BCFCO and B2bCO products are compared to discuss the positioning convergence, positioning accuracy, and timing performance of the two products. The observation data of 16 receivers in China were used in the assessment from 22 March to 4 April 2022.

4.1. Convergence Performance of Kinematic Positioning

In order to obtain more samples to evaluate the convergence performance, the observation from 1:00 to 23:00 (BDT) each day is divided into 11 segments (2 h for each segment) for kinematic positioning, and the positioning accuracy (RMS) is counted according to the convergence time. In Figure 7, compared with B2bCO, better convergence performance could be achieved using BCFCO products. In the horizontal direction, the 15 min positioning error decreased from 0.38 m to 0.29 m, and the 30-min positioning error decreased from 0.30 m to 0.23 m; In the vertical direction, the 15-min positioning error decreased from 0.40 m to 0.36 m, and the 30-min positioning error decreased from 0.29 m to 0.25 m. The convergence performance in the horizontal direction is improved more significantly than in the vertical direction, which are 25% and 12%, respectively.

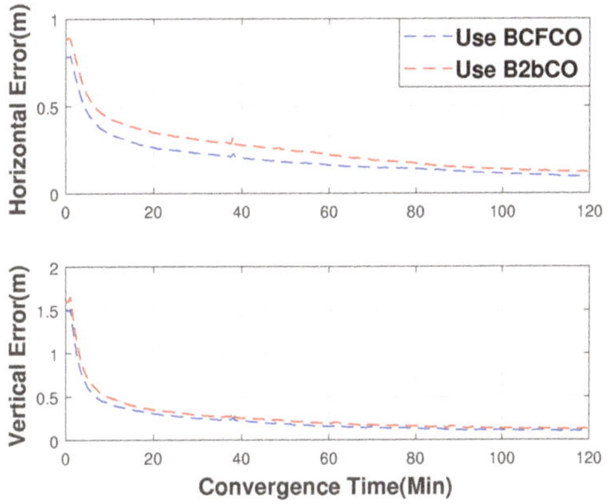

Figure 7. Convergence performance of BCFCO and B2bCO.

4.2. Accuracy of Kinematic Positioning

Observation data from 1:00 to 23:00 (BDT) is used in BDS-3 single system kinematic positioning, and 95% positioning error is counted from 3:00 to 23:00 each day. Considering that PPP-B2b is a region enhancement system, the positioning performance of receivers is discussed according to their region: Northeast, Northwest, and South of China. In addition, different types of receivers are deployed in some locations to compare the positioning improvement of different types of receivers. The statistical results are shown in Table 4.

Table 4. The 95% kinematic positioning accuracy after convergence.

RcvID	Horizontal Position Error			Vertical Position Error		
	BCFCO	B2bCO	Enhance	BCFCO	B2bCO	Enhance
NE1A	0.084	0.104	19%	0.137	0.160	15%
NE2A	0.061	0.086	29%	0.105	0.116	9%
NE3A	0.066	0.102	35%	0.122	0.141	14%
AVE_{NEA}	0.070	0.097	28%	0.121	0.139	13%
NW1A	0.179	0.232	23%	0.307	0.334	8%
NW2A	0.101	0.135	25%	0.164	0.260	37%
NW3A	0.078	0.107	27%	0.112	0.144	22%
AVE_{NWA}	0.119	0.158	24%	0.194	0.246	21%
SC1A	0.067	0.145	54%	0.184	0.256	28%
SC2A	0.045	0.074	39%	0.150	0.206	27%
SC3A	0.060	0.078	23%	0.118	0.159	26%
AVE_{SCA}	0.057	0.099	42%	0.151	0.207	27%
NE1B	0.151	0.175	13%	0.162	0.205	21%
NE2B	0.084	0.101	17%	0.129	0.169	24%
AVE_{NEB}	0.104	0.137	24%	0.156	0.206	24%
NW1B	0.159	0.158	−1%	0.342	0.329	−4%
NW2B	0.095	0.104	9%	0.153	0.163	6%
NW3B	0.179	0.213	16%	0.244	0.274	11%
AVE_{NWB}	0.144	0.158	9%	0.246	0.255	4%
SC1B	0.124	0.187	34%	0.219	0.277	21%
SC2B	0.109	0.115	5%	0.173	0.190	9%
AVE_{SCB}	0.126	0.153	18%	0.213	0.241	12%

The first two digits of the RcvID represent the region (NE is the northeast, NW is the northwest, and the SC is the South). The third digit represents the distance between the receiver and the core region of China (1 represents the receiver in the region farthest away from the core region). The receivers with the same first three letters have the same position. The fourth letter represents the type of receiver ('A' represents the receiver of the same type as the estimation receiver, and 'B' is another type).

In general, the receivers in the northeast can obtain the best positioning accuracy, followed by the south, and the worst in the northwest, which is related to the moving direction of BDS-3 satellites. Most MEO satellites enter China from the West, while IGSO satellites and few MEO satellites will enter from the south. It usually takes dozens of minutes for a satellite from reaching observation conditions to providing services, which results in a greater PDOP in the entry border area. Furthermore, the closer the receiver is to the core area of China, the more obvious the improvement on positioning accuracy is with BCFCO products, except the SC1 location. In all regions, compared with B-type receivers, the A-type receivers can obtain better positioning performance and greater performance improvement using BCFCO, which may be due to the code bias. The average improvement on the positioning accuracy of BCFCO is 22% and 17% in horizontal and vertical directions, respectively.

4.3. Timing Performance

With two calibrated timing receivers TM1A and TM1B, the timing performance of BCFCO and B2bCO can be evaluated by the static PPP. Observation data from 1:00 to 23:00 (BDT) is used and the error of clock offset and clock rate is counted from 3:00 to 23:00 in each day.

The timing performance of two receivers using different products is given in Table 5. Using BCFCO product, the clock offset error of receivers is about 0.9 ns (0.25 m), while it is 2.3 ns (0.69 m) and 3.7 ns (1.17 m) using B2bCO product, which is consistent with the bias of clock offset mentioned in Section 3.5. The clock rate error of the two products is basically the same, about 0.012 ns (0.004 m) per 30 s (step of PPP). This error is generally consistent with receiver observation noise of the carrier phase that determines the upper limit of the clock offset precision. The accurate clock rate of receivers can be obtained with both two products.

Table 5. Statistical results of timing performance of TM1A and TM1B.

STAID	Clock Offset Error (ns)		Clock Rate Error (ns/30 s)	
	BCFCO	B2bCO	BCFCO	B2bCO
TM1A	0.76	2.27	0.010	0.012
TM1B	0.88	3.74	0.013	0.013

5. Summary and Conclusions

This paper focuses on the bias that exists in PPP-B2b precision clock offset products (B2bCO). Firstly, the estimation process of PPP-B2b clock offset product is introduced, and then bias in B2bCO is analyzed with long-term historical data and post-processing clock offset (PPCO). According to the analysis, we propose a regional clock offset estimation strategy using broadcast clock offset for a priori constraint and get the new clock offset product (BCFCO). By comparing with the B2bCO, the performance improvement of BCFCO on the accuracy of clock offset, dynamic positioning, and timing is discussed.

The bias in B2bCO can be considered in two parts. The biases of all satellites have the same long-term variation trend, which ranges between ±3 m. After removing the long-term trend, the biases of satellites are not consistent, with the STD of 0.315 m (1.05 ns). In PPP, the consistent part of the bias will be absorbed into the receiver clock offset, while the inconsistent part will affect the convergence time and positioning accuracy. BCFCO has better accuracy and similar precision than B2bCO. The consistent part of BCFCO is almost eliminated, and the inconsistent part was reduced by 45%. The precision of the two clock offset products is both about 0.02 m.

The BCFCO and B2bCO are contrasted in terms of speed of convergence, positioning accuracy, and timing. Using BCFCO, the convergence speed increases by about 25% and 10% in horizontal and vertical directions, respectively. For the positioning accuracy, the average improvement is 22% and 17%. The closer it is to the core service area of China, the more obvious the improvement is. The improvement on the receivers of the same type is more obvious than that of receivers of another type. For timing, the error of the receiver clock offset can be reduced by 60% with the BCFCO product, while the error of clock rate is basically the same.

Author Contributions: Conceptualization, C.T., J.Y. and Y.L.; Data curation, Y.Y.; Formal analysis, J.L.; Resources, Y.Y.; Software, J.L.; Supervision, S.Z. and X.H.; Writing—original draft, J.L.; Writing—review & editing, C.T. and S.Z. All authors have read and agreed to the published version of the manuscript.

Funding: The work is supported by the National Natural Science Foundation of China (No. 12173072).

Data Availability Statement: Not applicable

Conflicts of Interest: The authors declare no conflict of interest.

References

1. Kouba, J.; Héroux, P. Precise Point Positioning Using IGS Orbit and Clock Products. *GPS Solut.* **2001**, *5*, 12–28. [CrossRef]
2. Liu, R.; Guo, B.; Zhang, A.; Yimwadsana, B. Research on GPS precise point positioning algorithm with a Sea Surface Height Constraint. *Ocean Eng.* **2020**, *197*, 106826. [CrossRef]
3. Geng, J.; Bock, Y.; Melgar, D.; Crowell, B.W.; Haase, J.S. A new seismogeodetic approach applied to GPS and accelerometer observations of the 2012 Brawley seismic swarm: Implications for earthquake early warning. *Geochem. Geophys. Geosyst.* **2013**, *14*, 2124–2142. [CrossRef]
4. Steigenberger, P.; Montenbruck, O. Consistency of MGEX Orbit and Clock Products. *Engineering* **2020**, *6*, 898–903. [CrossRef]
5. Bock, H.; Hugentobler, U.; Beutler, G. Kinematic and Dynamic Determination of Trajectories for Low Earth Satellites Using GPS. In *First CHAMP Mission Results for Gravity, Magnetic and Atmospheric Studies*; Reigber, C., Lühr, H., Schwintzer, P., Eds.; Springer: Berlin/Heidelberg, Germany, 2003; pp. 65–69._10. [CrossRef]
6. Grayson, B.; Penna, N.T.; Mills, J.P.; Grant, D.S. GPS precise point positioning for UAV photogrammetry. *Photogramm. Rec.* **2018**, *33*, 427–447. [CrossRef]
7. Dousa, J.; Vaclavovic, P. Real-time zenith tropospheric delays in support of numerical weather prediction applications. *Adv. Space Res.* **2014**, *53*, 1347–1358. [CrossRef]
8. China Satellite Navigation Office. BeiDou Navigation Satellite System Signal In Space Interface Control Document, Precise Point Positioning Service Signal PPP-B2b (Version 1.0). 2022. Available online: http://en.beidou.gov.cn/SYSTEMS/ICD/ (accessed on 12 August 2022).
9. China Satellite Navigation Office. The Application Service Architecture of BeiDou Navigation Satellite System. 2022. Available online: http://en.beidou.gov.cn/SYSTEMS/ICD/ (accessed on 12 August 2022).
10. Xu, Y.; Yang, Y.; Li, J. Performance evaluation of BDS-3 PPP-B2b precise point positioning service. *GPS Solut.* **2021**, *25*, 1–14. [CrossRef]
11. Ren, Z.; Gong, H.; Peng, J.; Tang, C.; Huang, X.; Sun, G. Performance assessment of real-time precise point positioning using BDS PPP-B2b service signal. *Adv. Space Res.* **2021**, *68*, 3242–3254. [CrossRef]
12. Hauschild, A.; Steigenberger, P.; Montenbruck, O. Inter-Receiver GNSS Pseudorange Biases and Their Effect on Clock and DCB Estimation. In Proceedings of the 32nd International Technical Meeting of the Satellite Division of The Institute of Navigation, Miami, FL, USA, 16–20 September 2019; pp. 3675–3685. [CrossRef]
13. Pan, J.; Hu, X.; Zhou, S.; Tang, C.; Guo, R.; Zhu, L.; Tang, G.; Hu, G. Time synchronization of new-generation BDS satellites using inter-satellite link measurements. *Adv. Space Res.* **2018**, *61*, 145–153. [CrossRef]
14. Pan, J.; Hu, X.; Zhou, S.; Tang, C.; Wang, D.; Yang, Y.; Dong, W. Full-ISL clock offset estimation and prediction algorithm for BDS3. *GPS Solut.* **2021**, *25*, 140. [CrossRef]
15. Zhou, S.S.; Hu, X.G.; Liu, L.; Guo, R.; Zhu, L.F.; Chang, Z.Q.; Tang, C.P.; Gong, X.Q.; Li, R.; Yu, Y. Applications of two-way satellite time and frequency transfer in the BeiDou navigation satellite system. *Sci. China Phys. Mech. Astron.* **2016**, *59*, 109511. [CrossRef]
16. Zhou, S.; Hu, X.; Zhou, J.; Chen, J.; Gong, X.; Tang, C.; Wu, B.; Liu, L.; Guo, R.; He, F.; et al. Accuracy analyses of precise orbit determination and timing for COMPASS/Beidou-2 4GEO/5IGSO/4MEO constellation. In Proceedings of the China Satellite Navigation Conference (CSNC), Wuhan, China, 15–17 May 2013; Volume 245, pp. 89–102._8. [CrossRef]
17. Yang, Y.; Yang, Y.; Hu, X.; Chen, J.; Guo, R.; Tang, C.; Zhou, S.; Liqian, Z.; Xu, J. Inter-Satellite Link Enhanced Orbit Determination for BeiDou-3. *J. Navig.* **2019**, *73*, 1–16. [CrossRef]
18. Zhou, S.; Hu, X.; Wu, B.; Liu, L.; Qu, W.; Guo, R.; He, F.; Cao, Y.; Wu, X.; Zhu, L.; et al. Orbit determination and time synchronization for a GEO/IGSO satellite navigation constellation with regional tracking network. *Sci. China Phys. Mech. Astron.* **2011**, *54*, 1089–1097. [CrossRef]
19. Xie, X.; Geng, T.; Zhao, Q.; Cai, H.; Zhang, F.; Wang, X.; Meng, Y. Precise orbit determination for BDS-3 satellites using satellite-ground and inter-satellite link observations. *GPS Solut.* **2019**, *23*, 1–12. [CrossRef]
20. Liqian, Z.; Hu, X.; Tang, C.; Zhou, S.; Cao, Y.; Wang, Q.; Su, R. Inter-satellite link augmented BeiDou-3 orbit determination for Precise Point Positioning. *Chin. J. Aeronaut.* **2021**, *35*, 332–343. [CrossRef]

remote sensing

Article

High-Accuracy Clock Offsets Estimation Strategy of BDS-3 Using Multi-Source Observations

Jianhua Yang [1,2], Chengpan Tang [1], Sanshi Zhou [1], Yezhi Song [1], Jinhuo Liu [1,2], Yu Xiang [3], Yuchen Liu [1,2], Qiuning Tian [4], Yufei Yang [3], Zuo Yang [3] and Xiaogong Hu [1,*]

1 Shanghai Astronomical Observatory, Chinese Academy of Sciences, Shanghai 200030, China
2 University of Chinese Academy of Sciences, Beijing 100049, China
3 Beijing Satellite Navigation Center, Beijing 100094, China
4 Beijing Research Institute of Telemetry, Beijing 100076, China
* Correspondence: hxg@shao.ac.cn

Abstract: Satellite clock offsets are the critical parameters for The Global Navigation Satellite Systems (GNSSs) to provide position and timing (PNT) service. Unlike other GNSSs, BDS-3 uses the two-way superimposition strategy to measure satellite clock offsets. However, affected by some deficiencies of the two-way superimposition strategy, the accuracy of BDS-3 clock offsets parameters is 1.29 ns (RMS), which is the main bottleneck for BDS-3 to improve its space signal accuracy. After analyzing problems in the clock offsets measurement process of BDS-3, the paper proposes a new strategy to real-time estimate high-accuracy satellite clock offsets. The clock offsets estimated by the new strategy show a good consistency with GBM clock offsets. The averaged STD of their differences in MEO is 0.14 ns, and the clock offsets estimated by the new strategy present less fluctuation in the 1-day fitting residuals. Applying the new clock offsets to prediction, BDS-3 can reduce its clock offsets errors from 1.05 ns to 0.29 ns (RMS), about 72%. The above results indicate that the new clock offsets estimated strategy can improve the accuracy of clock offsets parameters of BDS-3 effectively.

Keywords: BDS-3; clock offsets estimated; inter-satellite links; two-way satellite time and frequency transfer

Citation: Yang, J.; Tang, C.; Zhou, S.; Song, Y.; Liu, J.; Xiang, Y.; Liu, Y.; Tian, Q.; Yang, Y.; Yang, Z.; et al. High-Accuracy Clock Offsets Estimation Strategy of BDS-3 Using Multi-Source Observations. *Remote Sens.* **2022**, *14*, 4674. https://doi.org/10.3390/rs14184674

Academic Editors: Giuseppe Casula, Zhetao Zhang and Wenkun Yu

Received: 17 August 2022
Accepted: 15 September 2022
Published: 19 September 2022

1. Introduction

Because GNSSs measure distances and clock offsets by comparing clocks, GNSSs need to provide accurate messages, including their space and time. For BDS-3 open service, BDS-3 calculates the position and velocity of satellites in the Beidou Coordinate System (BDCS) [1], and the satellite clock offsets with respect to Beidou Time (BDT) [2,3] at first. After that, BDS-3 parameterizes the satellite's orbit and clock offsets to ephemeris and clock offset parameters and uploads them to the satellite for broadcasting [4,5]. The update frequency of the above messages is often every hour and the predicted time for users to use them is less than 2 h. The influence of their errors on ground users is presented as signal-in-space errors (SISRE) [6,7]. SISRE is one of the most significant error sources in PNT, which consist of broadcast ephemeris errors and clock offsets parameters errors. Galileo performs the lowest SISRE in "GNSS Big 4" at present [8]. The BDS-3 constellation consists of 24 MEOs, 3 IGSOs, and 3 GEOs, whose SISRE is second to Galileo [9,10]. Statistical results show that the broadcast ephemeris errors can reach centimeter levels [11,12], which is at the top level in "GNSS Big 4" [8–10]. However, the clock offsets parameters errors of BDS-3 are 0.35 m (RMS) with respect to BDT, and evaluated by IGS precise clock offsets are more than 0.51 ns, which is larger than the errors of 0.14 m of Galileo. Therefore, the main reason why SISRE of BDS-3 is larger than Galileo is that the clock offsets parameters of BDS-3 with larger errors [9,10].

Generally, GNSS satellite clock offsets are the differences between the satellite atomic clock onboard and GNSS system time in GNSS conventional reference system. Their measurement is the most crucial step in GNSS generating satellite clock offsets parameters. Orbit determination and time synchronization (ODTS) is GNSS's most proven and widely used clock offsets estimation technology [13,14]. The ODTS can estimate satellite clock offsets simultaneously when estimating satellite orbit using GNSS downlinks pseudorange and carrier phase measurements from dozens to hundreds of global GNSS stations. For BDS-3, some results suggest that the clock offsets differences between real-time and batch-processed solutions in ODTS are less than 0.15 ns, which is as the same level as other GNSS [15]. The ODTS estimation technology has high requirements for GNSS station distribution and quantity. However, since the BDS monitor station are only distributed within China, it is difficult for BDS to realize the above requirements. Under the above circumstances, BDS-3 uses the two-way comparison technology to measure satellite clock offsets [16]. The fundamental principle of the two-way comparison is to subtract two-way measurements to eliminate most of the errors related to spatial information, and obtain their clock offsets. As shown in Figure 1, the BDS-3 time synchronization system owns the Master Control Station (MCS) and Time Synchronization Stations (TSS). All of them are equipped with C-band antennas, allowing BDS-3 to measure their clock offsets by two-way measurements [17]. Moreover, the satellite and BDS-3 ground station are also equipped with L-band two-way antennas and Ka-band phased-array antennas. The above antennas allow BDS-3 to calculate L-band two-way clock offsets and Ka-band two-way inter-satellite clock offsets [18]. According to the tracking status of the ground antenna to the satellite and the link plan of the inter-satellite link (ISL) [19], BDS-3 superimposes the above clock offsets to obtain satellite clock offsets with respect to BDT. This strategy we named the two-way superimposition strategy. Although the two-way superimposition strategy can measure satellite clock offsets, whether overseas or intraregional, it puts forward high requirements for the ability and quantity of ground antenna. More importantly, the final satellite clock offset is only obtained by several superimpositions of the above two-way clock offset. Each superimposition of the two-way clock offset is accompanied with accuracy loss and will decrease the accuracy of satellite clock offset parameters. As a result, some research finds that the clock offset parameters of BDS-3 are not accurate enough [9,10]. With the continuous improvements and developments in terms of performance, availability, modernization, application, and hybridizing for current GNSS [20–22], BDS must face up to the problem in its clock measurements and take it as the guide to study the new approach to improve its signal-in space accuracy.

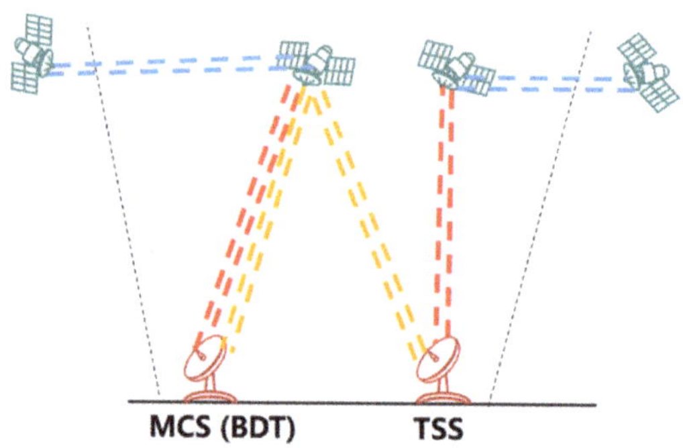

Figure 1. The construction of BDS time synchronization and the basic two-way measurements of BDS-3. Wherein, the red, blue, yellow curves are L-band satellite-station two-way measurements, Ka-band inter-satellite measurements, C-band inter-station two-way measurements, respectively.

For the above reason, we analyze the problems in the BDS-3 two-way superimposition strategy in the first part. Then, guided by solving the problems, we proposed a new strategy. The new strategy considers and combines BDS-3 constellation composition and multi-source measurements accuracy to real-time estimate BDS-3 satellite clock offsets with higher accuracy. We analyze the accuracy and improvements of the new strategy in the last part.

2. Materials and Methods

2.1. Two-Way Superimposition Strategy

2.1.1. Fundamental Principles Process

BDS-3 owns a time synchronization, which is independent of satellite orbit determination. The fundamental measurements of BDS-3 two measurements include L-band two-way satellite-station measurements, C-band two-way inter-station measurements, and Ka-band inter-satellite measurements.

The fundamental equation for using two-way measurements to calculate two-way clock offsets is given as Equation (1) [15]:

$$clk_{BA}(t_0) = [(\rho_{AB}(t_0) - r_{AB}(t_0)) - (\rho_{BA}(t_0) - r_{BA}(t_0))]/2 + \sum corr + Err + \varepsilon \quad (1)$$

where, $clk(t_0)_{BA}$ is the clock offsets of A with respect to B; $\rho_{BA}(t_0)$ and $\rho_{AB}(t_0)$ are the measurement values $r_{AB}(t_0)$ and $r_{BA}(t_0)$ are the signal propagation distance between two atomic clock A and B. $\sum corr$ is all kinds of errors correction in two-way measurements. They include phase center correction, relativity correction [23], time delay correction [24]. For L-band and C-band measurements, their signal passes through the ionosphere and troposphere, So, the errors correction of L-band and C-band measurements also include troposphere delay correction and ionosphere delay correction [25]. Err are the unknown residual errors. ε is two-way noise.

Considering all satellite clock offsets with respect to BDT. BDS-3 needs to use the two-way superimposition strategy to calculate final satellite clock offsets with respect to BDT. The procedure of the two-way superimposition strategy is given in Figure 2. For the intraregional satellites that are tracked by MCS antennas, BDS-3 can directly obtain the clock offsets by the L-band two-way satellite-station clock offsets of MCS. We present it as L in Figures 2 and 3. For the satellites that are tracked by TSS antennas, the two-way superimposition strategy subs the TSS's C-band inter-station clock offsets and L-band two-way satellite-station clock offsets to obtain the final satellite clock offsets. We present it as L + C in Figures 2 and 3. For the overseas satellites, the two-way superimposition strategy subs the two-way clock offsets of intraregional satellite and Ka-band inter-satellite clock offsets to obtain the overseas satellite two-way clock offsets. We present it as L + C + Ka or L + Ka in Figures 2 and 3. According to the antenna track messages and ISL plans [23,24], the two-way superimposition strategy can real-time calculate all satellite's two-way clock offsets [19]. Then, by 3-day comparison, BDS-3 estimate and deduct the bias between the two-way clock offsets and GNSS L-band downlinks offsets. Last, the final clock offset parameters of BDS-3 can be fitted by the linear model.

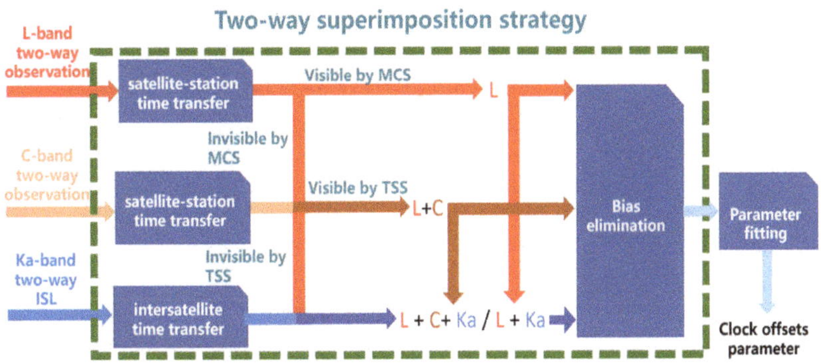

Figure 2. The procedure of the BDS-3 two-way superimposition strategy.

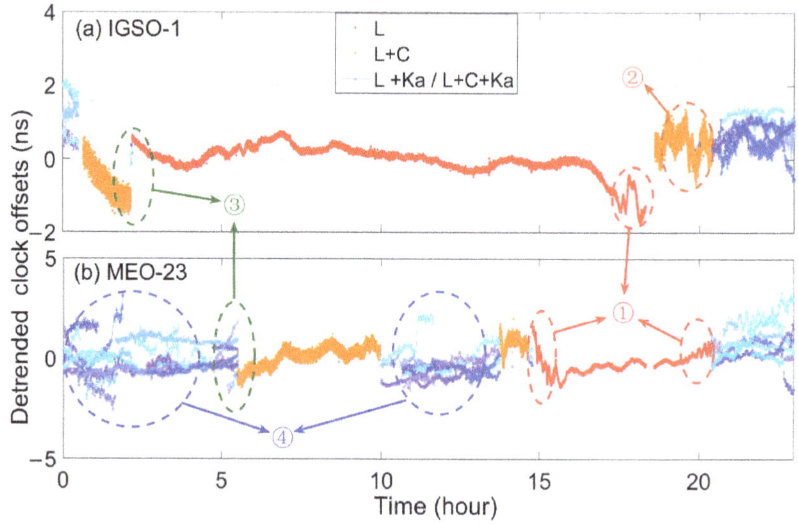

Figure 3. The typical one-day detrended two-way clock offsets of IGSO and MEO measured by BDS-3 two-way superimposition strategy. ①–④ describe the problems and abnormal phenomena in the two-way superimposition strategy.

2.1.2. Problems

Taking the IGSO-1 and MEO-1 as examples, the detrended 24 h two-way clock offsets calculated by the two-way superimposition strategy are plotted in Figure 3.

In Figure 3. the red curves are the clock offsets only calculated by L-band two-way clock. The orange curves are the clock offsets calculated by C-band inter-station two-way clock offsets and L-band two-way clock offsets. The blue curves are the clock offsets through Ka-band two-way inter-satellite clock offsets, and the blue curves with different depths represent the clock offsets calculated through different intraregional satellites. Figure 3 demonstrates that whatever satellite is or is not over China. The two-way superimposition strategy can always measure their clock offsets. However, this figure also presents its shortcoming and problems:

- Because there is little difference between the L-band uplinks and the L-band downlinks, Equation (1) cannot effectively eliminate the ionosphere delay and multipath effects. When the elevation of the L-band antenna is not low enough, the above residual errors will increase rapidly and decrease the accuracy of L-band two-way clock offsets (① in Figure 3).
- Because the noise of C-band measurements is far from other bands' measurements. The noise of clock offsets will become large when the satellite clock offsets are calculated through C-band two-way inter-stations clock offsets (② in Figure 3).
- Affected by temperature, antenna attitude, illumination, etc., the time delay of antennas has fluctuations, which are difficult to predict. Therefore, once the tracking antenna changes, the "step errors" will appear in the clock offsets (③ in Figure 3).
- The BDS-3 follows a time-division multiple access (TDMA) structure. One satellite connects with different visible satellites at different time in one connectivity cycle. Because an overseas satellite establishes links with different satellites frequently, the different "step errors" will be reflected in the clock offsets of overseas satellites, which are like the noise (④ in Figure 3).

The above reasons lead to poor accuracy of BDS-3 clock offsets, which is the main bottleneck limiting BDS-3 from improving itself signal-in-space accuracy. Therefore, based on the BDS-3 multi-source measurements, BDS-3 needs to urgently design a new strategy to measure the satellite clock offsets.

2.2. The New Strategy

From the above analysis, it can be seen that the reason why the accuracy of two-way clock offsets is poor is that the traditional strategy totally believes in the single two-way measurement and ignores that the errors of each measurement will also be superimposed in the final two-way clock offsets. All BDS-3 satellites are equipped with high-performance onboard atomic clocks [26–28], and the output signal has high stability in the short timescale. Therefore, the new strategy should believe in the ability of onboard atomic clocks instead of signal measurement. More importantly, unlike the traditional strategy, the new strategy abandons the C-band inter-station two-way clock offsets and L-band two-way clock offsets of MEOs and IGSOs, instead only using Ka-band inter-satellite between all satellites and L-band two-way clock offsets of GEOs. After the above changes, the new clock offsets estimated strategy includes three steps, shown in Figure 4.

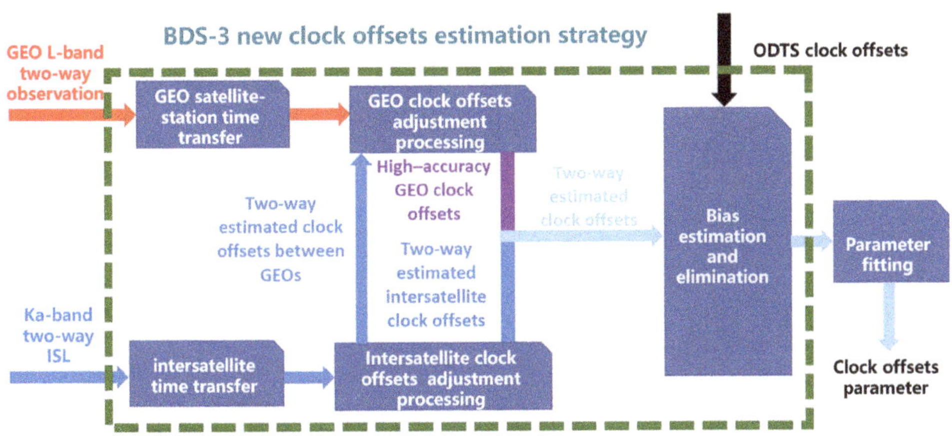

Figure 4. The procedure of the BDS-3 new clock offsets estimation strategy.

Firstly, the new strategy takes redundant Ka-band two-way inter-satellite clock offsets as the inputs and uses least square estimation to filter the noise and errors of the two-way clock offsets to estimate inter-satellite clock offsets with higher accuracy between all satellites, which is recorded as two-way estimated inter-satellite clock offsets. Then, taking the two-way estimated inter-satellite clock offsets between GEOs and their L-band two-way clock offsets as inputs, the new strategy further uses least square estimation to estimate one GEO's clock offsets with respect to BDT. Last, by making the differences between the ODTS clock offsets and two-way estimated clock offsets and averaging their differences, the bias between two kinds of clock offsets can be calibrated and thereby realize the alignment of the two-way estimated clock offsets to the ODTS clock offsets. The following subsection will introduce each step of the new clock offsets estimation strategy of BDS-3.

2.2.1. Estimation of Inter-Satellite Clock Offsets

As shown in Figure 5a, BDS-3 follows the TDMA structure. One satellite connects with different satellites at different intervals to perform dual, one-way ranging measurements. In one connectivity cycle whose duration is 60 s, multiple Ka-band two-way inter-satellite clock offsets between one specific satellite and those on other satellites can be obtained by processing measurements using Equation (1). In fact, BDS-3 onboard atomic clocks perform high stability in the short timescale. The above redundant Ka-band two-way inter-satellite clock offsets can be further reduced to the same time (see Figure 5b) [29]. After that, the redundant Ka-band inter-satellite two-way clock can be used to estimate the clock offsets with higher accuracy (see Figure 5c) [19].

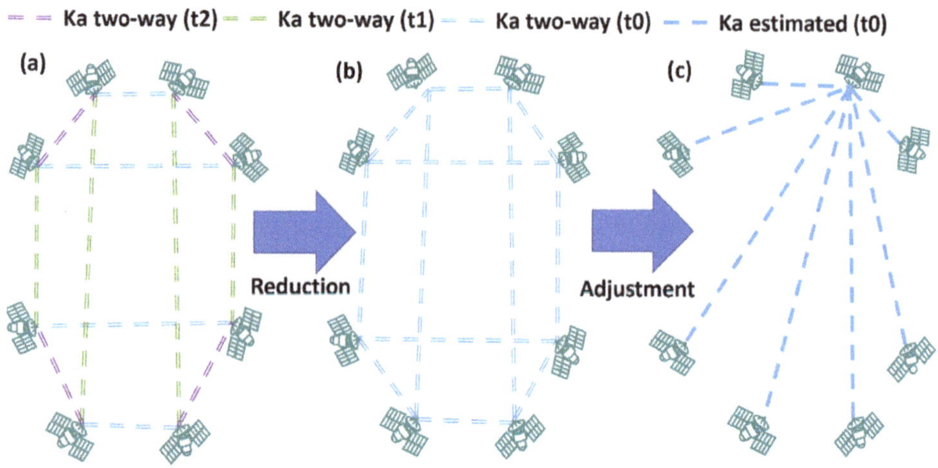

Figure 5. The procedure of the inter-satellite clock offset estimation. Wherein, t0, t1, t2 are different epoch within 60 s. Ka two-way is the Ka-band two-way clock offsets. Despite their epoch are different (as (**a**)), they can be reduced to the same epoch by prior message (as (**b**)) for further adjustment processing in the new strategy (as (**c**)).

After selecting a satellite as a reference satellite, "z." In the process of indirect adjustments, the observations are redundant Ka-band two-way inter-satellite clock offsets, and the parameters to be estimated are the Ka-band two-way inter-satellite clock offsets with respect to the reference satellite. The derivatives of observations of parameters to be estimated are given in Equation (2).

$$\begin{cases} \frac{\partial {Clk^{Ka}}_{AB}}{\partial \vec{x}} = [1, -1, 0 \ldots 0] \\ \ldots \\ \frac{\partial {Clk^{Ka}}_{Az}}{\partial \vec{x}} = [1, 0, 0 \ldots 0] \\ \ldots \\ \frac{\partial {Clk^{Ka}}_{zB}}{\partial \vec{x}} = [0, -1, 0 \ldots 0] \\ \ldots \end{cases} \tag{2}$$

where, $\vec{x}$ are the clock offsets of all satellites with respect to a reference satellite, ${Clk^{Ka}}_{AB}$ are the Ka-band two-way inter-satellite clock offsets between the A and B. The results from the trial suggest that if the Ka-band two-way inter-satellite clock offsets within 1 min of the whole constellation are used to reduce, there will be more than 300 two-way links, and the reduced errors will be less than 7 mm. The comparison of the clock offsets before and after indirect adjustment is shown in Figure 6. The dark blue curves present the clock offsets after indirect adjustments, and the azury blue present the original Ka-band two-way inter-satellite clock offsets. Both of them deduct the same trends. From Figure 6. the advantages of the further indirect adjustments processing are that it not only can estimate inter-satellite clock offsets when they are not linked. Moreover, compared with Ka-band two-way inter-satellite clock offsets, the clock offsets, by further estimated, have higher accuracy and less noise.

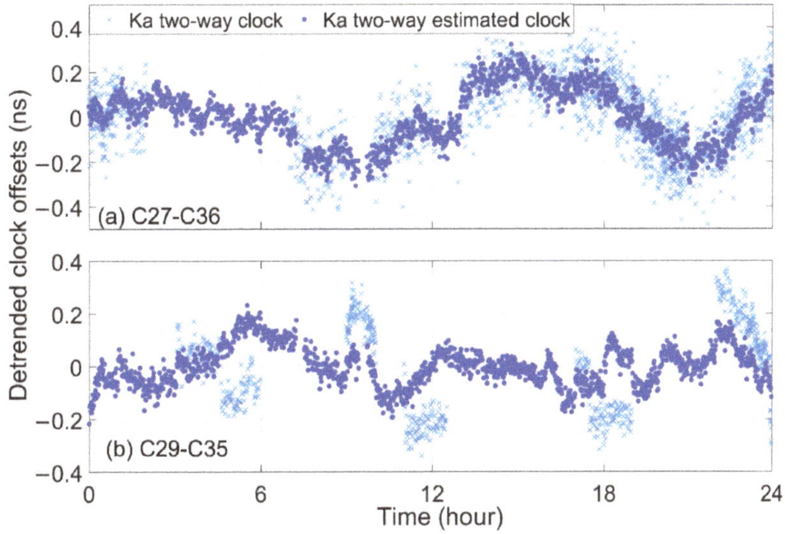

Figure 6. The improvements after using redundant Ka-band inter-satellite clock offsets to estimate Ka-band inter-satellite clock offsets (since 19 May 2022).

2.2.2. Estimation of Clock Offsets with Respect to BDT

The new strategy can estimate high-accuracy inter-satellite clock offsets by least square estimation processing for redundant Ka-band two-way inter-satellite clock offsets. However, the reference of satellite clock offsets parameter is BDT. The inter-satellite two-way estimated clock offsets need to be further traced to BDT. Despite no good ideas to directly solve the problems shown in Figure 1, we find that these problems can be effectively circumvented with some changes in the estimated clock offsets strategy. Firstly, the BDS-3 GEO is always over China, and MCS's antennas can track BDS-3 GEO anytime. BDS-3 can estimate GEOs clock offsets without passing through Ka-band two-way inter-satellite clock offsets and C-band two-way inter-station clock offsets. Then, the velocity of GEO related to antenna ground is slow, and the altitude angle and azimuth angle of tracked antenna nearly remain unchanged. Lastly, GEOs need to broadcast the messages of other services, including Satellite Based Augmentation System Service (SBAS) [30,31] and Precise Point Position Service in B2b signal (PPP-B2b) [32]. The above services require MCS to use antennas with a larger diameter and greater signal transmission power to track GEOs full time, which will help suppress the multipath effect in two-way observation. Therefore, in the satellite-ground clock offsets estimation, only the L-band two-way clock offsets of the large-diameter antennas following GEOs are considered the satellite-ground clock offsets estimation.

Similar to the inter-satellite clock offsets estimation, in order to filter the noise and errors, the GEO's clock offsets estimation also adopts the least square estimation method. However, in the estimation, the zero value of different L-band antennas should be calibration and aligned. Set a GEO satellite as a reference "Q" the parameter to be estimated includes Q's clock offsets with respect to BDT, and others are the differences of zero value of L-band larger-diameter antennas. After that, the observation equations in the least square estimation are given as Equation (3).

$$\begin{cases} \rho_Q(t) = clk_Q{}^L(t) = clk_Q(t) \\ \rho_A(t) = clk_A{}^L(t) - clk_{AQ}{}^{\text{Two-way estimated}}(t) = clk_Q(t) + Bias_A{}^L \\ \rho_B(t) = clk_B{}^L(t) - clk_{BQ}{}^{\text{Two-way estimated}}(t) = clk_Q(t) + Bias_B{}^L \end{cases} \tag{3}$$

where the derivatives of observation of the parameters to be estimated are given as Equation (4).

$$\begin{cases} \frac{\partial \rho_Q}{\partial \vec{x}} = [(0,0)_{\vec{x}_0}, (1)_{\vec{x}_1}] \\ \frac{\partial \rho_A}{\partial \vec{x}} = [(1,0)_{\vec{x}_0}, (1)_{\vec{x}_1}] \\ \frac{\partial \rho_B}{\partial \vec{x}} = [(0,1)_{\vec{x}_0}, (1)_{\vec{x}_1}] \end{cases} \quad (4)$$

wherein, $clk_A{}^L(t)$, $clk_B{}^L(t)$, $clk_Q{}^L(t)$ are the L-band two-way clock offsets of three GEO satellites. $clk^{Two-way\ estimated}$ are the Ka-band two-way inter-satellite estimated clock offsets. are the parameters to be estimated, which include the global parameter $\vec{x}_0$ and local parameter $\vec{x}_1$. Global parameters are the zero value differences, including the $Bias_A{}^L$ and $Bias_B{}^L$. Local parameters include the clock offsets of the reference satellite at different epochs. In the least square estimation processing, the global and local parameters can be effectively separated by operating the normal equation to eliminate parameters. Thus, the update of the antenna's bias and the estimation of higher-accuracy clock offsets of BDS-3 GEOs can be both real-time realized. The RMS of the 9-day residual for the least square estimation is 0.1 ns, which are given in Figure 7. Figure 7 preliminary suggests that the observations of different large-diameter antennas present a good consistency in short timescale. Their long-time consistency needs to be further verification in the future.

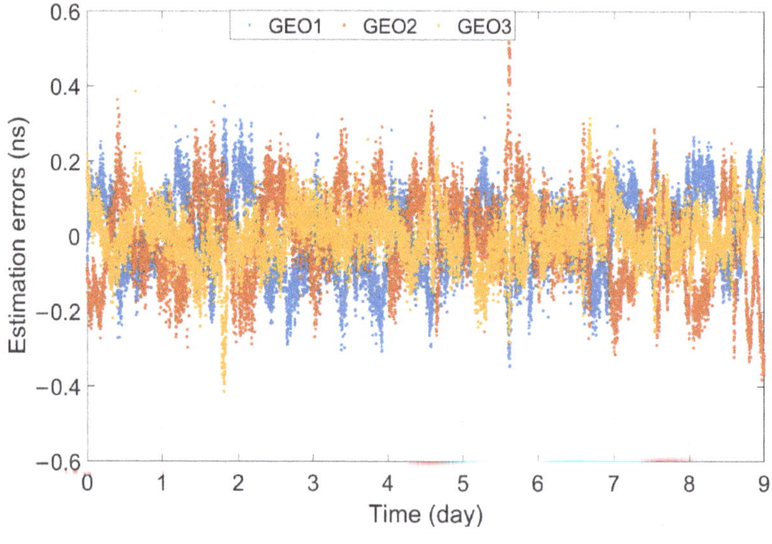

Figure 7. The estimation errors of the second least square estimation (since 19 May 2022), where the reference satellite is GEO-3.

Considering the high-accuracy inter-satellite clock offsets already obtained by 3.1. the clock offsets of all other satellites can be further traced to BDT by Equation (5).

$$clk_A(t_0) = clk_Q(t_0) + clk^{Two-way\ estimated}{}_{AQ}(t_0) \quad (5)$$

2.2.3. Estimation and Elimination of the Bias of Two-Way Clock Offsets

By the method in Sections 2.2.1 and 2.2.2, the new strategy can estimate the two-way clock offsets with higher accuracy in real-time. However, bias still exists between the two-way estimated clock offsets and ODTS clocks offsets. This bias includes the zero value of signal transmitting and receiving of Ka-band antennas and L-band antennas. However, BDS-3 users use one-way downlinks observation for positioning and timing. Therefore, to align the two-way estimated clock offsets with the one-way downlinks signal, the new strategy should finally correct the bias in the two-way estimated clock offsets.

The effective way to estimate the bias is using ODTS clock offsets. The duration of BDS-3 orbit determination is three days, and the specific strategy of BDS-3 orbit determination can refer to Tang et al. In order to suppress the influence of ODTS clocks offset errors on the bias estimation. The new strategy uses multiple-day ODTS clock offsets to estimate the bias. As shown in Figure 8, the estimation and elimination of the bias of the two-way estimated clock offsets include three steps. Firstly, as shown in Figure 8, splice the original ODTS clock offsets of multiple days to obtain the ODTS clock offsets (at present, the number of days is 3, which are consistent with the two-way superimposition strategy). Secondly, make the difference between the ODTS clock offsets and two-way estimated clock offsets, and average their differences to estimate the bias. Thirdly, deduct the bias in two-way estimated clock offsets so as to realize the alignment of the two-way clock offsets with the one-way downlink signal.

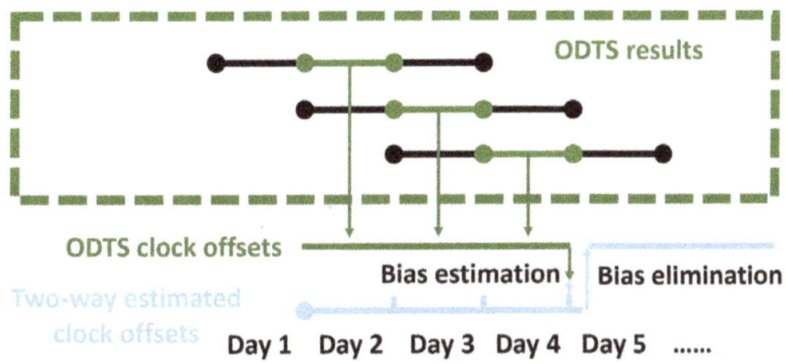

Figure 8. The procedure of the estimation and elimination of the bias of the two-way estimated clock offsets.

3. Results

In the new strategy, only using redundant ISL two-way observations and first least square estimation processing, the variations of inter-satellite clock offsets between other satellites can be real-time estimated. The two-way inter-satellite clock offsets are independent of L-band two-way measurements. Taking the clock offsets between the MEO-1 and MEO-2 as the example, the differences in the 1-day quadratic fitting residual between the two-way inter-satellite estimated clock offsets and the ODTS inter-satellite clock offsets from the German research center for geosciences Beidou multi-GNSS (GBM) [33] are plotted in Figure 9. Wherein, the blue curves of Figure 9a are the two-way estimated clock offsets between the MEO-1 and MEO-2. The green curves of Figure 9a are their ODTS clock offsets. Both of them are deducted with the same trends, and their differences are plotted in Figure 9b. Although the STD of their differences is less than 0.1 ns in the timescale of 1 day, due to the day-boundary discontinuities in the ODTS clock offsets [34], their differences exist in the obvious "hop" at the boundary between two days, which are given in Figure 10.

In order to suppress the influence of day-boundary discontinuities in the ODTS clock offsets, we deducted the same trends per day in the following comparisons between the ODTS clock offsets and the two-way clock offsets. Referring to BDS-3 M1, Tab1 lists 9-day averaging statistical results of their 1-day quadratic fitting residual and the STD of their differences for all BDS-3 satellites. Except for BDS-3 C61, the statistical results show that the averaged STD of the differences between the two-way estimated clock offsets and GBM clock offsets is 0.19 ns. Among them, the averaged STD for BDS-3 MEOs is 0.14 ns and 0.40 ns, 0.36 ns for IGSOs and GEOs, respectively.

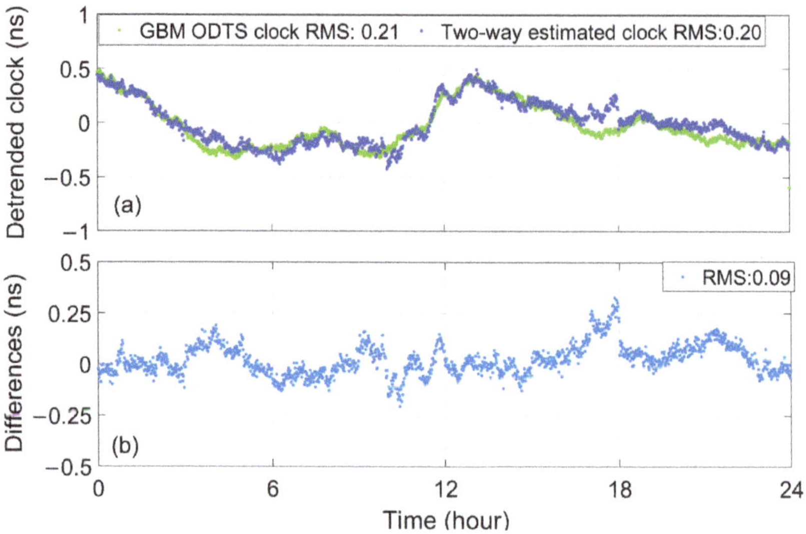

Figure 9. 1-day Satellite clock offset differences between the Ka-band inter-satellite estimated clock offsets and GBM products (since 19 May 2022). Wherein, (**a**) describes the two kinds of clock offsets after deducting the same trend. (**b**) describes their differences.

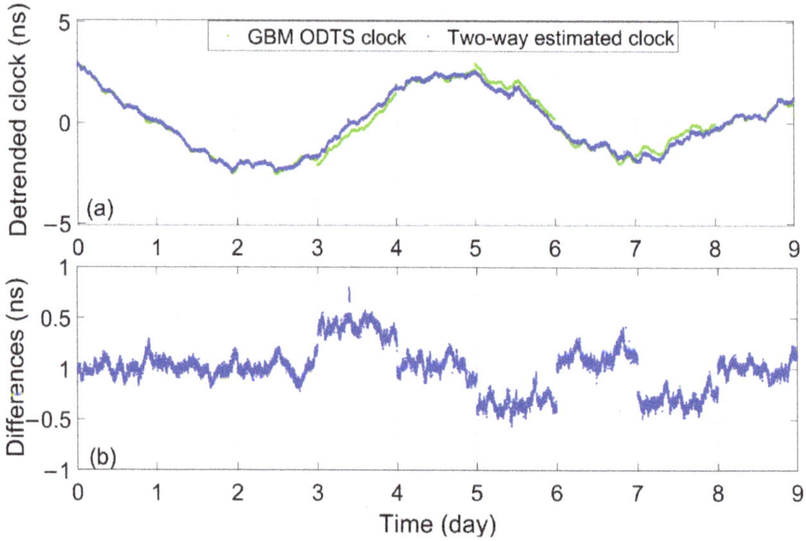

Figure 10. 9-day Satellite clock offset differences between the Ka-band inter-satellite estimated clock offsets and GBM products. Wherein, (**a**) describes the two kinds of clock offsets after deducting the same trend. (**b**) describes their differences. The reason why "hop" exist in the bottom figure is that the day-boundary discontinuities exist in the ODTS clock offsets (since 19 May 2022).

As shown in Table 1, In most cases, the large differences between the two kinds of clock offset with the larger fitting residuals in the GBM ODTS clock offsets. Epically for the GEOs and IGSOs. Taking the two kinds of clock offset between the IGSO-1 and MEO-1 as an example, their comparisons are given in Figure 11. Both of them are deducted the same trends per day. From Figure 10, It can be clearly found that their larger differences are caused by the large fluctuation in GBM ODTS clock offsets. We believe that the IGSOs and GEOs clock offsets with large fluctuations are that there exist larger orbit errors in GEOs and IGSOs. Some results suggest it is probably due to the not accurate enough solar radiation pressure model adopted in ODTS processing [33,35].

Table 1. Averaging statistical results of the two kinds of 1-day clock offsets comparison.

Satellite PRN	The STD of the Differences between Two Kinds of Clock Offsets (ns)	24 h Fitting Residual (ns)	
		ODTS Clock (GBM)	Two-Way Estimated Clock
C38	0.25	0.26	0.12
C39	0.36	0.37	0.09
C40	0.39	0.39	0.08
C25	0.16	0.18	0.11
C26	0.16	0.14	0.09
C27	0.11	0.09	0.11
C28	0.13	0.13	0.12
C29	0.19	0.13	0.22
C30	0.15	0.17	0.10
C20	0.07	0.18	0.16
C21	0.09	0.21	0.21
C22	0.07	0.18	0.17
C23	0.14	0.18	0.14
C24	0.12	0.16	0.17
C32	0.11	0.12	0.12
C33	0.12	0.18	0.14
C34	0.15	0.13	0.08
C35	0.19	0.19	0.10
C36	0.14	0.13	0.07
C37	0.11	0.15	0.10
C41	0.08	0.10	0.10
C42	0.11	0.12	0.11
C43	0.15	0.24	0.15
C44	0.12	0.20	0.16
C45	0.23	0.26	0.08
C46	0.23	0.25	0.11
C59	0.50	0.65	0.32
C60	0.61	1.00	0.17
MEO	0.14	0.19	0.13
IGSO	0.40	0.71	0.11
GEO	0.36	1.10	0.36
Average of all satellites	0.19	0.23	0.13

Taking the high-accuracy inter-satellite clock offsets and L-band two-way clock offsets of GEOs as inputs, the clock offsets of all satellites can be further traced to the BDT so as to estimate the final clock offsets, whose accuracy should be better than the clock offsets calculated by two-way superimposition strategy, which can be shown in Figure 9. The top panel of Figure 9 describes the GEO clock offsets. The middle panel of Figure 9 describes the typical IGSO clock offsets, and the bottom panel of Figure 12 describes the typical MEO clock offsets curves are the two-way estimated clock offsets. Wherein the red curves are the clock offsets directly calculated by L-band two-way clock offsets or superimposed by L-band two-way clock offsets and C-band inter-satellite clock offsets.

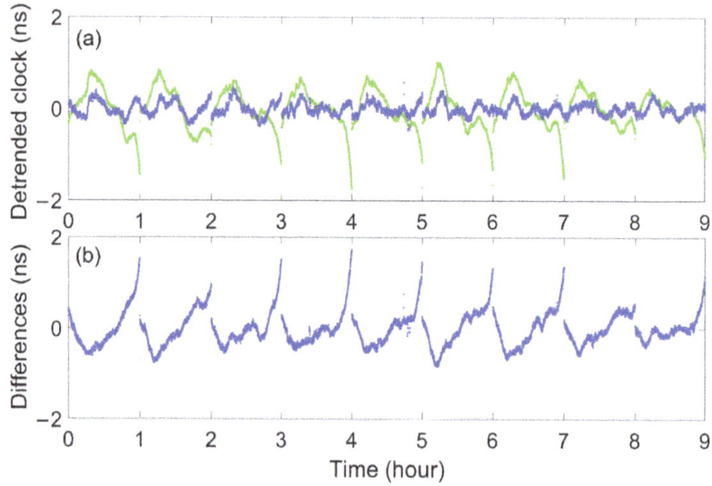

Figure 11. Satellite clock offset differences between the Ka-band inter-satellite estimated clock offsets and GBM products for 9 days. All of them are deducted the same trends per day (since 19 May 2022). Wherein, (**a**) are their detrended clock offsets, (**b**) are their differences.

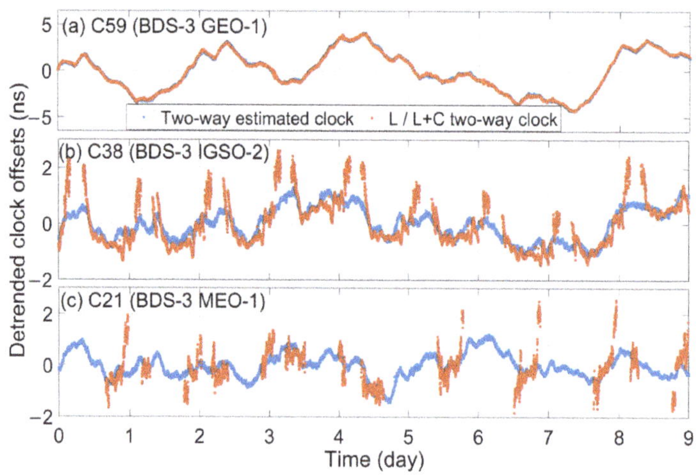

Figure 12. Satellite clock offset differences between the two-way estimated clock offsets and the two-way intraregional satellite clock offsets measured by superimposition strategy (since 19 May 2022).

By comparing the two curves, it can be found that although the red curves are in good agreement with the blue curves in most cases, the red curves become more divergent than the blue curves at the beginning and end of each segment, which indicate the two-way estimated clock offsets have more advantages than two-way clock offsets. The above phenomenon is because the new strategy effectively solves problems in the two-way superimposition strategy:

- Because the new strategy uses the least square estimation method to estimate clock offsets and will not consider C-band two-way inter-station clock offsets with large noise, the clock offsets can always keep higher precision.
- The final two-way clock offsets between the satellite atomic clock onboard and BDT are only realized through the two-way observation between the MCS larger diameter

antennas and GEOs, and the elevation of antennas tracking to GEOs is unchanged all the time. The problem of antennas with low elevation no longer exists.

- The new strategy only introduces three larger-diameter L-band antennas, each of which tracks to the specific GEO anytime. On this basis, the new strategy introduces the bias parameters to absorb the inconsistencies of observations of larger-diameter L-band antennas. All of the above considerations are able to suppress the "step errors."
- In the new strategy, the concept of intraregional satellite and overseas satellite no longer exist. All satellites play the same role in the inter-satellite clock offsets estimation. Furthermore, their clock offsets with respect to BDT are not dependent on ISL plans and antenna status but can be calculated through the same clock offsets (reference GEO). Therefore, even if IGSOs and MEOs are not able to be observed by ground antenna, the accuracy of their clock offsets will not be lost.

After obtaining the satellite clock offsets with higher accuracy, the BDS-3 is able to generate the satellite clock offsets parameters with higher accuracy. In most cases, BDS-3 satellite clock offsets parameters are generated by the satellite clock offsets of 2 h in history. Moreover, their prediction duration is less than 2 h. The Table 2. shows the statistical results of 2-h prediction errors of two-way estimated clock offsets for 9-day. From Table 2, the averaged RMS of all satellites is 0.29 ns, which is 29% of those calculated by the BDS-3 two-way superimposition strategy. Considering the clock offsets parameters errors of Galileo is 0.14 m (0.42 ns), the above results indicate that after using the new clock offsets estimation strategy, the accuracy of BDS-3 clock offsets parameters could reach or exceed those of Galileo.

Table 2. Statistical results of 2-h prediction errors of two-way estimated clock offsets for 10 days.

PRN	RMS (ns)	95% Error (ns)
C19	0.27	0.55
C20	0.27	0.58
C21	0.31	0.66
C22	0.31	0.67
C23	0.30	0.61
C24	0.29	0.62
C25	0.28	0.59
C26	0.29	0.65
C27	0.26	0.56
C28	0.28	0.62
C29	0.30	0.64
C30	0.27	0.59
C32	0.27	0.57
C33	0.28	0.59
C34	0.28	0.59
C35	0.28	0.59
C36	0.27	0.56
C37	0.25	0.53
C38	0.28	0.62
C39	0.26	0.53
C40	0.26	0.56
C41	0.29	0.65
C42	0.26	0.54
C43	0.28	0.57
C44	0.37	0.77
C45	0.27	0.58
C46	0.3	0.62
C59	0.33	0.71
C60	0.36	0.86
C61	0.38	0.66
Average of all satellites	0.29	0.61

4. Conclusions

In order to improve the signal-in-space accuracy of the BDS-3 satellite, the paper introduces the two-way superimposition strategy and analyzes its problems in satellite clock offsets measured. The analysis shows that BDS-3 uses the two-way superimposition strategy that can effectively calculate the clock offsets, both including intraregional satellite and overseas, with respect to BDT in real-time. Nevertheless, suffering from antenna zero value error, the low elevation angle of antennas, and the large noise of C-band inter-station clock offsets, the clock offsets calculated by BDS-3 are not accurate enough. This problem is the main factor limiting the improvement of space-in-signal accuracy for BDS-3.

Based on the above analysis, the paper optimizes and updates the BDS-3 strategy in BDS-3 satellite clock offsets estimation. Unlike the simple superposition of two-way observations, the new strategy from the perspective of least square estimation takes full advantage of the BDS-3 multi-source observation and the BDS-3 constellation characteristics to circumvent problems in the two-way superimposition strategy so as to improve the accuracy of the BDS-3 satellite clock offsets. The satellite clock offsets estimated by the new strategy have the following characteristic:

- The two-way estimated clock offsets retain the advantages of two-way clock offsets, which are less affected by propagation and position errors than one-way downlink clock offsets.
- All ISLs maintain the inter-satellite relative variation of two-way estimated clock offsets. The inter-satellite relative variation of two-way estimated clock offsets is independent of satellite-ground links.
- All satellite clock offsets are traced to BDT by the same clock offsets, which are only affected by the observations of three large-diameter L-band antennas.

By optimizing and updating, the BDS-3 is able to improve the accuracy of not only BDS-3 inter-satellite clock offsets but also the final BDS-3 satellite clock offsets with respect to BDT. Making the differences between the clock offsets estimated by the new strategy and the GBM products, the average STD of their clock offsets is 0.19 ns, wherein MEO is 0.14 ns. Furthermore, compared with the inter-satellite clock offsets of GBM products, the 24-h fitting residual of two-way estimated clock offsets is more stable. The above results indicate that: Only relying on ISLs and a few L-band large-diameter antennas, the BDS-3 can realize the global satellite clock offsets real-time estimation with the same accuracy level as the clock offsets estimated by more than one hundred global-distributed GNSS stations [33,35].

After adopting the new strategy, the accuracy of the satellite clock offsets parameter can be improved from 1.05 ns (0.35 m) [8] to 0.29 ns (0.09 m), about 71%. This result indicates the ISL can not only solve the problems of insufficient observation in overseas satellite clock offsets estimation but also can enhance the accuracy of satellite clock offsets parameters of intraregional satellites. The signal-in-space errors of BDS-3 broadcast ephemeris are 0.09 m. Considering that the SISRE are less than the sum of the broadcast ephemeris errors and clock offsets parameter errors [8], after adopting the proposed strategy in the BDS-3 time synchronization system, the signal-in-space accuracy of the BDS-3 navigation message can be less than 0.2 m at least. This result is of great significance to the performance improvement of BDS.

As shown in Table 2, after BDS-3 used the new strategy to estimate the clock offsets, a puzzling discovery is that despite the performance of all BDS-3 clock offsets are different, there is little difference between their prediction ability. The reason is plotted in Figure 13. Multi-day fitting residuals of all satellites with the highest-level-performance atomic clock present the almost common fluctuation, and the same fluctuation is found in the output signals of BDT. Although the common fluctuation will not affect the final position accuracy, it will affect the timing accuracy. It indicates that the common fluctuation in Figure 13 is caused by the fluctuation of BDT itself [2]! Therefore, the stability of the BDT's output signal limits the ability for BDS-3 to improve further the accuracy of BDS-3 satellite's clock offset parameters.

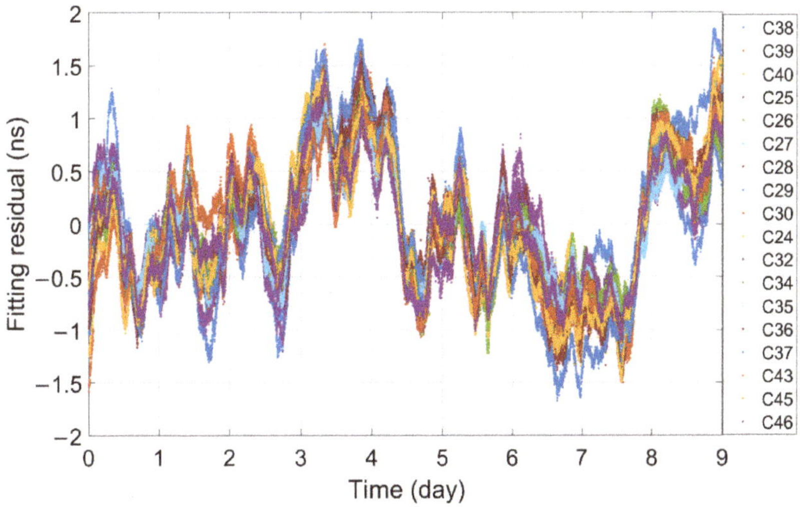

Figure 13. Multi-day fitting residuals of BDS-3 satellite with highest-level-performance atomic clock onboard with respect to BDT (since 19 May 2022). The common fluctuation is caused by the fluctuation in BDT output signals.

It should be noticed that, despite the above results are inspiring for BDS-3, from the perspective of service stability, BDS-3 still needs longer time to check this strategy. What is more, in the bias estimation of the two-way estimation clock offsets, the inputs ODTS clock offsets are dual-frequency ionospheric free combination clock offsets (usually B1I/B3I). Therefore, the final two-way clock offsets are aligned with the one-way downlinks clock offsets with the specific combination frequency. However, for the users of other BDS-3 frequencies. They need to use the broadcast DCB and TGD, and ionospheric parameters to further correct their observation. However, because of their strong correlation and poor distribution of BDS-3 GNSS stations, it is also tricky for BDS-3 to realize their high-precision calculation under the condition that GNSS stations of BDS-3 are only distributed within China. Therefore, after optimizing and updating the clock offsets estimation strategy, the accuracy of the parameters of ionospheric and DCB/TGD will become the main point of growth of BDS-3 accuracy. Improving their accuracy is another key technology that BDS-3 needs to further breakthrough.

Finally, we can see a good agreement between the one-way downlink clock offsets and the two-way estimated clock offsets, especially for the MEO satellites with higher ODTS clock offsets accuracy. It probably means that the unknown system errors between the ISL and L-band downlinks are not large. Considering that the two-way estimated clock offsets with less fluctuation than the current ODTS clock offsets in the timescale from the few hours to 1 day. We believe that if BDS-3 takes the two-way estimated clock offsets as the new observation in the GNSS orbit determination, BDS-3 will be able to enhance the accuracy of orbit determination further and accelerate the decouple and convergence time of GNSS parameters, which will be conducive to the study in other GNSS fields [20–22] and the improvement of the ability on BDS-3 PPP-B2b [32]. All of them deserve further study.

Author Contributions: J.Y., C.T., Y.Y. and X.H.; Data curation, Y.X. and Z.Y.; Formal analysis, J.L.; Funding acquisition, S.Z.; Investigation, Y.S., Y.L. and Z.Y.; Methodology, J.Y., S.Z. and Y.S.; Project administration, X.H.; Resources, X.H.; Software, J.Y. and Y.S.; Supervision, Y.X., Y.Y. and X.H.; Validation, J.L., Y.X., Y.L. and Z.Y.; Visualization, J.Y.; Writing—original draft, J.Y.; Writing—review & editing, J.Y., Q.T. and X.H. All authors have read and agreed to the published version of the manuscript.

Funding: This research was supported by Shanghai Observatory's key cultivation project (N20210601003); National Natural Science Foundation of China (No. 12173072); Civil Aerospace "14th Five-Year" Technology Pre-research Project (KJSP2020020203).

Data Availability Statement: The data used in this contribution, including the ISL measurements, L-band two-way measurements, C-band two-way measurements, are provided by the Beijing Satellite Navigation Canter. The above data are available from the corresponding author upon request. The other data used in this contribution include the BDS-3 broadcast message provided by the Test and Assessment Research Center of the China Satellite Navigation Office and the precise clock data provided by the GBM, which can be downloaded from the internet.

Conflicts of Interest: The authors declare no conflict of interest.

References

1. Wu, F.; Wei, Z.; Ming, F.; Zeng, A. The first realization of BeiDou coordinate system and Its assessment. *J. Surv. Eng.* **2021**, *147*, 04021004. [CrossRef]
2. Beidou Timescale Description, 2017, 01. Available online: http://en.beidou.gov.cn/SYSTEMS/Officialdocument/201912/P020191202322231024298.pdf (accessed on 1 January 2017).
3. Han, C.; Liu, L.; Cai, Z.; Lin, Y. The space–time references of BeiDou navigation satellite system. *Satel. Navig.* **2021**, *2*, 1–10. [CrossRef]
4. Guo, S.; Cai, H.; Meng, Y. BDS-3 RNSS technical characteristics and service performance. *Acta Geod. Et Cartogr. Sin.* **2019**, *48*, 810.
5. Zhou, S.; Wu, B.; Hu, X.; Cao, Y.; Yu, Y. Signal-Space Accuracy BeiDou navigation satellite system: Challenges and solutions. *Sci. China Phys. Mech. Astron.* **2017**, *2017*, 019531. [CrossRef]
6. Montenbruck, O.; Steigenberger, P.; Hauschild, A. Broadcast versus precise ephemerides: A multi-GNSS perspective. *GPS Solut.* **2015**, *19*, 321–333. [CrossRef]
7. Warren, D.L.M.; Raquet, J.F. Broadcast vs. Precise GPS ephemerides: A historical perspective. *GPS Solut.* **2003**, *7*, 151–156. [CrossRef]
8. Montenbruck, O.; Steigenberger, P.; Hauschild, A. Comparing the 'Big 4'-A User's View on GNSS Performance. In Proceedings of the 2020 IEEE/ION Position, Location and Navigation Symposium (PLANS), Portland, OR, USA, 20–23 April 2020; pp. 407–418.
9. Yang, J.; Tang, C.; Song, Y.; Hu, X.; Zhou, S.; Chang, Z. Analysis of signal-in-space ranging error of GNSS navigation message. *Sci. Sin. Phys. Mech. Astron.* **2021**, *51*, 019508. (In Chinese) [CrossRef]
10. Lv, Y.; Geng, T.; Zhao, Q.; Xie, X.; Zhou, R. Initial assessment of BDS-3 preliminary system signal-in-space range error. *GPS Solut.* **2019**, *24*, 1–13. [CrossRef]
11. Yang, Y.; Mao, Y.; Sun, B. Basic performance and future developments of BeiDou global navigation satellite system. *Satell. Navig.* **2020**, *1*, 8. [CrossRef]
12. Chen, J.; Hu, X.; Tang, C.; Zhou, S.; Yang, Y.; Pan, J.; Ren, H.; Ma, Y.; Tian, Q.; Wu, B.; et al. SIS accuracy and service performance of the BDS-3 basic system. *Sci. China Phys. Mech. Astron.* **2020**, *63*, 105–116. [CrossRef]
13. Pan, J.; Hu, X.; Zhou, S.; Tang, C.; Guo, R.; Zhu, L.; Tang, G.; Hu, G. Time synchronization of new-generation BDS satellites using inter-satellite link measurements. *Adv. Space Res. Off. J. Comm. Space Res.* **2018**, *61*, 145–153. [CrossRef]
14. Steigenberger, P.; Montenbruck, O. Galileo status: Orbits, clocks, and positioning. *GPS Solut.* **2016**, *21*, 319–331. [CrossRef]
15. Li, X.; Ge, M.; Dai, X.; Ren, X.; Fritsche, M.; Wickert, J.; Schuh, H. Accuracy and reliability of multi-GNSS real-time precise positioning: GPS, GLONASS, BeiDou, and Galileo. *J. Geod.* **2015**, *89*, 607–635. [CrossRef]
16. Formichella, V.; Galleani, L.; Signorile, G.; Sesia, I. Time–frequency analysis of the Galileo satellite clocks: Looking for the J2 relativistic efect and other periodic variations. *GPS Solut.* **2021**, *25*, 1–14. [CrossRef]
17. Liu, L.; Tang, G.; Han, C.; Shi, X.; Guo, R.; Zhu, L. The method and experiment analysis of two-way common-view satellite time transfer for compass system. *Sci. China Phys. Mech. Astron.* **2015**, *58*, 1–7. [CrossRef]
18. Zhou, S.; Hu, X.; Liu, L.; Guo, R.; Zhu, L.; Chang, Z.; Tang, C.; Gong, X.; Li, R.; Yu, Y. Applications of two-way satellite time and frequency transfer in the BeiDou navigation satellite system. *Sci. China Phys. Mech. Astron.* **2016**, *59*, 109511. [CrossRef]
19. Pan, J.; Hu, X.; Zhou, S.; Tang, C.; Wang, D.; Yang, Y.; Dong, W. Full-ISL clock offset estimation and prediction algorithm for BDS3. *GPS Solut.* **2021**, *25*, 140. [CrossRef]
20. Dow, J.M.; Neilan, R.E.; Rizos, C. The International GNSS Service in a changing landscape of Global Navigation Satellite Systems. *J. Geod.* **2009**, *83*, 689. [CrossRef]
21. Jin, S.; Wang, Q.; Dardanelli, G. A Review on Multi-GNSS for Earth Observation and Emerging Applications. *Remote Sens.* **2022**, *14*, 3930. [CrossRef]
22. Montenbruck, O.; Steigenberger, P.; Prange, L.; Deng, Z.; Zhao, Q.; Perosanz, F.; Romero, I.; Noll, C.; Stürze, A.; Weber, G.; et al. The Multi-GNSS Experiment (MGEX) of the International GNSS Service (IGS)-Achievements, prospects and challenges. *Adv. Space Res. Off. J. Comm. Space Res.* **2017**, *59*, 1671–1697. [CrossRef]
23. Wang, C.; Zhao, Q.; Guo, J.; Liu, J.; Chen, G. The contribution of intersatellite links to BDS-3 orbit determination: Model refinement and comparisons. *Navigation* **2019**, *66*, 71–82. [CrossRef]

24. Tang, C.; Hu, X.; Zhou, S.; Liu, L.; Pan, J.; Chen, L.; Guo, R.; Zhu, L.; Hu, G.; Li, X.; et al. Initial results of centralized autonomous orbit determination of the new-generation BDS satellites with inter-satellite link measurements. *J. Geod.* **2018**, *92*, 1155–1169. [CrossRef]
25. Pan, J.Y.; Xiaogong, H.U.; Tang, C.P.; Zhou, S.; Li, R.; Zhu, L.; Tang, G.; Hu, G.; Chang, Z.; Wu, S.; et al. System error calibration for time division multiple access inter-satellite payload of new-generation Beidou satellites. *Chin. Sci. Bull.* **2017**, *62*, 2671–2679. [CrossRef]
26. Yang, Y.; Yang, Y.; Hu, X.; Chen, J.; Guo, R.; Tang, C.; Zhou, S.; Zhao, L.; Xu, J. Inter-Satellite link enhanced orbit determination for BeiDou-3. *J. Navig.* **2020**, *73*, 115–130. [CrossRef]
27. Ma, Y.; Tang, C.; Hu, X.; Zhu, X.; Sun, L.; Yang, Y.; Chang, Z.; Yang, J. Long-Term Frequency Stability Analysis for RRFM and Power-Law Noise Determination of the BeiDou-3 Satellite On-Board Atomic Clocks. *IEEE Trans. Instrum. Meas.* **2022**, *71*, 1002515. [CrossRef]
28. Han, C.; Cai, Z. Relativistic effects to the onboard BeiDou satellite clocks. *Navigation* **2019**, *66*, 49–53. [CrossRef]
29. Wu, Z.; Zhou, S.; Hu, X.; Liu, L.; Shuai, T.; Xie, Y.; Tang, C.; Pan, J.; Zhu, L.; Chang, Z. Performance of the BDS3 experimental satellite passive hydrogen maser. *GPS Solut.* **2018**, *22*, 44. [CrossRef]
30. Liu, Y.; Cao, Y.; Tang, C.; Chen, J.; Zhao, L.; Zhou, S.; Hu, X.; Tian, Q.; Yang, Y. Pseudorange Bias Analysis and Preliminary Service Performance Evaluation of BDSBAS. *Remote Sens.* **2021**, *13*, 4815. [CrossRef]
31. Tian, Q.; Cao, Y.; Hu, X.; Tang, C.; Zhou, S.; Guo, R.; Li, X.; Tian, Y.; Yang, Y.; Yang, J. A Study on Pseudorange Biases in BDS B1I/B3I Signals and the Impacts on Beidou Wide Area Differential Services. *Remote Sens.* **2022**, *14*, 432. [CrossRef]
32. Yang, Y.; Ding, Q.; Gao, W.; Li, J.; Xu, Y.; Sun, B. Principle and performance of BDSBAS and PPP-B2b of BDS-3. *Satell. Navig.* **2022**, *3*, 1–9. [CrossRef]
33. Zhiguo, D.E.N.G.; Jungang, W.A.N.G.; Maorong, G.E. The GBM rapid product and the improvement from the undifferenced ambiguity. Resolution. *Acta Geodeatica Et Cartogr. Sin.* **2022**, *51*, 544–555. [CrossRef]
34. Zhang, X.; Guo, J.; Hu, Y.; Sun, B.; Wu, J.; Zhao, D.; He, Z. Influence of Precise Products on the Day-Boundary Discontinuities in GNSS Carrier Phase Time Transfer. *Sensors* **2021**, *21*, 1156. [CrossRef] [PubMed]
35. Xia, F.; Ye, S.; Chen, D.; Tang, L.; Wang, C.; Ge, M.; Neitzel, F. Advancing the Solar Radiation Pressure Model for BeiDou-3 IGSO Satellites. *Remote Sens.* **2022**, *14*, 1460. [CrossRef]

remote sensing

MDPI

Communication

An Observation Density Based Method for Independent Baseline Searching in GNSS Network Solution

Tong Liu [1], Yujun Du [2,*], Wenfeng Nie [3,4], Jian Liu [1], Yongchao Ma [1] and Guochang Xu [1,5]

1 Laboratory of Navigation and Remote Sensing, Institute of Space Science and Applied Technology (ISSAT), Harbin Institute of Technology, Shenzhen 518055, China
2 The Industrial Sciences Group, Sydney, NSW 2060, Australia
3 Institute of Space Sciences, Shandong University, Weihai 264209, China
4 State Key Laboratory of Geo-Information Engineering, Xi'an 710000, China
5 Key Laboratory of Urban Land Resources Monitoring and Simulation, Ministry of Natural Resources, Shenzhen 518000, China
* Correspondence: eugene@industrialsciences.com.au; Tel.: +61-411-644-071

Abstract: With applications such as precise geodetic product generation and reference frame maintenance, the global GNSS network solution is a fundamental problem that has constantly been a focus of concern. Independent baseline search is a prerequisite step of the double-differenced (DD) GNSS network. In this process, only empirical methods are usually used, i.e., the observation-max (OBS-MAX), which allows for obtaining more redundant DD observations, and the shortest-path (SHORTEST), which helps to better eliminate tropospheric and ionospheric errors between stations. Given the possible limitations that neither of the methods can always guarantee baselines of the highest accuracy to be selected, a strategy based on the 'density' of common satellites (OBS-DEN) is proposed. It takes the number of co-viewing satellites per unit distance between stations as the criterion. This method ensures that the independent baseline network has both sufficient observations and short baselines. With single-day solutions and annual statistics computed with parallel processing, the method demonstrates that it has the ability to obtain comparable or even higher positioning accuracy than the conventional methods. With a clearer meaning, OBS-DEN can be an option alongside the previous methods in the independent baseline search.

Keywords: GNSS; independent baseline; GNSS network solution; observation-max; shortest; observation density; minimum spanning tree

Citation: Liu, T.; Du, Y.; Nie, W.; Liu, J.; Ma, Y.; Xu, G. An Observation Density Based Method for Independent Baseline Searching in GNSS Network Solution. *Remote Sens.* **2022**, *14*, 4717. https://doi.org/10.3390/rs14194717

Academic Editors: Zhetao Zhang, Wenkun Yu and Giuseppe Casula

Received: 16 August 2022
Accepted: 16 September 2022
Published: 21 September 2022

1. Introduction

The global GNSS network solution plays an important role in geodesy, especially geodetic parameter estimation [1], high-precision product generation [2,3], datum maintenance [4], and geodynamics applications [5,6]. As a well-developed method, double differencing (DD) [7] is widely used in well-known GNSS data processing software such as Bernese 5.2 developed by Rolf Dach et al. at the Astronomical Institute of the University of Bern (AIUB), Switzerland [8] and GAIMIT/GLOBK 10.7 developed by T. A. Herring et al. from MIT, Scripps Institution of Oceanography and Harvard University in America [9]. How to improve the accuracy of the GNSS DD network is a topic that has been continuously explored.

In the implementation of the GNSS network solution, in order to reduce the computational load while not affecting the overall positioning accuracy, the independent baseline solution of multiple stations should be used before the entire network adjustment [7]. The principle of independent baseline selection is that only one path exists between any two stations, while all the stations should be connected. For a network with n stations, a total of $n(n+1)/2$ baselines exist, only $n-1$ of which are independent. The objective of the independent baseline selection is to optimize the overall accuracy of the baseline solutions in order to facilitate the subsequent network adjustment. Mathematically, this process can be described by the minimum spanning tree (MST) [10,11] problem.

In the process of MST, the criteria for selecting baselines can be defined according to the user's needs. One of the most easily conceived solutions is to make the total length of $n-1$ baselines the shortest, which is known as the shortest path (SHORTEST) method [7,8]. This is because the shorter the distance between stations, the greater the number of co-viewing satellites, thus more redundant observations are involved to facilitate the network adjustment. More importantly, the tropospheric and ionospheric delays of neighboring stations are similar. The shorter baselines help to better eliminate these errors. Since the original intention of SHORTEST is to improve positioning accuracy by increasing the number of co-viewing satellites, a more straightforward solution is to use the maximum common satellites as the criterion. This method is known as the maximum observation method (OBS-MAX) [8].

Both methods mentioned above have been investigated; for example, SHORTEST was used in a massive GNSS network of more than 2000 globally distributed stations [12], while OBS-MAX is shown to be beneficial in the tropospheric delay estimation [13]. In the ideal situation, the shorter the baseline, the more common observations there are. Then, SHORTEST and OBS-MAX should be fully equivalent. However, the statistics show that they are not consistent [14], i.e., on various days, the baselines generated by different methods could ultimately lead to different solution precision, which violates the assumption that the two methods are equivalent. This means that the number of observations does not necessarily increase as the baseline becomes shorter. This is due to the fact that the satellites are usually not evenly distributed across the sky, e.g., sparse observations in local areas and sufficient co-viewing satellites for some long baselines. In a word, the search for optimal independent baselines is still an open question to be further investigated.

A scheme of setting up weights (WEIGHT) between SHORTEST and OBS-MAX has been proposed [14]. The WEIGHT method was demonstrated to be of higher positioning precision than that of SHORTEST and OBS-MAX. However, how to set up weights lacks theoretical support and can only be empirical. For instance, the weights can be determined based on the posterior accuracy of the final baseline solutions using each of the two methods; on an a priori basis, the Bernese software could use a weight of 30% for the SHORTEST in addition to OBS-MAX as an option [8].

To avoid setting up empirical weights or doubling the computational load brought by a posteriori precision-based weighting, a new method called "observation-density" (OBS-DEN) is proposed here. It takes the ratio of baseline length and the number of observations between two stations as the criterion of the MST. The physical interpretation of this criterion is the number of common satellites per unit distance, which overcomes the degradation of baseline accuracy by seeking only maximum observations or the shortest baselines. The advantage of OBS-DEN is that it provides an explainable weighting scheme that can overcome the downsides of SHORTEST and OBS-MAX. This method can be used in various types of GNSS network solution-related software, alongside existing options for users to choose from.

The rest of the paper is organized as follows. In Section 2, the datasets and products, the principle of MST, and the criterion with OBS-DEN are introduced. Then the flowchart for generating independent baselines using various methods and the parallelization of network processing is presented in Section 3. Afterward, the results of single-day solutions and annual statistics are analyzed and discussed in Section 4. Finally, the paper is concluded in Section 5.

2. Data and Method

2.1. Data

Observation data from about 100 IGS stations distributed worldwide, was used to test the proposed method. The data was in Receiver Independent Exchange Format (RINEX) and can be accessed through (https://cddis.nasa.gov/archive/gnss/data/daily/2012/,

Weihai, China, 1 June 2018). First, data from 13 January 2012, was used to show the accuracy of a single-day solution. After that, data from the whole year 2012 was used for statistical analysis. We chose a data span over the year 2012, as it was the first peak of the last solar cycle [15], which helps to investigate the performance of the proposed method under various ionospheric situations.

Both GPS and GLONASS observations were included in the processing. The sampling interval of the data was 30 s. Precise products, including the precise orbit and clocks (*.SP3 and *.CLK), Earth rotation parameters (*.ERP), ionospheric parameters (*.ION), differential code biases (*.DCB), reference coordinates (*.CRD), antenna phase center corrections, receiver types, and tidal loading corrections (*.BLQ and *.ATL) were adopted to enable high-precision GNSS network solution. The data was processed with the Bernese software developed at the Astronomical Institute of the University of Bern (AIUB). Detailed descriptions or flow charts of data processing with Bernese can be found in the Bernese manual [8] or other publications [12,14,16–18]. The products were downloaded and used following the website (https://cddis.nasa.gov/Data_and_Derived_Products/GNSS/GNSS_product_holdings.html, Weihai, China, 1 June 2018) and the instruction of Bernese [8].

2.2. MST

MST is the mathematical basis for GNSS DD independent baseline solutions given the baseline length and the number of observations that are calculated and counted. Let V denote all possible Vertexes and E denote all possible edges. Let e denote an edge and $w(e)$ denote the criterion of that edge. This criterion $w(e)$ can be distance or any other weighting factors. MST is defined as follows. In an undirected graph $G = (V, E)$, if there exists a subset T of E, such that the sum of $w(e)$ of e that can connect all nodes (V) is minimal, then that subset T is called the minimum spanning tree of E, or the minimum weight spanning tree.

In geometry, the brief explanation of MST is that it is the shortest path that connects all n nodes. It is easy to understand that there are always $n-1$ edges in an MST, and these edges are independent, i.e., the path between any of two nodes is unique. The schematic diagram is shown in Figure 1. Usually, the solution of MST can be obtained using the Kruskal method [19,20] or the Prim method [21,22].

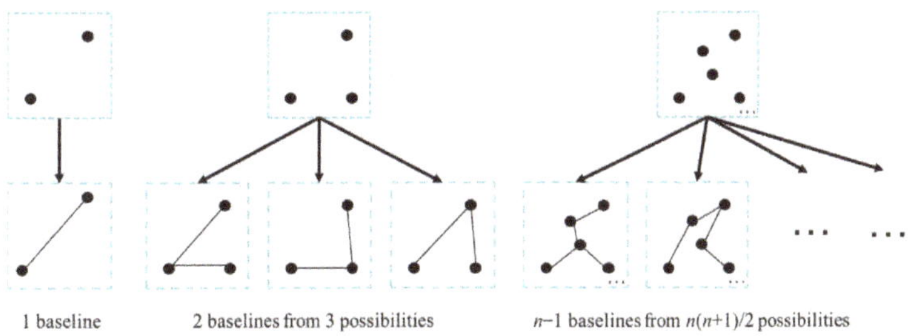

Figure 1. The schematic diagram for selecting the minimum spanning tree from all possibilities.

2.3. The Criteria—Distance, Observation, and Others

The goal of the baseline selection is to optimize the overall accuracy of baseline solutions. For the SHORTEST method, only the length of the path is used as the criterion. Mathematically, the weight $w_{\_SHORTEST}(e)$ in MST is proportional to the baseline length, i.e., the smaller the baseline length, the smaller $w_{\_SHORTEST}(e)$ is. This is based on the assumption that the closer two stations, the more thoroughly common errors can be eliminated or reduced by DD, thus the higher overall accuracies can be achieved for baseline solutions. For the OBS-MAX method, on the other hand, only the number of DD observations is used as the criterion, i.e., $w_{\_OBS\text{-}MAX}(e)$ is considered to be inversely proportional to the number of observations. The more the DD observations, the smaller $w_{\_OBS\text{-}MAX}(e)$ is. This strategy is based on the fact that more observations can bring higher redundancy in parameter estimation. WEIGHT is a synthesis of these two strategies. In this process, a normalization is introduced since the dimensions of OBS-MAX (number) and SHORTEST (meter) are not consistent [14]. Unfortunately, WEIGHT is still an empirical operation lacking theoretical support.

Since both shorter baselines and more observations can result in higher accuracies for baseline solutions, it is reasonable to adopt a special weight for MST which is both proportional to the length of the path and inversely proportional to the number of observations. The proposed criterion, i.e., the number of DD observations per unit distance, can be interpreted as the density of the observations over baselines. Thus, the proposed method is named observation-density (OBS-DEN).

Equation (1) represents the definition of $w(e)$ in different baseline selection methods.

$$\begin{cases} w_{\text{SHORTEST}}(e) = \mathrm{m_{sho}} \\ w_{\text{OBS-MAX}}(e) = 1/\mathrm{m_{obs}} \\ w_{\text{WEIGHT}}(e) = x_1/norm(\mathrm{m_{obs}}) + x_2 \times norm(\mathrm{m_{sho}}), \quad (x_1 \geq 0, x_2 \geq 0, x_1 + x_2 = 1) \\ w_{\text{OBS-DEN}}(e) = \mathrm{m_{sho}}/\mathrm{m_{obs}} \end{cases} \tag{1}$$

where $\mathrm{m_{obs}}$ denotes the number of co-viewing satellites observed by every two stations, $\mathrm{m_{sho}}$ denotes the geodetic distance of each two stations; x_1 and x_2 denote the weights applied to the $\mathrm{m_{obs}}$ and $\mathrm{m_{sho}}$ factors, respectively, which can be obtained empirically or based on the a posteriori accuracies of the solutions of the two methods. Note that in this paper, the number of observations is counted on a daily basis.

2.4. The Calculation Process of the Independent Baseline

The calculation process is illustrated in Figure 2. First, the observation files of all stations are loaded. After that, the co-observations between every two stations are retrieved per epoch. The total co-observations of each station pair of a day are aggregated, respectively. In this way, the common observation matrix $\mathbf{M_{obs}}$ is formed, and each element $\mathrm{m_{obs}}$ represents the number of common observations between every station pair. At the same time, the geodetic distance between every station pair is calculated to form the distance matrix $\mathbf{M_{sho}}$. The unit of element $\mathrm{m_{sho}}$ is meter. $\mathbf{M_{obs}}$ and $\mathbf{M_{sho}}$ are shown in Equation (2).

Then, the MST is applied to the matrix $\mathbf{M_{sho}}$, which chooses the solution that lets the sum of $\mathrm{m_{sho}}$ be the smallest. The main diagonal elements of the $\mathbf{M}$ matrix represent all available observations of individual stations or 0 distances, which are not involved in the MST generation.

$$\mathbf{M}_{\text{obs}} = \begin{bmatrix} \mathrm{m}_{\text{obs}}^{1,1} & \mathrm{m}_{\text{obs}}^{1,2} & \mathrm{m}_{\text{obs}}^{1,3} & \cdots & \mathrm{m}_{\text{obs}}^{1,n} \\ & \mathrm{m}_{\text{obs}}^{2,2} & \mathrm{m}_{\text{obs}}^{2,3} & \cdots & \mathrm{m}_{\text{obs}}^{2,n} \\ & & \ddots & \ddots & \vdots \\ & & & \ddots & \mathrm{m}_{\text{obs}}^{n-1,n} \\ & & & & \mathrm{m}_{\text{obs}}^{n,n} \end{bmatrix}, \quad \mathbf{M}_{\text{sho}} = \begin{bmatrix} \mathrm{m}_{\text{sho}}^{1,1} & \mathrm{m}_{\text{sho}}^{1,2} & \mathrm{m}_{\text{sho}}^{1,3} & \cdots & \mathrm{m}_{\text{sho}}^{1,n} \\ & \mathrm{m}_{\text{sho}}^{2,2} & \mathrm{m}_{\text{sho}}^{2,3} & \cdots & \mathrm{m}_{\text{sho}}^{2,n} \\ & & \ddots & \ddots & \vdots \\ & & & \ddots & \mathrm{m}_{\text{sho}}^{n-1,n} \\ & & & & \mathrm{m}_{\text{sho}}^{n,n} \end{bmatrix} \tag{2}$$

For the matrix $\mathbf{M_{obs}}$, the reciprocal of each element or the maximum spanning tree should be used, since the largest observation needs to be chosen instead of the smallest. Correspondingly, the WEIGHT matrix $\mathbf{M_{wei}}$ and the OBS-DEN matrix $\mathbf{M_{den}}$ can be computed as follows:

$$\begin{cases} \mathbf{M}_{\text{wei}} = x_1 \times \mathbf{M}_{\text{sho}} + x_2/\mathbf{M}_{\text{obs}}, (x_1 \geq 0, x_2 \geq 0, x_1 + x_2 = 1) \\ \mathbf{M}_{\text{den}} = \mathbf{M}_{\text{sho}}/\mathbf{M}_{\text{bos}} \end{cases} \tag{3}$$

In the subsequent experiment and data analysis, both x_1 and x_2 of the WEIGHT were set to 0.5. In order to apply the above methods with Bernese, the generated baseline file can be used to replace the baseline file generated by Bernese's default scheme. Except for the independent baseline option, all other processing sessions and parameter settings follow Bernese's default options [8].

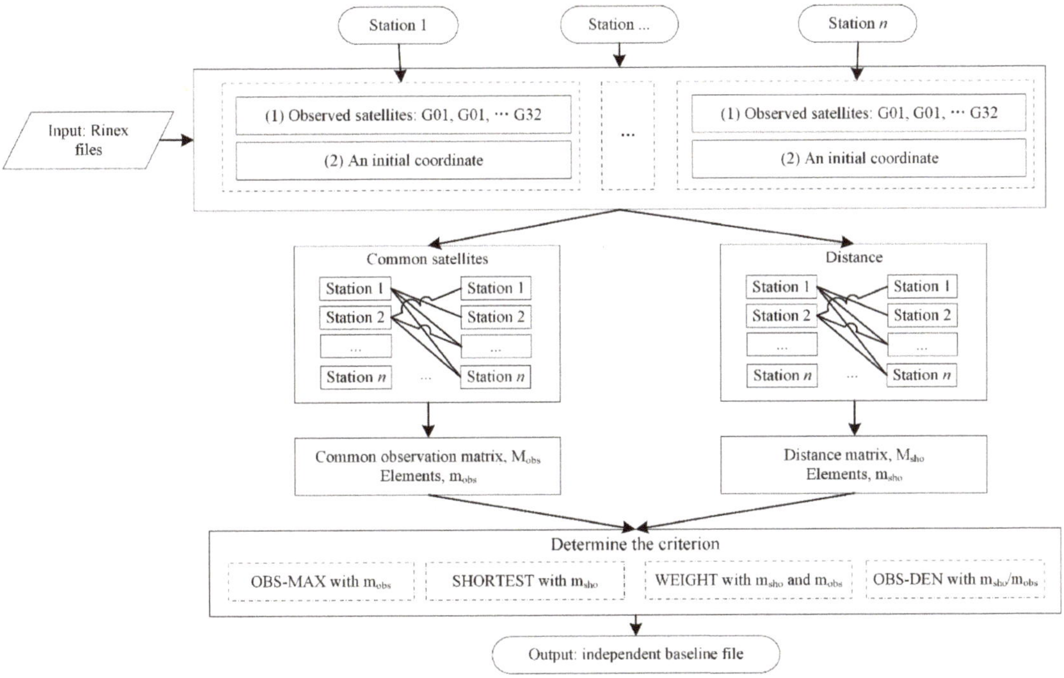

Figure 2. Flow chart for independent baseline selection, starting from reading the RINEX files of all stations, to generate different independent baseline files according to different baseline selection strategies.

2.5. Parallel Computation

Since the data of a global GNSS network may be very large, parallel computation [12,16–18,23] is recommended for such data processing. This can be done by utilizing the Bernese Processing Engine (BPE). As shown in Figure 3, first, the CPU file that comes with Bernese needs to be defined. After that, commands are submitted to the supercomputing platform, the main server of which accepts the commands and performs parallel computation according to the settings in the CPU file. In this experiment, the parallel computation was performed in two different layers, one was the parallelization of the BPEs for a single-day solution of independent baselines, and the other was the parallelization of multiple daily solutions.

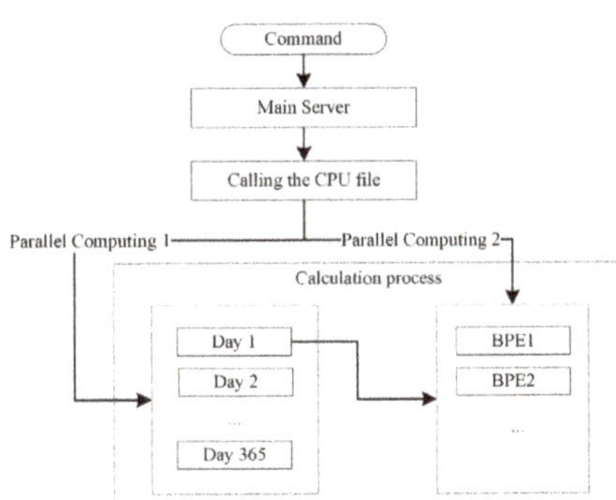

Figure 3. Flowchart of parallel computing.

3. Results

This section first shows the results of a single-day solution, including a comparison of the precision of the different methods, and the generated baseline map for our proposed method. The number of observations versus baseline length is also analyzed. After that, statistical results for one year using different methods are shown.

3.1. Single-Day Solution

At first, an experiment was conducted using globally distributed stations on 13 January 2012. The GNSS network solution was performed in the ITRF08 (International Terrestrial Reference Frame 2008). After the network adjustment, the results were converted to the local ENU (East-North-Up) coordinate system using the final coordinate products in SINEX format (Solution Independent Exchange Format) provided by CODE (Center for Orbit Determination in Europe) as a reference.

As is shown in Table 1, OBS-DEN has the smallest 3D RMS error of 7.30 mm, followed by SHORTEST. For SHORTEST, the large error in the E direction drags down its RMS. Compared with the commonly used OBS-MAX and SHORTEST, OBS-DEN has mainly improved the East and North accuracies.

Table 1. Accuracy comparison of single-day solutions of different methods. The statistics of station accuracies are calculated in the local coordinate system. The three axes of the local coordinate frame are East (E), North (N), and Up (U). The left, middle and right columns show the mean, the standard deviation (STD), and the root mean square (RMS) of the station coordinate errors of each method, respectively.

	MEAN (mm)			STD (mm)			RMS (mm)			
	E	N	U	E	N	U	E	N	U	3D
SHORTEST	−0.99	0.89	−0.53	3.66	3.15	5.97	3.79	3.28	6.00	7.81
OBS-MAX	−0.70	−1.09	0.13	2.83	3.53	7.28	2.92	3.69	7.28	8.67
WEIGHT	−1.45	0.29	0.00	2.96	3.07	6.95	3.30	3.08	6.95	8.29
OBS-DEN	−0.70	0.38	0.19	2.78	2.41	6.25	2.86	2.44	6.25	7.30

STD represents the degree of dispersion of the error for all stations. Although OBS-DEN has the lowest 3D RMS error, it has a larger STD compared to SHORTEST in the U direction. The STD of N and U components of OBS-MAX are the largest, which indicates that some individual stations may have large errors with OBS-MAX. Generally, it can be seen that the positioning errors of OBS-DEN are less discrete compared to other approaches.

Histograms of the single-day solution showing the coordinate error distributions in the E, N, and U directions of different methods are presented in Figure 4. In the East direction, the error distribution of OBS-DEN is closest to 0 and has rare discrete bars (> 10 mm), followed by OBS-MAX; the center of the error distribution of WEIGHT deviates from 0 at about −1.4 mm, and there are agminated bars around −5 mm, which makes its precision worse than OBS-DEN and OBS-MAX in East. In the N direction, the errors of OBS-DEN are more concentrated, while OBS-MAX has the most discrete values. For the U component, the results of the various methods are broadly similar, with the errors of OBS-MAX being slightly dispersed.

Figure 5 shows the baseline map of OBS-DEN. Since OBS-MAX emphasizes more DD observations, a 'STAR'-like shape, i.e., a central station with plenty of observations connected with multiple nearby stations [24,25], will exist in many regions. For example, some stations in South America, Australia, and Europe shown in [14] could connect more than 5 baselines. SHORTEST, on the other hand, has fewer baselines clustered towards the central stations, i.e., most stations are connected to only two or three baselines. As a result of a combination of the above two methods, some of the central stations of OBS-DEN, such as POVE in South America and ALIC in Australia, are connected to four baselines, while other stations are mostly connected to two or three baselines.

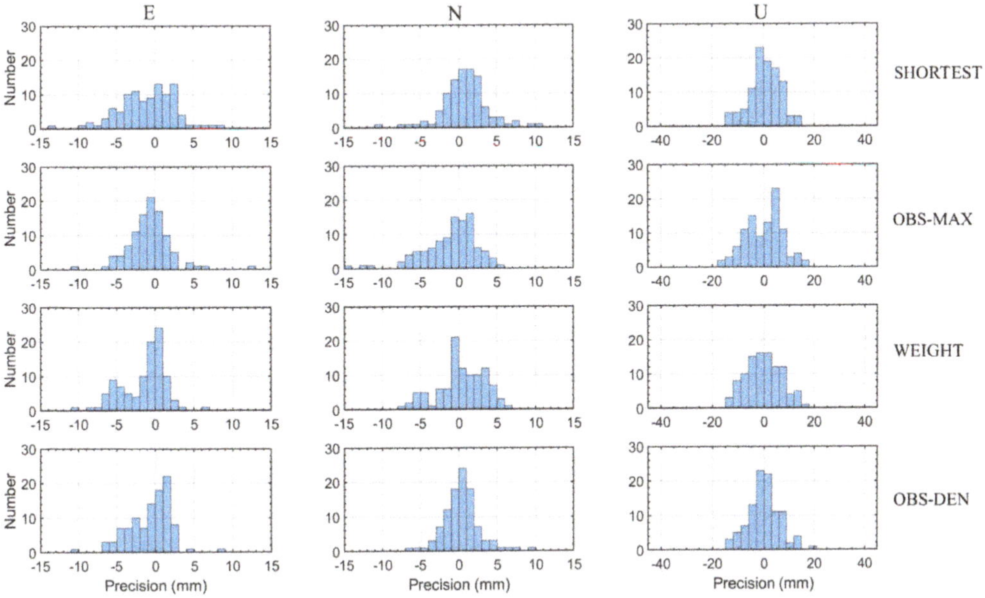

Figure 4. Histograms of single-day solutions. The x-axis of each subplot is the final station coordinate accuracy in millimeters, and the interval of each bin of North and East is 1 mm (3 mm for Up). The y-axis of each subplot represents the number of stations accommodated in each bin. The columns from left to right denote the East (E), North (N), and Up (U) component, respectively. The four methods from top to bottom are SHORTEST, OBS-MAX, WEIGHT, and OBS-DEN, respectively.

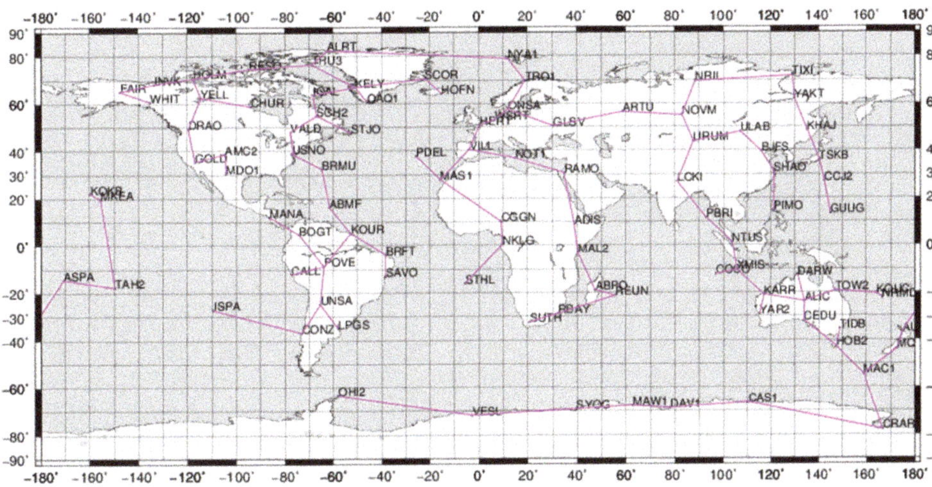

Figure 5. Independent baseline network diagram of about 100 stations generated using OBS-DEN.

The number of DD observations versus the baseline length of each baseline is plotted in Figure 6. It can be seen that the co-viewing satellites decrease roughly linearly with increasing distance. When the station spacing is greater than 17,000 km, the co-viewing satellites are almost absent. However, at distances of several thousand kilometers, there are still large numbers of common observations between stations. It is obvious that the baseline selection dominated by the number of DD observations, which is applied in OBS-MAX, is no longer applicable at this point. This is due to the long distances resulting in different tropospheric and ionospheric conditions, especially the baselines from mid-latitude to low-latitude/equatorial regions.

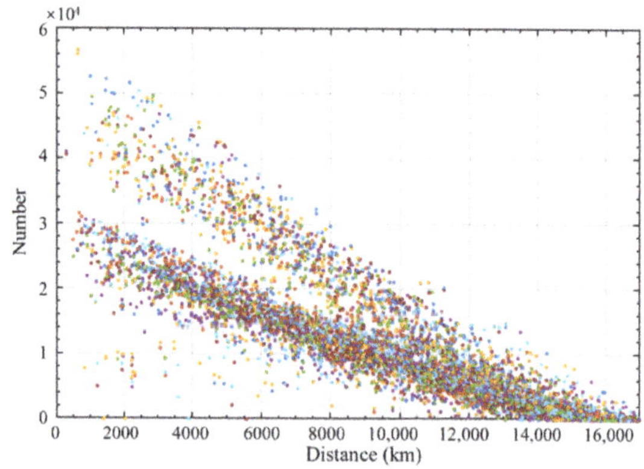

Figure 6. Variation of the number of DD observations between every two stations with distance. This is based on a single-day solution. Some of the stations have only GPS observations while others have both GPS and GLONASS observations, leading to two linear patterns in the plot.

In the data analysis, we found that some stations have both GPS and GLONASS observations while others can receive only GPS signals. That is why there are two linear aggregations presented in Figure 6. In addition to the two obvious linear aggregations, one can see some scattered dots to the lower left. These dots indicate that although the stations are close to each other, there are not many common observations. This may be due to a long-time loss of signal lock or bad observations being excluded. In this case, the baselines are short but with fewer observations. Thus, the SHORTEST method can possibly degrade the accuracy of the baseline solutions due to the introduction of these stations with a small number of observations, while OBS-DEN would avoid such baselines and instead choose baselines with sufficient satellites, but which are slightly longer, i.e., those from the upper-left region of Figure 6.

3.2. One-Year Statistical Results

To better evaluate the performance of various methods, we have tabulated the statistical results for a year. The RMS errors in each direction and the distribution are summarized in Table 2 and Figure 7. Generally, the RMS and the distribution of the methods are comparable. In more detail, the probabilities that 3D errors exceed ε, 2ε, and 3ε mm are presented, respectively, in the right column of Table 2. The threshold ε is set as 9.67 mm, which is the average 3D RMS value of the four methods. From the statistical results, we can see that OBS-DEN has the most stations with accuracies within one ε, and WEIGHT the least. However, the probability that WEIGHT is larger than 2ε and 3ε is the smallest. That is, the coordinate errors of WEIGHT lie more in the interval from ε to 3ε. OBS-DEN and OBS-MAX have more 3D errors larger than 3ε, which pulls down the performance of OBS-DEN and OBS-MAX somewhat.

In addition to the tails of the distributions explored on the right side of Table 2, the histograms showing the coordinate error distributions can be seen in Figure 7. Overall, the distribution of the four methods is similar. However, SHORTEST has fewer burrs for errors greater than 30 mm, especially in the North direction. The distributions of the four methods in the East direction seem to be a little fatter than that in the North, which shows that the STD is minimal in N. For the Up direction, there are large discrete errors around or larger than 50 mm for all four methods.

Table 2. Statistics of one-year solutions. The left side represents the RMS, and the right side represents the probability that the 3D errors for each method exceed certain thresholds. The threshold ε is set as 9.67 mm, which is the average 3D RMS value of the four methods.

	RMS				Probability		
	E (mm)	N (mm)	U (mm)	3D	<ε	<2ε	<3ε
SHORTEST	4.38	4.21	7.63	9.75	71.89%	96.17%	99.38%
OBS-MAX	3.92	3.94	7.79	9.57	71.96%	96.54%	99.16%
WEIGHT	4.14	3.92	7.78	9.64	71.82%	96.77%	99.41%
OBS-DEN	4.31	4.15	7.68	9.73	72.49%	96.45%	99.33%

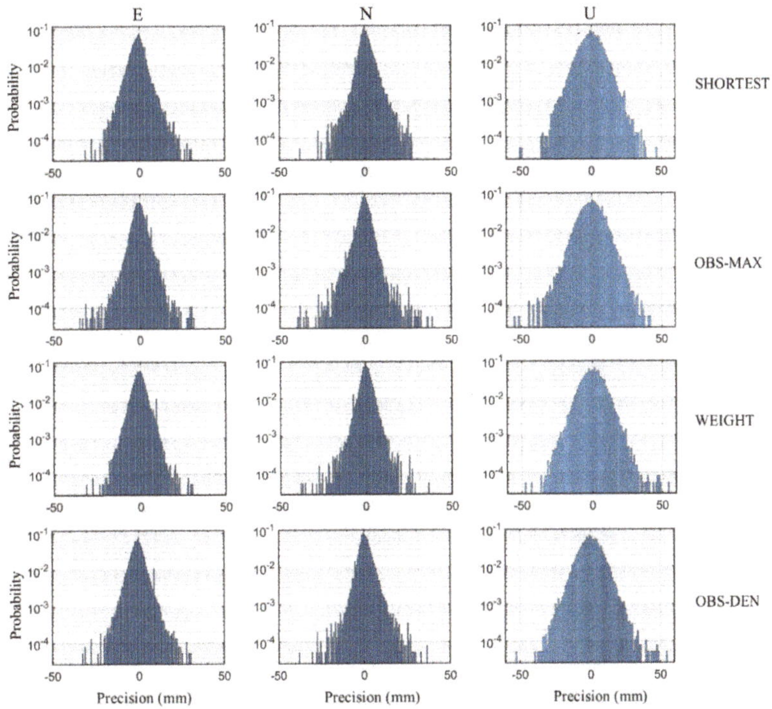

Figure 7. Histograms of one-year solutions. The x-axis of each subplot is the final station coordinate accuracy in millimeters, and the interval of each bin is 1 mm. The y-axis of each subplot, which is on a logarithmic scale, represents the quotient of the number of stations accommodated in each bin and the total number. The columns from left to right denote the East (E), North (N), and Up (U) component, respectively. The four methods from top to bottom are SHORTEST, OBS-MAX, WEIGHT, and OBS-DEN, respectively.

4. Discussion

Overall, OBS-DEN achieves the desired precision in terms of the RMS 3D of station coordinates and shows its capability to get comparable or even better precision than other methods. OBS-MAX is overly focused on the number of observations, but it may include some long baselines with low precision, while SHORTEST is excessively focused on baselines' length and may have incorporated some short baselines with less co-viewed satellites. OBS-DEN excludes these two extreme conditions by both pursuing high observation numbers and also emphasizing short baselines. When compared with WEIGHT, although the accuracy improvement of OBS-DEN is limited, it provides a rational option rather than determining weights empirically.

In theory, with the same information obtained, the final results should be equivalent, but the different ways of data processing led to inconsistent information or data involved. The advantage of OBS-MAX is that it absorbs more redundant observations involved in the adjustment. However, from the above results, especially in Figure 6, there are still a considerable number of observations at a certain range with baselines getting too long. In this case, OBS-MAX may pick some long baselines and make the results worse. In addition to the impacts of the tropospheric and ionospheric delays, the DD ambiguity is more difficult to deal with when the baseline becomes longer [26]. The advantage of SHORTEST is that it uses stations from short baselines whose atmospheric delays are basically the same. However, the shortest baseline could not necessarily exclude the baselines with few co-viewing satellites. As a synthesis of the above two methods, the ratio of the baseline length to the number of observations can be used to overcome the respective shortcomings of the previous individual methods, resulting in a better baseline solution in certain scenarios.

In addition to these most common methods, there are the maximum-ambiguity-fixed-rate method [27] and the STAR method [8]. However, the former uses the outcome of the solution as a basis for selection and cannot provide a pre-defined option for the independent baseline solution as other methods. The STAR method is commonly used for local networks rather than global ones. Therefore, only OBS-MAX and SHORTEST from the traditional methods are involved in the comparison. In future work, the performance of different constellations including positioning accuracy, number of observations, and signal quality could also be used as another baseline searching criteria.

The baseline solution precision is closely related to the station location and density, the shape of the network, and the local atmospheric environment. Different baseline search strategies can be adapted to specific situations. For example, baseline solutions at low latitudes, equatorial and polar regions are usually affected more heavily by ionospheric effects [28,29], especially during a solar maximum period. Thus, more consideration should be given to making the baselines shorter during such periods.

It should be noted that the stations selected for this experiment are globally distributed. The results of these methods may be less different in a local area network where all stations have comparable observations. For example, for a local area network [6] or network RTK (Real-Time Kinetic) [30–32], the different baseline selection methods are theoretically close to being equivalent, especially with a large number of observations of multiple systems [33]. While all stations are close to each other, the number of co-viewing satellites between them is also similar. The baselines selected by different methods may differ from each other, but the total length of the baseline and the total number of satellite observations will not vary significantly.

5. Conclusions

In light of the limitations of current independent baseline selection methods, such as OBS-MAX and SHORTEST, an alternative optimized scheme named OBS-DEN is proposed for GNSS network solutions. It is characterized by maximum co-viewing satellites per unit distance. Since the SHORTEST pursues only short baselines, there is a risk of introducing low-precision baselines with small co-observations numbers; OBS-MAX aims only for more observations and will potentially introduce baselines with large tropospheric and ionospheric differences. OBS-DEN considers both shorter paths and more DD observations in an independent baseline network. It compensates for the shortcomings of SHORTEST and OBS-MAX and does not require empirical weighting. It can be a new independent baseline search strategy for baseline selection in GNSS software, e.g., Bernese.

In both the single-day and annual solutions, OBS-DEN demonstrates its ability to obtain comparable or even higher 3D accuracies. In the single-day solution, the distribution of OBS-DEN is more concentrated. The RMS is smaller than OBS-MAX and SHORTEST. In the statistical results of annual solutions, the 3D RMS of OBS-DEN has the highest probability to be less than 9.67 mm, i.e., the average 3D RMS of all the four methods, compared to other methods.

Due to the uncertainty of the error distribution, OBS-DEN would not be better than other methods in all cases. Different network types and application scenarios correspond to different optimal baseline schemes. In scenarios where the traditional methods are both limited, OBS-DEN can be considered as the preferred scheme.

Author Contributions: Conceptualization, Y.D.; methodology, T.L.; software, T.L. and W.N.; validation, T.L., Y.M., and J.L.; formal analysis, T.L., Y.M., and J.L.; writing—original draft preparation, T.L.; writing—review and editing, Y.D.; visualization, T.L.; funding acquisition, G.X. and W.N. All authors have read and agreed to the published version of the manuscript.

Funding: This research was funded by the Open Fund of the Key Laboratory of Urban Land Resources Monitoring and Simulation, Ministry of Natural Resources, grant number (KF-2021-06-104); Guangdong Basic and Applied Basic Research Foundation, grant number (2021A1515012600); The Opening Project of Guangxi Wireless Broadband Communication and Signal Processing Key Laboratory (No. GXKL06200217); The National Nature Science Foundation of China (No. 42004012); The Natural Science Foundation of Shandong Province (No. ZR2020QD048); Wenhai Program of the S&T Fund of Shandong Province for Pilot National Laboratory for Marine Science and Technology (Qingdao) (NO. 2021WHZZB1004, 2021WHZZB1004_01).

Data Availability Statement: The data and products are downloaded from (www.igs.org, Weihai, China, 1 June 2018).

Acknowledgments: Thanks for the support from the Navigation and Remote Sensing Group of Shandong University. Mowen Li and Zhenlong Fang provided part of the code for data batch downloading; Tianhe Xu, Chunhua Jiang, and Yan Xu provided valuable discussions. This experiment was conducted at the Supercomputing Center of Shandong University in Weihai, China. We also thank Ta-Kang Yeh of Taipei University and Baoqi Sun of the National Time Service Center-Chinese Academy of Sciences, who provided assistance in using and compiling the Bernese software.

Conflicts of Interest: The authors declare no conflict of interest.

References

1. Rodriguez-Solano, C.; Hugentobler, U.; Steigenberger, P.; Bloßfeld, M.; Fritsche, M. Reducing the draconitic errors in GNSS geodetic products. *J. Geod.* **2014**, *88*, 559–574. [CrossRef]
2. Stępniak, K.; Bock, O.; Bosser, P.; Wielgosz, P. Outliers and uncertainties in GNSS ZTD estimates from double-difference processing and precise point positioning. *GPS Solut.* **2022**, *26*, 74. [CrossRef]
3. Sun, M.; Liu, L.; Yan, W.; Liu, J.; Liu, T.; Xu, G. Performance analysis of BDS B1C/B2a PPP using different models and MGEX products. *Surv. Rev.* **2022**, 1–12. [CrossRef]
4. Yang, Y.; Mao, Y.; Sun, B. Basic performance and future developments of BeiDou global navigation satellite system. *Satell. Navig.* **2020**, *1*, 1. [CrossRef]
5. Zajdel, R.; Sośnica, K.; Dach, R.; Bury, G.; Prange, L.; Jäggi, A. Network effects and handling of the geocenter motion in multi-GNSS processing. *J. Geophys. Res. Solid Earth* **2019**, *124*, 5970–5989. [CrossRef]
6. Specht, C.; Specht, M.; Dąbrowski, P. Comparative analysis of active geodetic networks in Poland. *Int. Multidiscip. Sci. GeoConf. SGEM* **2017**, *17*, 163–176.
7. Xu, G.; Xu, Y. *GPS: Theory, Algorithms and Applications*; Springer: Berlin, Germany, 2016.
8. Dach, R.; Walser, P. *Bernese GNSS Software*, Version 5.2; University of Bern: Bern, Switzerland, 2015.
9. Herring, T.; King, R.; McClusky, S. *Introduction to Gamit/Globk*; Massachusetts Institute of Technology: Cambridge, MA, USA, 2010.
10. Gilad, E.-T. Graph theory applications to GPS networks. *GPS Solut.* **2001**, *5*, 31–38.
11. Graham, R.L.; Hell, P. On the history of the minimum spanning tree problem. *Ann. Hist. Comput.* **1985**, *7*, 43–57. [CrossRef]
12. Cui, Y.; Chen, Z.; Li, L.; Zhang, Q.; Lu, Z. An efficient parallel computing strategy for the processing of large GNSS network datasets. *GPS Solut.* **2021**, *25*, 36. [CrossRef]
13. Stepniak, K.; Bock, O.; Wielgosz, P. Reduction of ZTD outliers through improved GNSS data processing and screening strategies. *Atmos. Meas. Tech.* **2018**, *11*, 1347–1361. [CrossRef]
14. Liu, T.; Xu, T.; Nie, W.; Li, M.; Fang, Z.; Du, Y.; Jiang, Y.; Xu, G. Optimal Independent Baseline Searching for Global GNSS Networks. *J. Surv. Eng.* **2021**, *147*, 5020010. [CrossRef]
15. Krypiak-Gregorczyk, A. Ionosphere response to three extreme events occurring near spring equinox in 2012, 2013 and 2015, observed by regional GNSS-TEC model. *J. Geod.* **2019**, *93*, 931–951. [CrossRef]
16. Chen, Z.; Zhiping, L.; Yang, C.; Hao, L. Parallel computing of GNSS data based on Bernese processing engine. *J. Geod. Geodyn.* **2013**, *33*, 79–82.

17. Li, L.; Lu, Z.; Chen, Z.; Cui, Y.; Sun, D.; Wang, Y.; Kuang, Y.; Wang, F. GNSSer: Objected-oriented and design pattern-based software for GNSS data parallel processing. *J. Spat. Sci.* **2019**, *66*, 27–47. [CrossRef]
18. Jiang, C.; Xu, T.; Du, Y.; Sun, Z.; Xu, G. A parallel equivalence algorithm based on MPI for GNSS data processing. *J. Spat. Sci.* **2019**, *66*, 513–532. [CrossRef]
19. Naidoo, K. MiSTree: A Python package for constructing and analysing Minimum Spanning Trees. *arXiv* **2019**, arXiv:1910.08562. [CrossRef]
20. Kruskal, J.B. On the shortest spanning subtree of a graph and the traveling salesman problem. *Proc. Am. Math. Soc.* **1956**, *7*, 48–50. [CrossRef]
21. Spira, P.M.; Pan, A. On finding and updating spanning trees and shortest paths. *SIAM J. Comput.* **1975**, *4*, 375–380. [CrossRef]
22. Prim, R.C. Shortest connection networks and some generalizations. *Bell Syst. Tech. J.* **1957**, *36*, 1389–1401. [CrossRef]
23. Cui, Y.; Lv, Z.; Li, L.; Chen, Z.; Sun, D.; Kwong, Y. A Fast Parallel Processing Strategy of Double Difference Model for GNSS Huge Networks. *Acta Geod. Cartogr. Sin.* **2017**, *46*, 48–56.
24. Paziewski, J.; Kurpinski, G.; Wielgosz, P.; Stolecki, L.; Sieradzki, R.; Seta, M.; Oszczak, S.; Castillo, M.; Martin-Porqueras, F. Towards Galileo+ GPS seismology: Validation of high-rate GNSS-based system for seismic events characterisation. *Measurement* **2020**, *166*, 108236. [CrossRef]
25. Fotiou, A.; Pikridas, C.; Rossikopoulos, D.; Chatzinikos, M. The effect of independent and trivial GPS baselines on the adjustment of networks in everyday engineering practice. In Proceedings of the International Symposium on Modern Technologies, Education and Professional Practice in Geodesy and Related Fields, Sofia, Bulgaria, 5–6 November 2009; pp. 201–212.
26. Geng, J.; Mao, S. Massive GNSS network analysis without baselines: Undifferenced ambiguity resolution. *J. Geophys. Res. Solid Earth* **2021**, *126*, e2020JB021558. [CrossRef]
27. Hua, C. Application Research of Method of Large GNSS Network Realtime Data Rapid Solution. Ph.D. Thesis, Wuhan University, Wuhan, China, 2010.
28. Liu, T.; Yu, Z.; Ding, Z.; Nie, W.; Xu, G. Observation of Ionospheric Gravity Waves Introduced by Thunderstorms in Low Latitudes China by GNSS. *Remote Sens.* **2021**, *13*, 4131. [CrossRef]
29. Dao, T.; Harima, K.; Carter, B.; Currie, J.; McClusky, S.; Brown, R.; Rubinov, E.; Choy, S. Regional Ionospheric Corrections for High Accuracy GNSS Positioning. *Remote Sens.* **2022**, *14*, 2463. [CrossRef]
30. Bakuła, M.; Przestrzelski, P.; Kaźmierczak, R. Reliable technology of centimeter GPS/GLONASS surveying in forest environments. *IEEE Trans. Geosci. Remote Sens.* **2014**, *53*, 1029–1038. [CrossRef]
31. Wang, P.; Liu, H.; Yang, Z.; Shu, B.; Xu, X.; Nie, G. Evaluation of Network RTK Positioning Performance Based on BDS-3 New Signal System. *Remote Sens.* **2021**, *14*, 2. [CrossRef]
32. Liu, J.; Zhang, B.; Liu, T.; Xu, G.; Ji, Y.; Sun, M.; Nie, W.; He, Y. An Efficient UD Factorization Implementation of Kalman Filter for RTK Based on Equivalent Principle. *Remote Sens.* **2022**, *14*, 967. [CrossRef]
33. Siejka, Z. Validation of the accuracy and convergence time of real time kinematic results using a single galileo navigation system. *Sensors* **2018**, *18*, 2412. [CrossRef]

remote sensing

MDPI

Article

Experimental Study of Accuracy of High-Rate GNSS in Context of Structural Health Monitoring

Xuanyu Qu [1,2], Bao Shu [3,4,*], Xiaoli Ding [1,2], Yangwei Lu [1], Guopeng Li [5] and Li Wang [3,4]

1 Department of Land Surveying and Geo-Informatics, The Hong Kong Polytechnic University, Hong Kong, China
2 Research Institute for Land and Space, The Hong Kong Polytechnic University, Hong Kong, China
3 College of Geological Engineering and Geomatics, Chang'an University, Xi'an 710054, China
4 Key Laboratory of Western China's Mineral Resources and Geological Engineering, Ministry of Education, Xi'an 710054, China
5 The First Geodetic Surveying Bridge of Ministry of Natural Resources, Xi'an 710054, China
* Correspondence: baos613@chd.edu.cn

Abstract: Global Navigation Satellite Systems (GNSS)-based technologies have become an indispensable part of current structural health monitoring (SHM) systems because of the unique capability of the GNSS technologies to provide accurate real-time displacement information. GNSS equipment with a data sampling rate of up to about 20 Hz has been widely used for this purpose. High-rate GNSS systems (typically up to about 100 Hz) offer additional advantages in structural health monitoring as some highly dynamic civil structures, such as some bridges, require high-rate monitoring data to capture the dynamic behaviors. However, the performance of high-rate GNSS positioning in the context of structural health monitoring is not entirely known, as studies on structural monitoring with high-rate GNSS positioning are very limited, especially considering that some of the satellite systems just reached their full constellations very recently. We carried out a series of experiments with the help of a shaking table to assess the SHM performance of a set of 100 Hz GNSS equipment and three commonly used GNSS positioning techniques, PPP (precise point positioning), PPP-AR (precise point positioning with ambiguity resolution), and RTK (real-time kinematic). We found that the standard deviations of the 100 Hz GNSS displacement solutions derived from PPP, PPP-AR, and RTK techniques were 5.5 mm, 3.6 mm, and 0.8 mm, respectively, when the antenna was in quasi-static motion, and about 9.2 mm, 6.2 mm, and 3.5 mm, respectively, when the antenna was in vibration (up to about 0.7 Hz), under typical urban observational conditions in Hong Kong. We also found that the higher a sampling rate is, the lower the accuracy of a measured displacement series is. On average, the 10 Hz and 100 Hz results are 5.5% and 10.3%, respectively, noisier than the 1 Hz results.

Keywords: high-rate GNSS; structural health monitoring; PPP; PPP-AR; RTK

Citation: Qu, X.; Shu, B.; Ding, X.; Lu, Y.; Li, G.; Wang, L. Experimental Study of Accuracy of High-Rate GNSS in Context of Structural Health Monitoring. *Remote Sens.* **2022**, *14*, 4989. https://doi.org/10.3390/rs14194989

Academic Editors: Zhetao Zhang, Wenkun Yu and Giuseppe Casula

Received: 31 August 2022
Accepted: 4 October 2022
Published: 7 October 2022

1. Introduction

Civil engineering structures, such as long-span bridges and tall buildings, are critical components of urban infrastructures. The conditions of these engineering structures are normally monitored closely as they may deteriorate over time, and the loadings may also change due to, e.g., increasing traffic loads on a bridge [1,2]. The failure of an engineering structure may disrupt transport and lead to casualties and economic losses [3]. Global Navigation Satellite Systems (GNSS)-based technologies have been widely used in monitoring displacements of engineering structures and ground motions (e.g., [4,5]). GNSS-based technologies have a distinct advantage in that they are able to provide accurate real-time displacement information. Two main GNSS data processing strategies have been in use in structural health monitoring: real-time kinematic (RTK) positioning and precise point positioning (PPP) [6]. Most of the GNSS positioning errors (e.g., clock errors, orbit errors, and tropospheric and ionospheric effects) can be removed in RTK positioning, resulting in up to millimeter-level displacement accuracy [7]. The RTK positioning technique has been widely used for monitoring engineering structures, including high-rise buildings, bridges, dams, and wind turbines [8–11]. The accuracy of RTK positioning decreases in general with the length of the baseline [12]. The PPP positioning technique obtains positioning information with a standalone GNSS receiver [13]. Aided with a precise ephemeris, the positioning accuracy of PPP can be up to the centimeter level [14,15]. The technique of PPP with ambiguity resolution (PPP-AR) can further enhance positioning accuracy [16]. PPP and PPP-AR techniques have also been applied in applications such as seismic waveform detection [17]. PPP and PPP-AR have been verified with a low-rate GNSS receiver, a shaking table, and field experiments for monitoring structural displacements [18].

Low data sampling rate GNSS receivers, typically 1–20 Hz, are the mainstream GNSS equipment deployed to date partly due to the wide availability of such GNSS equipment on the market. For example, Yu et al. [19] monitored a bridge using a 1 Hz GPS/GLONASS RTK system to identify the bridge's dynamic characteristics. Moschas et al. [20] evaluated the performance of 10 Hz PPP with the help of a shaking table. Many 10-Hz GNSS RTK systems have been deployed on several real-world long-span bridges, e.g., the Severn suspension bridge in the UK, the Tsing Ma suspension bridge, and the Stonecutters cable-stayed bridge in Hong Kong [21]. However, it is often more desirable to have higher sampling rate data to better define the dynamic characteristics of engineering structures, which may be more suitable for monitoring structures with very high-frequency vibrations, such as the short-span railway bridge, or the bridge with earthquake loading. For example, Cantieni et al. [22] reported that about one-quarter of the railway bridges in Switzerland exhibited natural frequencies above 10 Hz, with some of them reaching about 50 Hz, which is hard to detect based on the GNSS with a normal sampling rate. According to the Nyquist theorem, the Nyquist frequency of GNSS observations should be generally higher than the natural vibration frequency of the structures to avoid aliasing problems. Kaloop et al. [23] indicated that displacement information from very high-rate GNSS was useful for monitoring structural health conditions. Moreover, with the advance in GNSS hardware technologies, the high-sampling rate GNSS receivers (typically 20-100 Hz) have become widely available and emerged as a promising technique for monitoring structural deformation and health assessment. For example, Yi et al. [24] collected 50 Hz GPS data by a JAVAD TRIMMPH-1 receiver to monitor the displacement of a two-floor flexible steel frame model in real-time. Moschas and Stiros [25] used 100 Hz GNSS data (Javad Delta receiver) to study a stiff pedestrian bridge and then analyze the displacement noise characteristics [26]. Paziewski et al. [27] showed that 100 Hz PPP (recorded by Topcon Net-G5 receiver) derived displacements with the help of a filter. In addition, the high-rate GNSS receivers (model: Javad Sigma-G3TAJ) have also successfully been applied in monitoring seismic responses [28,29].

Since studies on high-rate GNSS systems in structural health monitoring are limited, it is not entirely clear how high-rate GNSS systems perform when different positioning techniques are used and different types of displacements are monitored. Although there are some early studies, it still feels necessary to carry out further investigations as some of the satellite constellations have changed. For example, the BeiDou system only reached its full constellation on 23 June 2020 [30]. We aim to evaluate the performance of a high-rate GNSS system when applied for structural health monitoring in a typical urban environment. The study tests the performance of the system when the three most commonly used positioning techniques, i.e., RTK, PPP, and PPP-AR, are used with different sampling rates and with different combinations of the GNSS satellite systems. A uniaxial shaking table and a laser transducer are deployed to simulate different motion signals for the tests. The positioning results are evaluated in both the frequency and time domains.

2. Methods and Materials

2.1. Real-Time Kinematic Positioning

RTK positioning is based on double-differenced (DD) GNSS observations between observational stations and the time to reduce the effects of the spatially correlated errors, such as the ionospheric and tropospheric delays and clock errors, over a short baseline. The DD equations of the pseudorange and carrier-phase observations are [31]

$$P_{br}^{ij,g} = \mu_r^{ij,g} s_r - \rho_{br}^{ij,g} + \delta_P \tag{1}$$

$$\lambda \Phi_{br}^{ij,g} = \mu_r^{ij,g} s_r - \rho_{br}^{ij,g} - \lambda N_{br}^{ij,g} + \delta_\Phi \tag{2}$$

where P and Φ are, respectively, the pseudorange and carrier phase DD observations; δ represents the noise in the pseudorange and carrier phase observations; μ is the LOS (line-of-sight) unit vector from a receiver to a satellite; s is the coordinate components; ρ denotes the geometric distance between the receiver and the satellite; N is the integer phase ambiguity; λ is the wavelength of the carrier phase observations; i and j are the two satellites tracked by both the base (b) and rover (r) stations, where i represents the reference satellite; and superscript g indicates the satellite system, i.e., Global Positioning System (GPS), BeiDou navigation system (BDS), or Galileo and GLONASS. Detailed computation steps for RTK are not given here for brevity. Interested readers may refer to [31].

2.2. Precise Point Positioning

The PPP determines the position of a point with a stand-alone GNSS receiver. With the help of good quality satellite orbit and clock products, the PPP-derived position can reach a centimeter level. Dual-frequency ionospheric-free (IF) pseudorange and carrier phase observations are usually used to reduce the ionospheric effects. After the clock, orbit, and DCB (differential code bias) corrections, the linearized PPP IF observations can be expressed as:

$$\begin{cases} P_{IF}^g = \mu_r^g x + c \cdot \overline{dt_r^g} + T^g + \delta_P \\ \Phi_{IF}^g = \mu_r^g x + c \cdot \overline{dt_r^g} + \overline{N_{IF}^g} + T^g + \delta_\Phi \end{cases} \tag{3}$$

with

$$\begin{cases} \overline{c \cdot dt_r^g} = c \cdot (dt_r^g + \overline{b}_{r,IF}^g / c) \\ \overline{N_{r,IF}^g} = N_{r,IF}^g + \overline{B}_{r,IF}^g - \overline{B}_{IF}^g \end{cases} \tag{4}$$

where P_{IF}^g and Φ_{IF}^g represent the corrected pseudorange and carrier phase observations; c is the speed of light in the vacuum; x is the 3D receiver coordinates; $\overline{dt_r^g}$ and $\overline{N^g}$ denote, respectively, the reformed receiver clock error and phase ambiguity; the slant tropospheric delay is marked as T^g, which contains the hydrostatic delay T_d^g and wet delay T_w^g; $\overline{b}_{r,IF}^g$ is the code hardware bias and grouped into the receiver clock offsets; $\overline{B}_{r,IF}^g$ and $\overline{B}_{IF}^g$ are the receiver and satellite phase hardware biases grouped into the ambiguity parameters, respectively. Based on the assumption that the multi-GNSS code observations share the same receiver clock, the code-specific inter-system bias (ISB) is considered for each system except for GPS.

To obtain a PPP ambiguity resolution, three steps can be followed. First, the fractional cycle biases (FCBs) of the receiver are eliminated by the between-satellite single-difference (BSSD) algorithm [32], while the FCBs of a satellite can be corrected by the WL/NL FCB products or the observable-specific signal bias (OSB) products. The WL float phase ambiguity can be fixed by rounding. Finally, the least-squares ambiguity de-correlation adjustment (LAMBDA) method [33] is adopted to resolve the NL phase ambiguities. When WL and NL ambiguities are fixed, the IF-combined fixed-phase ambiguity can be derived by the following equation.

$$\lambda_{IF} \widetilde{N}_{IF}^g = \frac{cf_2}{f_1^2 - f_2^2} \widetilde{N}_{WL}^g + \frac{c}{f_2 + f_2} \widetilde{N}_{NL}^g \tag{5}$$

where λ_{IF} denotes the wavelength of the IF combined observations; f_1 and f_2 are the frequency of the two-carrier phase; and $\widetilde{N}_{WL}^{g}$ and $\widetilde{N}_{NL}^{g}$ represent the WL and NL fixed phase ambiguity in cycles, respectively.

2.3. *Experiments*

2.3.1. Set Up of Shaking Table

A shaking table test was carried out on the 6th floor of Block Z, The Hong Kong Polytechnic University. Specifically, as shown in Figure 1a, the shaking table was equipped with a GNSS antenna and a MEMS triaxial accelerometer to record the dynamic displacement response of the shaking table. An LDT (Laser Displacement Transducer, model: SENPUM LM-S155) that can record linear displacement at up to 1,000 Hz with an accuracy of 0.01 mm was used to record the true displacement. Another GNSS receiver was set up in an open environment about 100 m away from the shaking table to serve as the base station for the RTK solution, and the location of the base and rover stations is shown in the Figure A1. A GNSS antenna was connected to the data logger (Figure 1c) to synchronize all the sensors to GPS time.

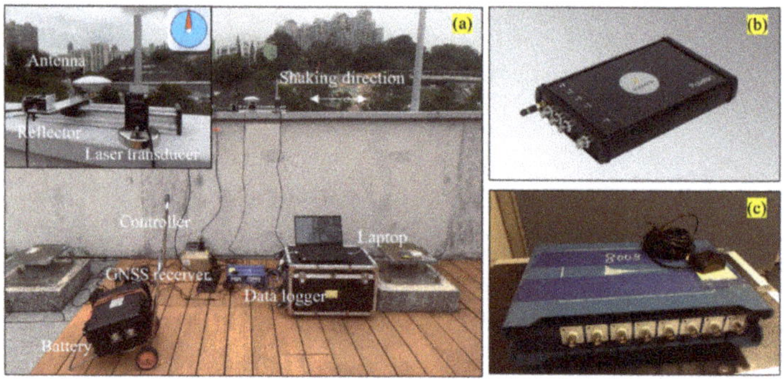

Figure 1. (**a**) Set up of experimental equipment; (**b**) Septentrio GNSS receiver (PolaRx5s) used; and (**c**) data logger with a GPS antenna.

At the start of the experiment, the shaking table remained stationary for over 1 h to obtain the initial coordinates of the antenna. Both dynamic and quasi-static motion signals were then loaded to the shaking table and recorded by the GNSS, accelerometer, and LDT. Specifically, the quasi-static motion (amplitude of about 6 cm) was firstly loaded with very low velocity to avoid possible oscillations. Subsequently, following a short stationary time (about one minute), three sets of sinusoidal motions with different amplitudes and peak frequencies were loaded successively, with a 50-s interval between any two sets.

2.3.2. Data Collection and Processing

Two Septentrio GNSS receivers (model: PolaRx5s, Figure 1b) were used in the experiments to record the GNSS observations at a sampling rate of 100 Hz (for both the rover and base stations). The Septentrio PolaRx5 can track multi-frequency and multi-constellations GNSS signals with 8 logging sessions (more technical parameters can be found at www.septentrio.com). The sampling rate of the LDT was also set to 100 Hz for easy comparison. The Phase Lock Loop bandwidth of the GNSS receiver was set to the default value of 15 Hz. The GNSS data were processed with GPSSHM [34] software and Net_Diff [35] software in RTK, PPP, and PPP-AR modes, respectively. Some details for the data processing are given in Table 1. The final ephemerids and 5-min clock products of the Multi-GNSS Experiment (MGEX) produced by the GeoForschungs Zentrum (GFZ) were used in the PPP processing [36]. The multi-GNSS rapid ephemerides and clock and phase bias products produced by Wuhan University (WUM) were adopted for PPP-AR processing [37]. Four computational schemes based on the different GNSS satellite systems were adopted. The GPS-only and BeiDou-only solutions were first considered (denoted by 'G' and 'C', respectively). The combined GPS/BeiDou and GPS/BeiDou/Galileo/GLONASS solutions were then computed (denoted by 'GC' and 'GCER', respectively). The accuracies of the estimated displacements were finally evaluated and compared.

Table 1. Details of PPP, PPP-AR, and RTK data processing.

Parameters	PPP	PPP-AR	RTK
Frequency	Dual	Dual	Dual
Sampling rate	0.01s (100 Hz)	0.01s (100 Hz)	0.01s (100 Hz)
Elevation mask	10°	10°	10°
Tropospheric model	Zenith: GPT2w Mapping function: VMF1 [38]	Zenith: GPT2w Mapping function: VMF1 [38]	
Ionospheric model	IF	IF	
Orbit	Final products from GFZ (5 min)	Rapid products from WUM (1 min)	Broadcast ephemerides
Clock	Final products from GFZ (30 s)	Rapid products from WUM (30 s)	Broadcast ephemerids
Estimator	Kalman filter	Kalman filter	Kalman filter

3. Results and Discussion

3.1. Displacement Time Series

The displacement time series computed from the 100-Hz GCER observations are shown in Figure 2. The displacements measured with the LDT (refers to True in Figure 2) are considered the truth and are also shown in the figure for comparison. The time series can be divided into three parts, namely stationary (07:00–08:26), quasi-static (08:26–08:28), and sinusoidal (08:28–08:38) motions. The first part also includes the time for the PPP convergence. The quasi-static motion simulated permanent deformation and the sinusoidal motion may correspond to vibrations.

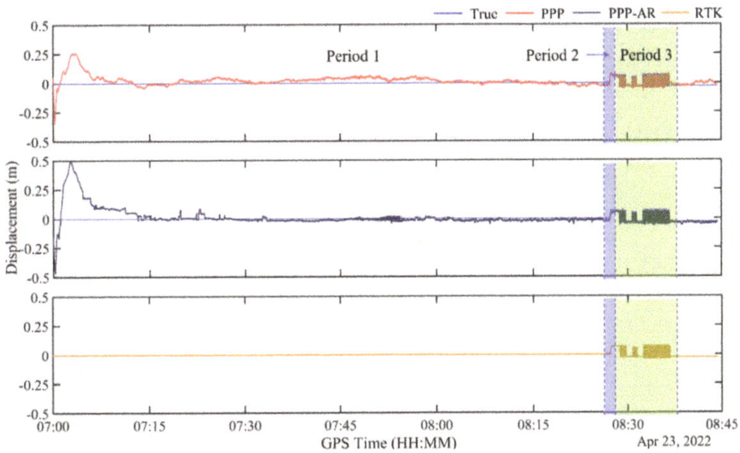

Figure 2. Displacement time series computed from the GCER observations and (top) PPP, (middle) PPP-AR, and (bottom) RTK methods.

Figures 2a and 2b show that the PPP and PPP-AR solutions took about 15 min to converge to a centimeter level. After that, the PPP, PPP-AR, and RTK techniques all captured both the stationary and dynamic displacement signals well. In general, as expected, the RTK solutions are the best. The PPP-AR and PPP results show a sub-centimeter to centimeter-level accuracy. Further analysis of the performance of the ultra-high-rate GNSS will be given in the following sub-sections.

3.2. *Evaluation of Accuracy of the 100-Hz Displacement Time Series*

3.2.1. Stationary Test

We first compared the 3D displacement results during the stationary period to study the noise level of different displacement time series computed with the PPP, PPP-AR, and RTK techniques based on different GNSS observations. The standard deviations (STDs) are calculated for quantifying the noise level of displacement and labeled in Figure 3. As depicted in Figure 3, the RTK displacement time series show more stability, followed by the PPP-AR solutions. Apparent oscillations can be observed in the displacement time series derived by the PPP technique, especially for upward direction, with the STDs being at 6.98 cm, 5.32 cm, 4.03 cm, and 2.92 cm for G, C, GC, and GCER observations, respectively. For all three positioning techniques, the GPS-only solutions have the highest noise level with STD values of up to 6.98 cm, 5.38 cm, and 0.60 cm for the PPP, PPP-AR, and RTK techniques, respectively. The main reason is that the limited GPS satellites are tracked during the experiment (about 5 G, Figure A1). With more satellite observations involved, the other three solutions (C, GC, GCER) show an improvement of up to 75% compared to the GPS-only solutions. When using four GNSS system observations, the RTK solutions can achieve a millimeter-level accuracy of 3D displacement (STD = 0.10 cm, 0.09 cm, and 0.22 cm for E, N, and U, respectively), followed by the PPP-AR that allowed millimeter-level accuracy in the horizontal direction and centimeter-level in the vertical direction.

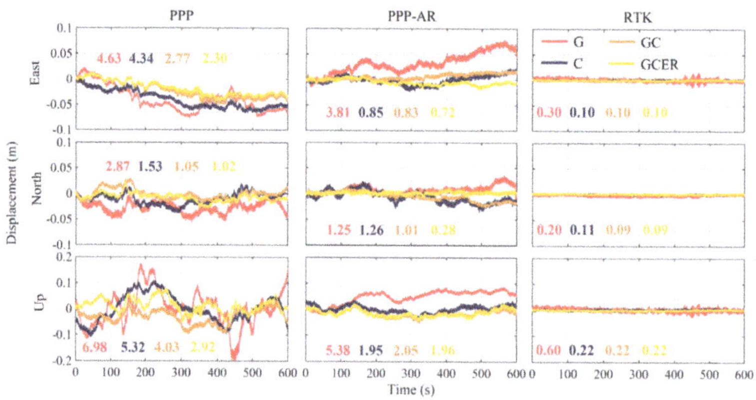

Figure 3. 100-Hz 3D displacement time series from G, C, GC, and GCER observations, respectively, and PPP, PPP-AR, and RTK techniques. The upper, middle, and lower panels represent displacements in the East, North, and Up directions, respectively, while the left, middle, and right columns represent results from PPP, PPP-AR, and RTK, respectively. The standard deviation (in cm) of each time series is given in the corresponding color.

We also calculated the amplitude spectra of the displacement for the stationary test calculated by the Fast Fourier Transform (FFT) to identify the noise spectral characteristics, shown in Figure 4. According to the FFT results of the background noise analysis, more significant attenuation of the noise level can be found in the RTK-derived displacement solutions, especially for using the GCER observations. In the low-frequency band, results computed with GCER show better performance for all three positioning techniques.

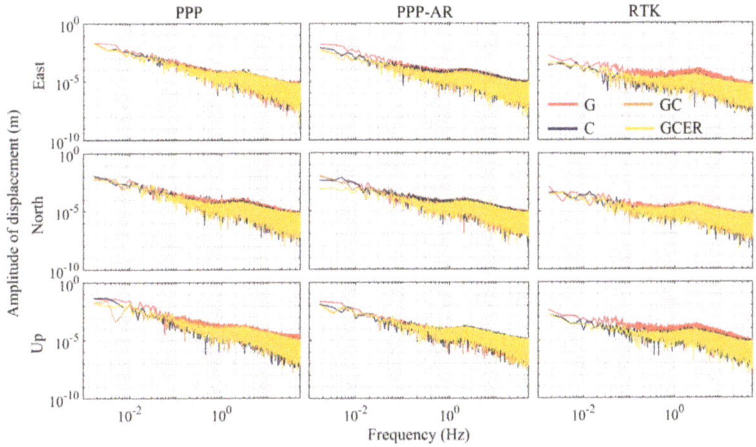

Figure 4. Amplitude–frequency spectra of 3D displacement results from G, C, GC, and GCER observations and PPP, PPP-AR, and RTK techniques in the stationary test. The upper, middle, and lower panels represent displacements in the East, North, and Up directions, respectively, while the left, middle, and right columns represent results from PPP, PPP-AR, and RTK, respectively.

3.2.2. Quasi-Static Motion Test

Figure 5 shows the 100-Hz displacement time series corresponding to the quasi-static motion test computed with the PPP, PPP-AR, and RTK techniques. Significant discrepancies can be seen among the different solutions. The GPS-only solutions have the lowest accuracy, with the RMSE value being 21.56 mm, 7.07 mm, and 6.51 mm for the PPP, PPP-AR, and RTK techniques, respectively. The BDS-only solutions show an improvement of up to 70% compared to the GPS-only solutions due mainly to the much smaller number of GPS satellites observed than that of BDS satellites (about 15 C and 5 G, Figure A2). The combined use of the GPS and BDS observations improved the accuracy by about 10% over the BDS-only solutions as judged by the RMSE value. When data from all four systems were used, the accuracy of the displacements was enhanced significantly for all three positioning techniques where the RTK technique achieved a sub-millimeter level of accuracy (RMSE = 0.75 mm), followed by the PPP-AR which allowed millimeter-level measurements of the quasi-static and abrupt motions.

The histograms of the displacement errors of the different processing techniques are shown in Figure 6 where the rows and columns represent the different processing techniques and the different combinations of GNSS observations, respectively. The range of the G and C solution errors from the PPP (first row in Figure 6) was from -4.62 to 2.27 cm, which was reduced significantly after the GC (−1.40–1.75 cm) and GCER (−0.83–1.48 cm) observations were used. The majority of the errors from the GCER observations were greater than 1 cm. When the PPP-AR technique was applied, the errors were reduced markedly to the range of −1.58–2.78 cm. The distribution pattern of the PPP-AR errors was similar to that of the PPP error; although, more errors were centered around 0 even for the GPS-only solutions. The GCER solutions achieved the best results with most of the errors within 1 cm. All the errors of the RTK results were within ±5 mm except for the GPS-only results that ranged between −0.40 and 1.40 cm.

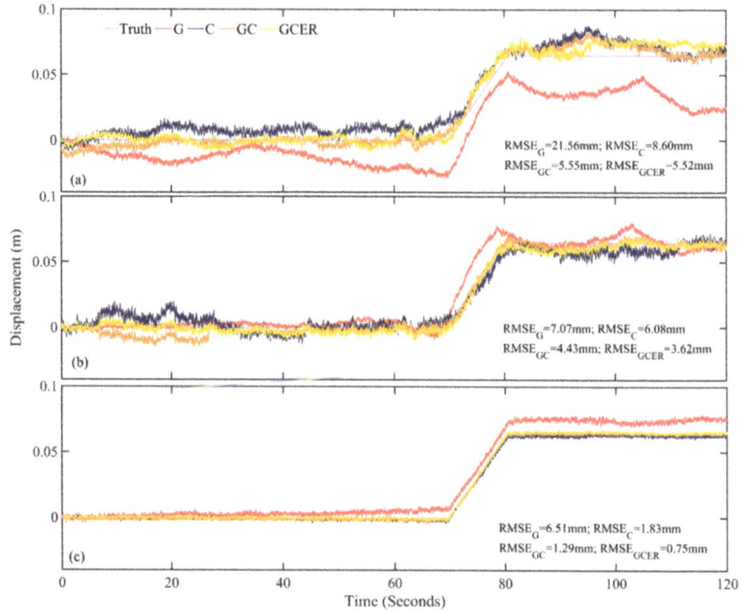

Figure 5. Displacement time series from the quasi-static motion test computed with observations from the different satellite systems and (**a**) PPP, (**b**) PPP-AR, (**c**) and RTK techniques (starting from 08:26:00 GPST).

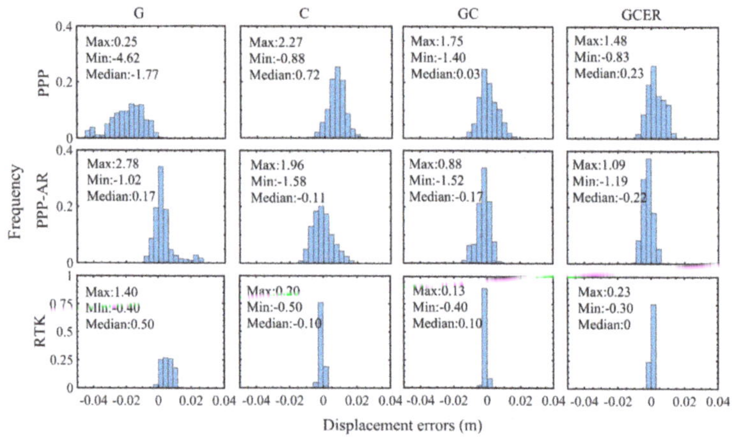

Figure 6. Histograms of displacement errors. From top to bottom, results from PPP, PPP-AR, and RTK; from left to right, results calculated by G, C, GC, and GCER, respectively. Statistics (maximum, minimum, and median) of each time series are given in each plot (in cm).

The amplitude–frequency spectra of the displacement errors from each positioning technique computed with Fast Fourier Transform (FFT) are shown in Figure 7 where the smaller amplitude indicates higher accuracy. The results indicate that the RTK results are more accurate than the other two positioning techniques, especially for those in the low-frequency band (less than 1 Hz). The displacement error spectra of PPP-AR are similar to that of PPP. The main discrepancies among different satellite system results are in the low-frequency band, where the GPS-only results have the worst performance for all positioning techniques due to limited satellites. The results suggest that using the combination of the multi-GNSS observations could improve the displacement estimation performance effectively.

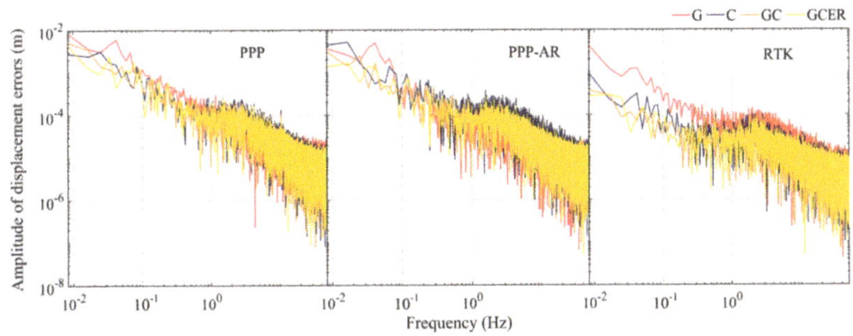

Figure 7. Amplitude–frequency spectra of displacement errors from quasi-static motion test computed with observations from the different satellite systems G, C, GC, and GCER, respectively. From left to right, results from PPP, PPP-AR, and RTK.

3.2.3. Sinusoidal Motion Test

As shown in Figure 2, a sinusoidal motion was simulated over three time periods. The motion had multiple vibration frequencies. Figure 8 shows the displacement times series from the different satellite systems and positioning techniques. The GPS-derived displacements from all the techniques are significantly biased due to the limited number of visible satellites, resulting in the RMSE being 4.59 cm, 3.60 cm, and 2.20 cm for PPP, PPP-AR, and RTK techniques, respectively. Similar to the quasi-static motion results, the BeiDou results were better than the GPS results due to much more satellites being tracked. Solutions from multi-GNSS observations (the last two rows in Figure 8) were improved noticeably compared to solutions from single satellite systems. A millimeter-level accuracy could be achieved with the RMSE of the solutions from the PPP, PPP-AR, and RTK techniques being 0.92 cm, 0.62 cm, and 0.35 cm, respectively.

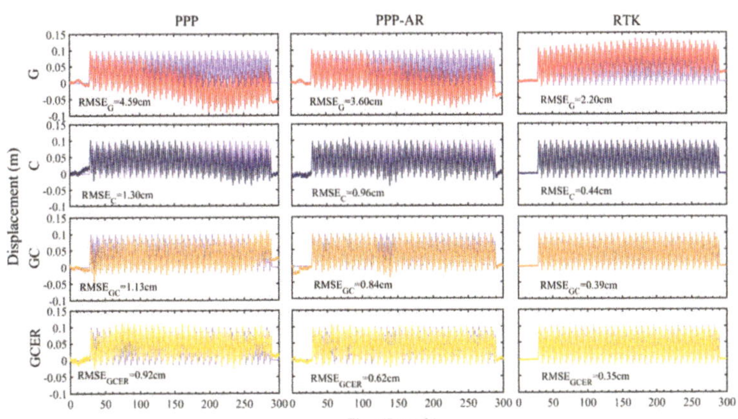

Figure 8. Sinusoidal displacement time series from the LDT (purple) and different satellite systems derived from (left column) PPP, (middle column) PPP-AR, and (right column) RTK techniques (starting from 08:32:00). From top to bottom, results, respectively, were computed from G (red), C (dark blue), GC (orange), and GCER (yellow) observations, respectively.

The histograms and spectra of the displacement errors from the different techniques are given in Figures 9 and 10, respectively. The displacement errors from PPP and GPS-only observations distribute fairly evenly over the range of −8.71–1.20 cm, while a uni-modal structure centered around ±0.5 cm can be observed for displacement errors from Beidou-only and multi-GNSS observations. The distribution patterns of the errors from PPP-AR and PPP are similar, although the error ranges of the PPP-AR results are narrower. The errors of the RTK results concentrate much more around 0 except for the GPS-only solutions that have a bias.

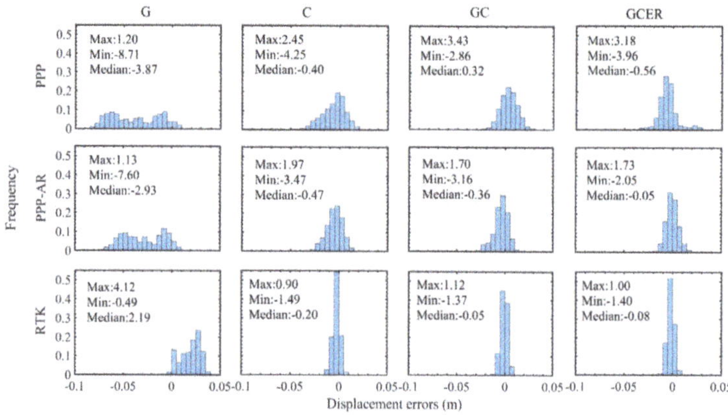

Figure 9. Histograms of displacement errors. From top to bottom, results correspond to PPP, PPP-AR, and RTK techniques while from left to right, results correspond to G, C, GC, and GCER observations, respectively. The maximum, minimum, and median values of each time series are given in each plot (in cm).

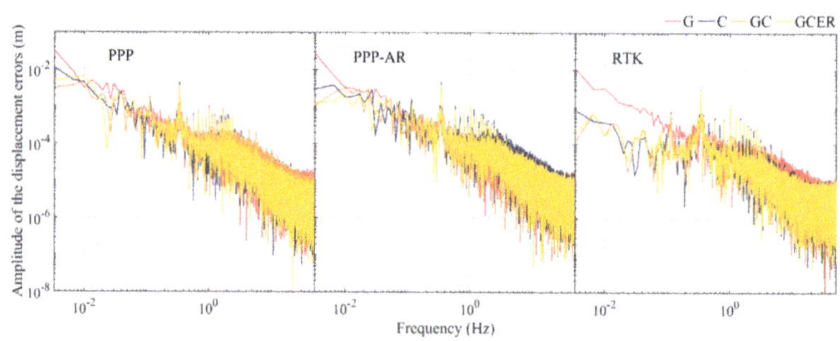

Figure 10. Amplitude–frequency spectra of displacement errors from sinusoidal motion test computed with observations from the different satellite systems G, C, GC, and GCER, respectively. From left to right, results from PPP, PPP-AR, and RTK.

The results in Figure 10 show the FFT spectra of the displacement errors. The RTK results have the best performance, especially for those in the low-frequency band (less than 1-Hz), followed by the PPP-AR technique. For different GNSS observation combinations, in terms of the low-frequency band, better results can be observed based on Beidou-only and multi-GNSS observations, while GPS-only solutions have the worst performance due to the biases (also see Figure 9). In addition, the results from different observations systems have similar accuracy in the high-frequency band.

The peak frequency and vibration amplitude of the results from each positioning technique computed with Fast Fourier Transform (FFT) are shown in Figure 11 and Table 2. The frequency and amplitude derived from the LDT solutions are regarded as benchmarks. The length and resolution of the FFT were 30,000 and 0.0033 Hz, respectively. The results indicate that the two peak frequencies could be well detected by all the positioning techniques, i.e., PPP, PPP-AR, and RTK; although, there are slight differences in the derived amplitudes.

The BeiDou-only results provided more accurate amplitude estimations than the GPS-only results, which is likely due to the BeiDou satellites being more available. The errors in the amplitudes estimated with RTK, PPP-AR, and PPP are within 0–0.3 mm, 0–0.6 mm, and 0.1–0.7 mm, respectively. In addition, the errors in the estimated amplitudes of the first peak are slightly larger than those in the second peak, which may be attributed to the fact that the second vibration frequency is more dominant and has a larger signal-to-noise ratio with respect to the first frequency [39].

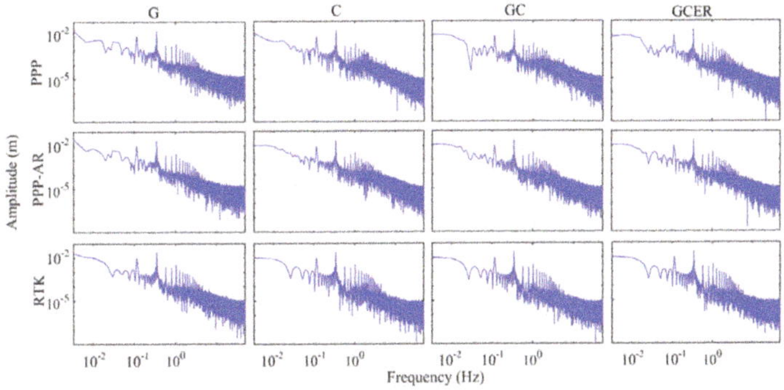

Figure 11. Amplitude–frequency spectra of displacement results from sinusoidal motion test. From top to bottom, results correspond to PPP, PPP-AR, and RTK techniques, while from left to right, results correspond to G, C, GC, and GCER observations (starting from 08:32:00 GPST).

Table 2. Peak frequencies and displacement amplitudes of the different data processing techniques. (in Hz and cm, respectively).

GNSS System	1st Peak (Ref: 0.11/1.04)			2nd Peak (Ref: 0.34/2.68)		
	PPP	PPP-AR	RTK	PPP	PPP-AR	RTK
G	0.11/0.97	0.11/0.98	0.11/1.07	0.34/2.63	0.34/2.64	0.34/2.65
C	0.11/1.02	0.11/1.03	0.11/1.03	0.34/2.64	0.34/2.66	0.34/2.67
GC	0.11/0.99	0.11/0.99	0.11/1.05	0.34/2.66	0.34/2.69	0.34/2.67
GCER	0.11/1.00	0.11/1.02	0.11/1.04	0.34/2.69	0.34/2.68	0.34/2.68

3.3. Comparison of Displacement Results from Data of Different Sampling Rates

There has been a general understanding that high-sampling rate GNSS measurements may be noisier (e.g., [26]). We first tested data acquired when the antenna was stationary to study the accuracies of the displacement time series computed from data of different sampling rates. The 3D displacement time series are given in Figure 12 together with the standard deviations of the results. In general, the higher-rate results are slightly noisier; although, the differences are not very significant, and there are also some exceptions. The results indicate that PPP and PPP-AR are more sensitive to the increase in the sampling rate. There is also a bias that can be observed in the eastern and vertical PPP-AR results, in addition to noisier data for higher-rate observations. The possible reason is that high-rate observations not only contain white noise but also have relatively high levels of color noise [28], which may degrade the accuracy of the PPP-AR solutions [40]. There are not any noticeable differences in the RTK results.

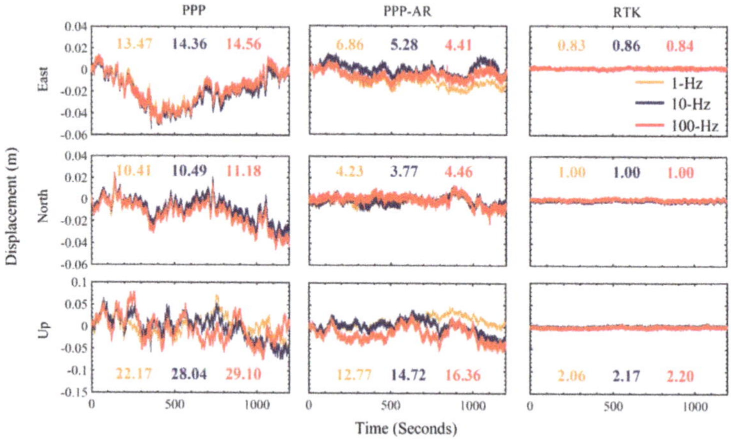

Figure 12. Three-dimensional displacement time series from multi-GNSS data of 100-, 10-, and 1-Hz, respectively, and PPP, PPP-AR, and RTK techniques. The upper, middle, and lower panels represent displacements in the East, North, and Up directions, respectively, while the left, middle, and right columns represent results from PPP, PPP-AR, and RTK, respectively. The standard deviation (in mm) of each time series is given in the corresponding color (Starting from 07:25:00 GPST).

The measured displacement time series when the antenna was in quasi-static and sinusoidal motions (starting from 08:26:00 (GPST) and lasting for 720 s) are shown in Figures 13 and A3. It can be seen from the results that the four simulated motion signals, starting at about 80 s and lasting about 570 s, can be clearly observable in all the results. Before testing the accuracies of the GNSS solutions with different sampling rates during the vibration period, we first validated whether there was any phase shift due to the time synchronization errors between the GNSS solutions and the LDT time series. Figures 14 and A5 show the 10-s comparison results between the GNSS and the LDT solutions from the different sampling rates, which are taken from the last two sinusoidal vibrations (also see Figure 13). In Figures 14 and A5, no obvious phase shift can be found between GNSS and LDT solutions for different sinusoidal oscillations.

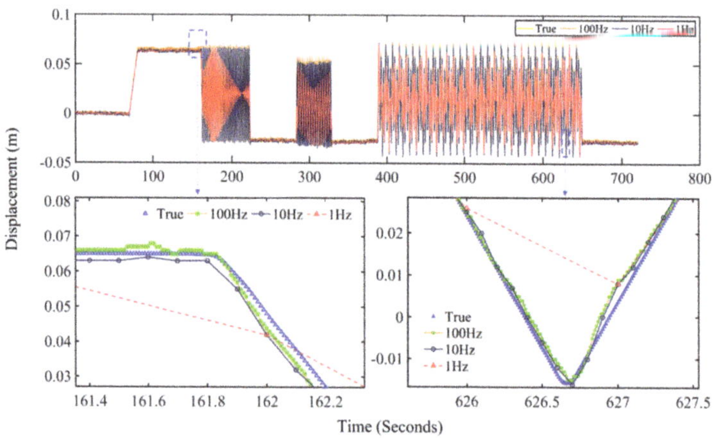

Figure 13. 100-Hz, 10-Hz, and 1-Hz displacement time series corresponding to quasi-static and sinusoidal motions derived from multi-GNSS observations and RTK technique (starting from 08:32:00 GPST).

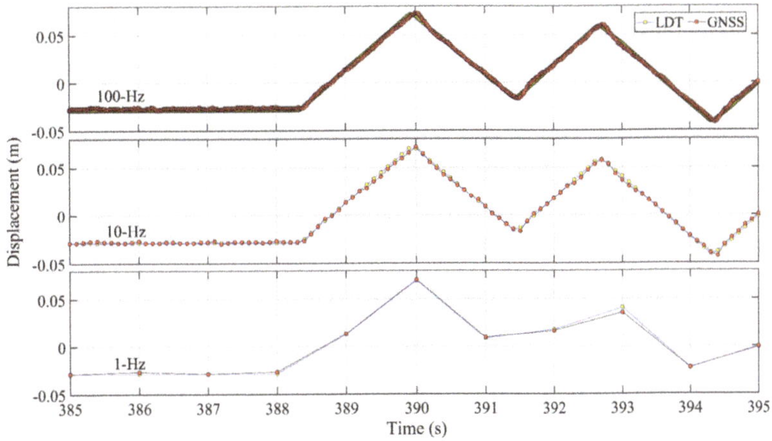

Figure 14. Comparison between 100-Hz, 10-Hz, and 1-Hz GNSS time series (red dots) and laser displacement transducer (yellow dots) for the third sinusoidal motion.

The RMSE of each of the time series in Figure 13 is given in Table 3. It is clear from the results that the higher the sampling rate is, the lower the accuracy. On average, the 10 Hz and 100 Hz results are 5.5% and 10.3% noisier than the 1 Hz results, respectively. Among the three positioning techniques, the accuracy of RTK is the highest and that of PPP is the lowest. The 1-Hz displacement results are significantly different in some parts of the time series, especially during the first two sinusoidal motion sections. This is considered due to the aliasing problem. The amplitude–frequency spectra of the first two sinusoidal motion sections calculated by the LDT solutions are shown in Figure A4. The peak frequencies of the two displacement signals were 0.50 Hz and 0.68 Hz, respectively, which could not be fully detected by the 1 Hz displacement series. Parts of the displacement time series were enlarged and shown in Figure 9b,c. The 10-Hz solutions were nearly the same as the 100-Hz solutions, although the 100-Hz solutions provided more detailed deformation. The advantage of 100 Hz solutions (i.e., the redundancy of data) is important in case of data filtering and data analysis, such as estimating the high-frequency parameters in certain loads (e.g., pre-dominant frequency of earthquakes loading [41] and accurate peak ground velocity in the near field [42]).

Table 3. RMSE of 100-, 10-, and 1-Hz displacement time series derived with PPP, PPP-AR, and RTK techniques (in mm). The values in the brackets are the percentage increases compared to the 1-Hz standard deviations.

Sampling Rate	PPP	PPP-AR	RTK
1-Hz	16.34	6.06	4.66
10-Hz	17.18 (5.1%)	6.58 (8.6%)	4.79 (2.8%)
100-Hz	17.96 (9.9%)	7.01 (15.6%)	4.91 (5.4%)

4. Conclusions

We have systematically assessed the performance of high-rate GNSS position observations computed with PPP, PPP-AR, and RTK techniques in monitoring structural deformation and dynamics based on a set of controlled experiments carried out in Hong Kong. When high-rate positioning was carried out, the RTK technique with observations from all the available GNSS offered the best accuracy among the tested positioning techniques and different combinations of satellite systems, followed by PPP-AR, which could potentially be an alternative approach to the commonly used RTK technique for millimeter-level SHM applications. The standard deviations of the 100 Hz GNSS displacement solutions derived from PPP, PPP-AR, and RTK techniques were 5.5 mm, 3.6 mm, and 0.8 mm, respectively, when the antenna was in quasi-static motion, and were about 9.2 mm, 6.2 mm, and 3.5 mm, respectively, when the antenna was in vibration (up to about 0.7 Hz), under typical urban observational conditions in Hong Kong. We have also shown that the higher a sampling rate was, the lower the accuracy of a measured displacement series was. On average, the 10 Hz and 100 Hz results were 5.5% and 10.3% noisier than the 1 Hz results, respectively. However, since the high-rate data series offered more detailed displacement information, it is still critical to deploy the high-rate systems to monitor highly dynamic civil infrastructures despite the lower accuracy.

Author Contributions: Conceptualization, X.Q. and B.S.; data curation, X.Q., B.S. and Y.L.; formal analysis, X.Q., B.S. and X.D.; funding acquisition, X.D. and L.W.; investigation, X.Q. and Y.L.; methodology, X.Q. and B.S.; project administration, X.D. and L.W.; resources, Y.L.; software, B.S.; supervision, X.D. and L.W.; validation, X.D., G.L. and L.W.; visualization, Y.L.; writing – original draft, X.Q.; writing—review and editing, B.S. and X.D. All authors have read and agreed to the published version of the manuscript.

Funding: The research was jointly supported by the National Key R&D Program of China (2021YFC3000501) and the National Natural Science Foundation of China (42004024; 41877289) and the Research Grants Council (RGC) of the Hong Kong Special Administrative Region (PolyU 152164/18E and PolyU 152233/19E).

Data Availability Statement: The phase-bias products can be obtained from ftp://igs.gnsswhu.cn/pub/whu/phasebias. The precise products of GFZ can be obtained from ftp://ftp.gfz-potsdam.de.

Acknowledgments: The first author is grateful to the Chinese National Rail Transit Electrification and Automation Engineering Technology Research Centre (Hong Kong Branch) for the PhD studentship provided. The authors thank the Wuhan University and IGS centers for providing the precise GNSS products for this study. We would also like to thank Yize Zhang of the Shanghai Observatory of the Chinese Academy of Sciences, author of the Net_Diff software, for making the Net_Diff software available.

Conflicts of Interest: The authors declare no conflict of interest.

Appendix A

Figure A1. Location of the rover and the base stations.

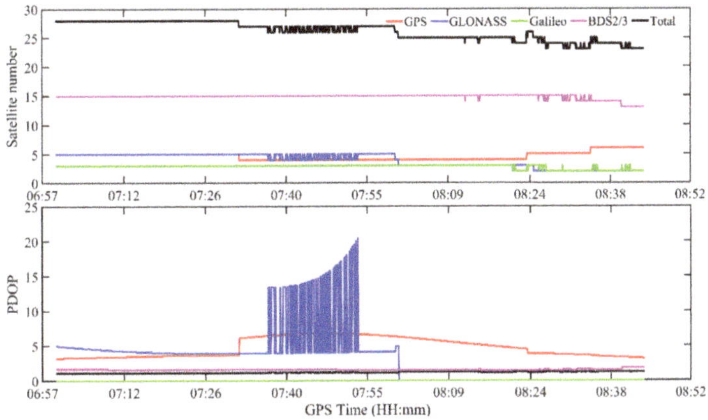

Figure A2. Satellite number (top) and Position Dilution of Precision (PDOP, bottom) during the experiment.

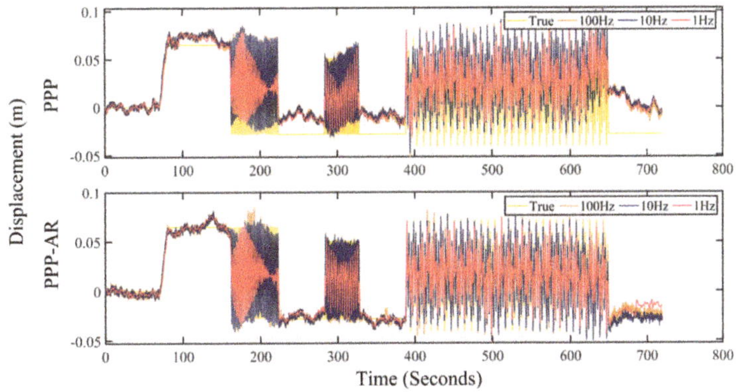

Figure A3. 100 Hz, 10 Hz and 1 Hz displacement time series corresponding to quasi-static and sinusoidal motion derived from multi-GNSS observations with PPP and PPP-AR techniques, respectively.

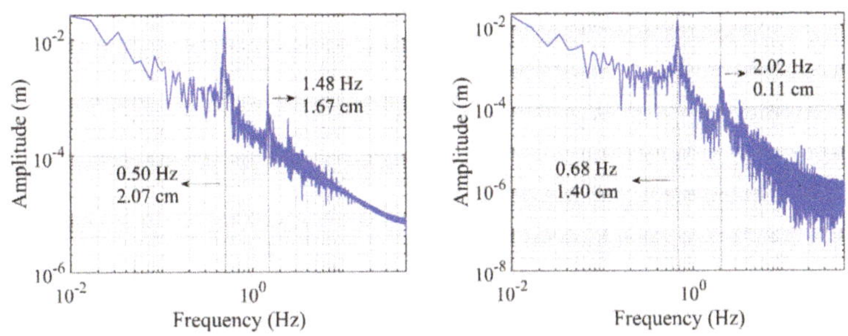

Figure A4. Amplitude-frequency spectra derived from the 100-Hz LDT data for the first (left) and second (right) sinusoidal motion sections.

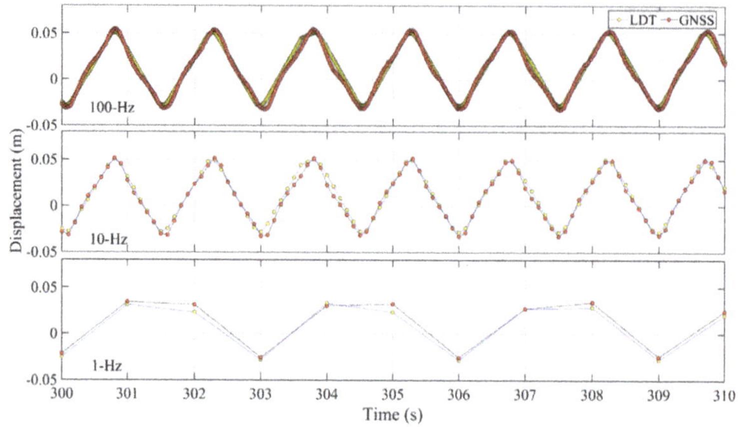

Figure A5. Comparison between 100-Hz, 10-Hz and 1-Hz GNSS time series (red dots) and laser displacement transducer (yellow dots) for the second sinusoidal motion.

References

1. Chan, W.S.; Xu, Y.L.; Ding, X.L.; Xiong, Y.L.; Dai, W.J. Assessment of dynamic measurement accuracy of GPS in three directions. *J. Surv. Eng.* **2006**, *132*, 108–117. [CrossRef]
2. Górski, P. Dynamic characteristic of tall industrial chimney estimated from GPS measurement and frequency domain decomposition. *Eng. Struct.* **2017**, *148*, 277–292. [CrossRef]
3. Xie, L.; Xu, W.; Ding, X. Precursory motion and deformation mechanism of the 2018 Xe Pian-Xe Namnoy dam Collapse, Laos: Insights from satellite radar interferometry. *Int. J. Appl. Earth Obs. Geoinf.* **2022**, *109*. [CrossRef]
4. Meng, X.; Wang, J.; Han, H. Optimal GPS/accelerometer integration algorithm for monitoring the vertical structural dynamics. *J. Appl. Geod.* **2014**, *8*, 265–272. [CrossRef]
5. Geng, J.; Pan, Y.; Li, X.; Guo, J.; Liu, J.; Chen, X.; Zhang, Y. Noise characteristics of high-rate multi-GNSS for subdaily crustal deformation monitoring. *J. Geophys. Res. Solid Earth* **2018**, *123*, 1987–2002. [CrossRef]
6. Zumberge, J.F.; Heflin, M.B.; Jefferson, D.C.; Watkins, M.M.; Webb, F.H. Precise point positioning for the efficient and robust analysis of GPS data from large networks. *J. Geophys. Res. Solid Earth* **1997**, *102*, 5005–5017. [CrossRef]
7. Zhong, P.; Ding, X.; Yuan, L.; Xu, Y.; Kwok, K.; Chen, Y. Sidereal filtering based on single differences for mitigating GPS multipath effects on short baselines. *J. Geod.* **2010**, *84*, 145–158. [CrossRef]
8. Chen, Q.; Jiang, W.; Meng, X.; Jiang, P.; Wang, K.; Xie, Y.; Ye, J. Vertical deformation monitoring of the suspension bridge tower using GNSS: A case study of the forth road bridge in the UK. *Remote Sens.* **2018**, *10*, 364. [CrossRef]
9. Dai, W.; Huang, D.; Cai, C. Multipath mitigation via component analysis methods for GPS dynamic deformation monitoring. *GPS Solut.* **2014**, *18*, 417–428. [CrossRef]
10. Xi, R.; Jiang, W.; Meng, X.; Chen, H.; Chen, Q. Bridge monitoring using BDS-RTK and GPS-RTK techniques. *Measurement* **2018**, *120*, 128–139. [CrossRef]
11. Xi, R.; Zhou, X.; Jiang, W.; Chen, Q. Simultaneous estimation of dam displacements and reservoir level variation from GPS measurements. *Measurement* **2018**, *122*, 247–256. [CrossRef]
12. Xu, P.; Shi, C.; Fang, R.; Liu, J.; Niu, X.; Zhang, Q.; Yanagidani, T. High-rate precise point positioning (PPP) to measure seismic wave motions: An experimental comparison of GPS PPP with inertial measurement units. *J. Geod.* **2013**, *87*, 361–372. [CrossRef]
13. Tang, X.; Liu, Z.; Roberts, G.W.; Hancock, C.M. PPP-derived tropospheric ZWD augmentation from local CORS network tested on bridge monitoring points. *Adv. Space Res.* **2022**, *69*, 3633–3643. [CrossRef]
14. Tu, R.; Wang, R.; Ge, M.; Walter, T.R.; Ramatschi, M.; Milkereit, C.; Dahm, T. Cost-effective monitoring of ground motion related to earthquakes, landslides, or volcanic activity by joint use of a single-frequency GPS and a MEMS accelerometer. *Geophys. Res. Lett.* **2013**, *40*, 3825–3829. [CrossRef]
15. Yigit, C.O.; El-Mowafy, A.; Anil Dindar, A.; Bezcioglu, M.; Tiryakioglu, I. Investigating Performance of high-rate GNSS-PPP and PPP-AR for structural health monitoring: Dynamic tests on shake table. *J. Surv. Eng.* **2021**, *147*, 05020011. [CrossRef]
16. Li, X.; Ge, M.; Zhang, X.; Zhang, Y.; Guo, B.; Wang, R.; Wickert, J. Real-time high-rate co-seismic displacement from ambiguity-fixed precise point positioning: Application to earthquake early warning. *Geophys. Res. Lett.* **2013**, *40*, 295–300. [CrossRef]
17. Yigit, C.O.; Gurlek, E. Experimental testing of high-rate GNSS precise point positioning (PPP) method for detecting dynamic vertical displacement response of engineering structures. *Geomat. Nat. Hazards Risk* **2017**, *8*, 893–904. [CrossRef]
18. Yu, W.; Peng, H.; Pan, L.; Dai, W.; Qu, X.; Ren, Z. Performance assessment of high-rate GPS/BDS precise point positioning for vibration monitoring based on shaking table tests. *Adv. Space Res.* **2022**, *69*, 2362–2375. [CrossRef]

19. Yu, J.; Meng, X.; Shao, X.; Yan, B.; Yang, L. Identification of dynamic displacements and modal frequencies of a medium-span suspension bridge using multimode GNSS processing. *Eng. Struct.* **2014**, *81*, 432–443. [CrossRef]
20. Moschas, F.; Avallone, A.; Saltogianni, V.; Stiros, S.C. Strong motion displacement waveforms using 10-Hz precise point positioning GPS: An assessment based on free oscillation experiments. *Earthq. Eng. Struct. Dyn.* **2014**, *43*, 1853–1866. [CrossRef]
21. Msaewe, H.A.; Psimoulis, P.A.; Hancock, C.M.; Roberts, G.W.; Bonenberg, L. Monitoring the response of Severn Suspension Bridge in the United Kingdom using multi-GNSS measurements. *Struct. Control. Health Monit.* **2021**, *28*, e2830. [CrossRef]
22. Cantieni, R. Transportation Research Board, National Research Council, Washington, DC, USA. In *Dynamic Load Testing of Highway Bridges*; National Research Council: Washington, DC, USA, 1984.
23. Kaloop, M.R.; Hu, J.W.; Elbeltagi, E. Adjustment and assessment of the measurements of low and high sampling frequencies of GPS real-time monitoring of structural movement. *ISPRS Int. J. Geo-Inf.* **2016**, *5*, 222. [CrossRef]
24. Yi, T.H.; Li, H.N.; Gu, M. Experimental assessment of high-rate GPS receivers for deformation monitoring of bridge. *Measurement* **2013**, *46*, 420–432. [CrossRef]
25. Moschas, F.; Stiros, S. Measurement of the dynamic displacements and of the modal frequencies of a short-span pedestrian bridge using GPS and an accelerometer. *Eng. Struct.* **2011**, *33*, 10–17. [CrossRef]
26. Moschas, F.; Stiros, S. PLL bandwidth and noise in 100 Hz GPS measurements. *GPS Solut.* **2015**, *19*, 173–185. [CrossRef]
27. Paziewski, J.; Sieradzki, R.; Baryla, R. Detection of structural vibration with high-rate precise point positioning: Case study results based on 100 Hz multi-GNSS observables and shake-table simulation. *Sensors* **2019**, *19*, 4832. [CrossRef]
28. Häberling, S.; Rothacher, M.; Zhang, Y.; Clinton, J.F.; Geiger, A. Assessment of high-rate GPS using a single-axis shake table. *J. Geod.* **2015**, *89*, 697–709. [CrossRef]
29. Shu, Y.; Fang, R.; Li, M.; Shi, C.; Li, M.; Liu, J. Very high-rate GPS for measuring dynamic seismic displacements without aliasing: Performance evaluation of the variometric approach. *GPS Solut.* **2018**, *22*, 121. [CrossRef]
30. Yang, Y.; Mao, Y.; Sun, B. Basic performance and future developments of BeiDou global navigation satellite system. *Satell. Navig.* **2020**, *1*, 1–8. [CrossRef]
31. Odolinski, R.; Teunissen, P.J. Best integer equivariant estimation: Performance analysis using real data collected by low-cost, single-and dual-frequency, multi-GNSS receivers for short-to long-baseline RTK positioning. *J. Geod.* **2020**, *94*, 1–17. [CrossRef]
32. Ge, M.; Gendt, G.; Rothacher, M.A.; Shi, C.; Liu, J. Resolution of GPS carrier-phase ambiguities in precise point positioning (PPP) with daily observations. *J. Geod.* **2008**, *82*, 389–399. [CrossRef]
33. Teunissen, P. The Least-Squares Ambiguity Decorrelation Adjustment: A Method for Fast GPS Ambiguity Estimation. *J. Geod.* **1999**, *73*, 587–593. [CrossRef]
34. Chan, W.S.; Xu, Y.L.; Ding, X.L.; Dai, W.J. An integrated GPS–accelerometer data processing technique for structural deformation monitoring. *J. Geod.* **2006**, *80*, 705–719. [CrossRef]
35. Zhang, Y.; Chen, J.; Gong, X.; Chen, Q. The update of BDS-2 TGD and its impact on positioning. *Adv. Space Res.* **2020**, *65*, 2645–2661. [CrossRef]
36. Montenbruck, O.; Steigenberger, P.; Prange, L.; Deng, Z.; Zhao, Q.; Perosanz, F.; Schaer, S. The Multi-GNSS Experiment (MGEX) of the International GNSS Service (IGS)–achievements, prospects and challenges. *Adv. Space Res.* **2017**, *59*, 1671–1697. [CrossRef]
37. Geng, J.; Chen, X.; Pan, Y.; Zhao, Q. A modified phase clock/bias model to improve PPP ambiguity resolution at Wuhan University. *J. Geod.* **2019**, *93*, 2053–2067. [CrossRef]
38. Böhm, J.; Möller, G.; Schindelegger, M.; Pain, G.; Weber, R. Development of an improved empirical model for slant delays in the troposphere (GPT2w). *GPS Solut.* **2015**, *19*, 433–441. [CrossRef]
39. Psimoulis, P.; Pytharouli, S.; Karambalis, D.; Stiros, S. Potential of Global Positioning System (GPS) to measure frequencies of oscillations of engineering structures. *J. Sound Vib.* **2008**, *318*, 606–623. [CrossRef]
40. Jiang, Y.; Gao, Y.; Zhou, P.; Gao, Y.; Huang, G. Real-time cascading PPP-WAR based on Kalman filter considering time-correlation. *J. Geod.* **2021**, *95*, 1–15. [CrossRef]
41. Psimoulis, P.A.; Houlié, N.; Behr, Y. Real-time magnitude characterization of large earthquakes using the predominant period derived from 1 Hz GPS data. *Geophys. Res. Lett.* **2018**, *45*, 517–526. [CrossRef]
42. Michel, C.; Kelevitz, K.; Houlié, N.; Edwards, B.; Psimoulis, P.; Su, Z.; Giardini, D. The potential of high-rate GPS for strong ground motion assessment. *Bull. Seismol. Soc. Am.* **2017**, *107*, 1849–1859. [CrossRef]

remote sensing

Article

Evaluation of BDS-3 B1C/B2b Single/Dual-Frequency PPP Using PPP-B2b and RTS SSR Products in Both Static and Dynamic Applications

Ruohua Lan, Cheng Yang *, Yanli Zheng, Qiaozhuang Xu, Jie Lv and Zhouzheng Gao

School of Land Science and Technology, China University of Geosciences Beijing, Beijing 100083, China
* Correspondence: c.yang@connect.polyu.hk; Tel.: +86-157-1284-7511

Abstract: The BeiDou Global Navigation Satellite System (BDS-3) can provide PPP-B2b satellite-based Precise Point Positioning (PPP) real-time service to the Asia–Pacific region via PPP-B2b signal transmitted from the three Geostationary Earth Orbit (GEO) satellites. This paper provides a comprehensive evaluation of the accuracies of the satellite's precise real-time orbit and clock products, including BDS-3 PPP-B2b precise products and the precise real-time products provided by four IGS centers (CAS, DLR, GFZ, and WHU). In addition, the influences of these real-time precise satellite products on the PPP positioning accuracy with single-frequency and dual-frequencies are also studied. Furthermore, the accuracies of broadcast ephemeris and IGS ultra-rapid products are studied, as well as their impact on PPP accuracies. Results illustrate that the orbits accuracies of PPP-B2b orbits are 9.42 cm, 21.26 cm, and 28.65 cm in the radial, along-track, and cross-track components, which are slightly lower than those of real-time orbits provided by the four IGS centers. However, the accuracy of PPP-B2b clock biases is 0.18 ns, which is higher than those provided by IGS Real-Time Service (RTS). In the static positioning test, the 3D positioning accuracy of B1I+B3I dual-frequencies PPP and B1C single-frequency PPP are centimeter-level while using PPP-B2b service, which is slightly lower in horizontal components compared to those obtained based on IGS RTS products. The results of the dynamic vehicle test indicate that the positioning accuracies of B1I+B2b dual-frequency PPP are about 50 cm and 120 cm in horizontal and vertical components, which are close to those of B2b single-frequency PPP using PPP-B2b service. Generally, the PPP-B2b orbit and clock accuracies on real-time PPP present similar performance to that based on IGS RTS.

Keywords: PPP-B2b service; BDS-3 new signals; RT-PPP; single-/dual-frequency RT-PPP

Citation: Lan, R.; Yang, C.; Zheng, Y.; Xu, Q.; Lv, J.; Gao, Z. Evaluation of BDS-3 B1C/B2b Single/Dual-Frequency PPP Using PPP-B2b and RTS SSR Products in Both Static and Dynamic Applications. *Remote Sens.* **2022**, *14*, 5835. https://doi.org/10.3390/rs14225835

Academic Editors: Giuseppe Casula, Zhetao Zhang and Wenkun Yu

Received: 18 October 2022
Accepted: 14 November 2022
Published: 17 November 2022

1. Introduction

The 3rd generation of the BeiDou Global Navigation Satellite System (BDS-3) officially provides services of Positioning, Navigation, and Timing (PNT) to global users as of 31 July 2020. The BDS-3 constellation consists of 30 satellites, including 24 Medium Earth Orbit (MEO) satellites, 3 Geostationary Earth Orbit (GEO) satellites, and 3 Inclined Geosynchronous Satellite Orbit (IGSO) satellites [1–3]. In addition to the standard PNT services provided by GPS, Galileo, and GLONASS, BDS-3 is the first system that provides short message communication services and Precise Point Positioning (PPP) real-time service. Real-Time Kinematic (RTK) and PPP are the most widely applied high-accuracy positioning techniques. The RTK positioning accuracy, however, declines along with the increasing baseline, and the conventional Real-time PPP (RT-PPP) based on the real-time orbit/clock products of the International GNSS Service (IGS) is limited by the internet connection. The RT-PPP service provided by BDS-3 uses the B2b (1207.14 MHz) signal to broadcast the orbit and clock corrections directly.

The PPP positioning accuracy is highly related to the satellite orbits and clocks accuracies, which are mainly obtained from the IGS [4–7]. However, IGS final products are generated with a two-week delay, and the IGS fast orbit and clock products are seventeen hours delayed. To fulfill the RT-PPP applications, IGS started providing ultra-rapid products on 3 November 2000, with centimeter to decimeter-level accuracy [8–12]. The IGS ultra-rapid files involve 24 h of properly arranged orbit and clock ephemerides. The first 24 h of each IGS ultra-rapid orbit and clock information is estimated from the latest observations from the IGS hourly tracking network. The second 24 h of data are the predicted orbits and clocks, which are extrapolated from the formerly observed orbits. To satisfy the growing demand for real-time high-precision applications (such as precision agriculture, earthquake warning, and tsunami warning) [13], IGS established a real-time working group to provide GPS/GLONASS orbit and clock corrections based on RTCM for the RT-PPP service [14,15]. Currently, IGS analysis centers (ACs), such as GFZ (Deutsches GeoForschungs Zentrum), WHU (Wuhan University), and CAS (Chinese Academic of Science), provide real-time orbits and clock products. The types of precise products and related constellations provided by different IGS centers are presented in Table 1.

Table 1. The IGS SSR message information and the corresponding interval.

IGS Center	Orbit	Clock	Code Bias	Phase Bias	VTEC	GNSS
BKG	60 s	5 s	60 s	60 s	-	G/R/E
CAS	5 s	5 s	10 s	-	60 s	G/R/E/C
CNES	5 s	5 s	5 s	5 s	60 s	G/R/E/C
DLR	30 s	5 s	30 s	30 s	-	G/R/E/C/J [1]
GFZ	5 s	5 s	5 s	-	-	G/R/E/C
WHU	5 s	5 s	-	-	-	G/R/E/C

[1] G, R, E, C, and J denote GNSS, GOLANSS, Galileo, BDS, and Quasi-Zenith Satellite System (QZSS), respectively.

Scholars have evaluated the accuracies of different IGS ACs' real-time products and their impact on the RT-PPP. Elsobeiey and Al-Harbi [16] analyzed the RT-PPP accuracies based on the global IGS stations using the IGS ultra-rapid products and real-time service (RTS). The results show that using the RT-PPP based on the IGS RTS can provide 50% position improvements compared to ultra-rapid products in terms of the Root Mean Square Error (RMSE). Li et al. [17] implemented a network RT-PPP model based on RTKLIB software and further evaluated the mode using both static and kinematic GPS experiments observations. According to the static positioning results, the horizontal and the vertical components achieved centimeter-level accuracy, while the kinematic RT-PPP achieved decimeter-level in those components. The quality of GPS/GLONASS/BDS-2/Galileo quad-system SSR products from eight different IGS analysis centers was evaluated by Wang et al. [18]. The study indicated that the RMSE between GPS real-time orbit and final precise orbit was centimeter level for all ACs, in which the Standard Deviation (STD) of the GPS clock was within 0.3 ns. The GLONASS orbit RMSE was also centimeter level, and the clock STD was twice large as that of GPS. The BDS and Galileo orbit accuracies from CNES were about 14.54 cm and 4.42 cm, respectively, and the clock accuracies of those two constellations were 0.32 ns and 0.18 ns, respectively. The study further studied the real-time product accuracy influences on the kinematic PPP with both simulation and field tests. The positioning results from the studied IGS ACs were centimeter level in horizontal and vertical components, respectively. The kinematic RT-PPP field test indicates that the positioning accuracies estimated by CNES' real-time products were better than those based on CAS precise products. Ouyang et al. [19] studied the performance of BDS-2 real-time products from four IGS centers, WHU, CNES, DLR, and GFZ. The availabilities of all these real-time products were more than 85%, and the real-time orbit accuracies of the GEO and IGSO/MEO satellites were meter level and decimeter level, respectively. Among all of these products, CNES provided the highest accuracy on 3D orbit and clock products. In static PPP mode, the convergence time on average was less than 1.5 h with 11.0 cm positioning accuracy, while in the kinematic mode, the convergence time was 2.11 h~9.84 h with 30.7 cm~68.0 cm positioning accuracy. Ge et al. [20] evaluated the accuracy availability of the BDS-3 real-time products from CENS using continuous data of 41 days. The results showed that satellite orbit accuracies were about 6 cm in radial and cross-track components but more than 10 cm in the along-track component, and the accuracy of the clock was 0.46 ns. The BDS-3 real-time product availability from CENS is over 80%. The PPP positioning errors in static were within 3 cm at East (E), North (N), and Up (U) components, which was comparable to that of GPS RT-PPP.

Different from the internet-based RT-PPP service provided by the IGS, BDS-3 broadcasts the correction of satellite orbit and clock for both BDS-3 and GPS via the PPP-B2b signal. Simultaneously, the BDS-3 Differential Code Bias (DCB) is broadcast to facilitate dual-frequency PPP positioning applications [3]. Table 2 lists the broadcast PPP-B2b correction messages [21–23].

Table 2. The PPP-B2b message information.

Information	Message Type	Sample Rate (s)	Nominal Validity (s)
Satellite mask	1	48	-
Orbit correction and User range accuracy	2	48	96
Differential code bias	3	48	86,400
Clock correction	4	6	12

The performance evaluations of PPP-B2b products have gained interest from researchers recently. Yang et al. [3] presented the positioning accuracy evaluation of dual-frequency B1C/B2a PPP with Ionosphere-Free (IF) combination based on the PPP-B2b data from 1 August to 30 November 2020. The results indicated that about 30 cm and 50 cm positioning accuracies in the horizontal and vertical components could be achieved with a 30 min convergence time. Lu et al. [21] analyzed the accuracy, integrity, and stability of the orbit clock and DCB corrections of both BDS and GPS provided by the BDS-3 PPP-B2b signal, which proved that the PPP-B2b signal could provide stable and accurate PPP services in China and surrounding areas. Nie et al. [22] compared the orbits and clocks accuracies between PPP-B2b corrections and GFZ final products with three days of observation. The BDS-3 satellite orbit accuracies in radial, along-track, and cross-track components are about ten centimeters, and the BDS-3 clock accuracy was centimeter-level. The static PPP with PPP-B2b correction messages achieved centimeter-level accuracy in the E, N, and U components, and the kinematic PPP of those also achieved centimeter-level accuracy. Tao et al. [23] illustrated that the real-time BDS-3 PPP-B2b correction messages presented better availability and integrity than CNES on BDS satellites. The positioning accuracy of PPP-B2b products-based BDS-3 kinematic PPP can achieve centimeter-level. Xu et al. [24] provided an extensive evaluation of the RT-PPP performance of the PPP-B2b products in terms of matching characteristics, product accuracies, and availability. The results showed that the RMSE of the MEO satellite orbit in radial, along-track, and cross-track components were 6.8 cm, 33.4 cm, and 36.6 cm, respectively, and the accuracy of the clock was 0.2 ns. The availability of BDS-3 PPP-B2b products is over 80%. The PPP test results presented that the accuracies of BDS-3 dual-frequency kinematic RT-PPP after convergence were 11 cm and 17 cm in horizontal and vertical components. In addition, the positioning accuracy of the B1C/B2a IF combination PPP was better than that of the B1I/B3I IF PPP. Ren et al. [25] evaluated the accuracy of PPP-B2b products and the corresponding RT-PPP, and the results showed that the real-time orbit accuracy is about 7.25, 24.79, and 25.87 cm in radial, along-track, and cross-track components for BDS-3 satellites, and 7.29, 30.98, and 21.93 cm for those of GPS satellites. The STD of the clock offset errors for BDS-3 and GPS are within 0.2 ns and 0.15 ns, respectively. The positioning accuracy of real-time BDS+GPS integrated static PPP was about 1.07 cm, 2.69 cm, and 2.25 cm in the N, E, and U directions, and the kinematic PPP of those were about 3.6 cm, 5.9 cm, and 9.4 cm in the N, E, and U directions.

Earlier studies evaluated the PPP-B2b performance using the dual-frequency PPP with static and simulated kinematic mode. However, the PPP-B2b RT-PPP based on the practical kinematic data has hardly been studied, especially using the BDS-3 new signals

(B1C and B2b). Therefore, this contribution will present a comprehensive performance evaluation on the real-time orbit and clock products provided by broadcast ephemeris, BDS-3 PPP-B2b, CAS, DLR, GFZ, and WHU, as well as the ultra-rapid products from WHU. The performance of the kinematic dual-and single-frequency RT-PPP is analyzed with B1C+B2b observation data collected in Beijing, China. The paper is arranged as follows. The recovery method of real-time orbit and clock offset and the mathematical models of dual-/single-frequency PPP are described in the next section. Then, the accuracy of the orbit and clock offset provided by broadcast ephemeris, PPP-B2b, CAS, DLR, GFZ, and WHU are evaluated separately. The positioning accuracy with dual-/single-frequency real-time dynamic PPP is afterward assessed, and the conclusions about the accuracy of real-time satellite orbit and clock products and their impacts on dual-/single-frequency PPP are drawn.

2. Methodologies

The recovery methods of PPP-B2b real-time orbit and clock, the evaluation principle, and the models of real-time dual-frequency and single-frequency PPP are presented in this section.

2.1. PPP-B2b Orbit Recovery

BDS-3 PPP-B2b orbit and clock products are corrections with respect to the broadcast ephemeris. The orbit correction coordinate is related to the satellite-fixed coordinate system in radial, along-track, and cross-track directions. The satellite positions used for PPP, however, are located in the ECEF frame. Thus, the orbit corrections should be transformed into the ECEF system to correct the satellite positions from the broadcast ephemeris [26]. The corrected real-time precise satellite position can be expressed as,

$$\begin{bmatrix} X \\ Y \\ Z \end{bmatrix}_t = \begin{bmatrix} X_{brdc} \\ Y_{brdc} \\ Z_{brdc} \end{bmatrix}_t - \boldsymbol{R} \cdot \begin{bmatrix} \Delta_r \\ \Delta_a \\ \Delta_c \end{bmatrix}_t \tag{1}$$

With

$$\boldsymbol{R} = \begin{bmatrix} \frac{\dot{\boldsymbol{r}}}{|\dot{\boldsymbol{r}}|} \times \frac{\boldsymbol{r}\times\dot{\boldsymbol{r}}}{|\boldsymbol{r}\times\dot{\boldsymbol{r}}|} & \frac{\dot{\boldsymbol{r}}}{|\dot{\boldsymbol{r}}|} & \frac{\boldsymbol{r}\times\dot{\boldsymbol{r}}}{|\boldsymbol{r}\times\dot{\boldsymbol{r}}|} \end{bmatrix} \tag{2}$$

where Δ_r, Δ_a and Δ_c represent the position corrections in the radial, along-track, and cross-track directions, respectively; $\begin{bmatrix} X & Y & Z \end{bmatrix}^T$ and $\begin{bmatrix} X_{brdc} & Y_{brdc} & Z_{brdc} \end{bmatrix}^T$ represent the corrected satellite coordinates in the EFEC system and the satellite coordinates calculated from the broadcast ephemeris, respectively; $\boldsymbol{R}$ is the transformation matrix from the satellite-fixed coordinate system to the ECEF system [26,27]; $\boldsymbol{r}$ and $\dot{\boldsymbol{r}}$ represent the satellite position and velocity calculated from the broadcast ephemeris, respectively.

It should be indicated that the orbit products provided by IGS ACs in this article are with respect to the Center of Mass (CoM) of satellites. In contrast, the orbits of PPP-B2b products are with respect to the satellite Antenna Phase Center (APC); the BDS is based on the APC of B3I frequency, while GPS is based on the APC of IF combination [26]. Therefore, the antenna phase center correction should be considered. The relationship between APC and CoM can be described by,

$$\begin{bmatrix} X_{APC} \\ Y_{APC} \\ Z_{APC} \end{bmatrix} = \begin{bmatrix} X_{CoM} \\ Y_{CoM} \\ Z_{CoM} \end{bmatrix} + \boldsymbol{A}^T \cdot \begin{bmatrix} N_{PCO} \\ E_{PCO} \\ U_{PCO} \end{bmatrix} \tag{3}$$

where $\begin{bmatrix} X_{APC} & Y_{APC} & Z_{APC} \end{bmatrix}^T$ and $\begin{bmatrix} X_{CoM} & Y_{CoM} & Z_{CoM} \end{bmatrix}^T$ represent the APC and CoM satellite position in the EFEC system; $\begin{bmatrix} N_{PCO} & E_{PCO} & U_{PCO} \end{bmatrix}^T$ represent the PCO correction of the satellite from the latest "igs14.atx" file released by IGS; $\boldsymbol{A}$ is the satellite attitude matrix.

2.2. PPP-B2b Clock Recovery

The clock corrections broadcast by PPP-B2b can be recovered to form the precise clock by [26],

$$t^s = t^s_{brdc} - C_0/c \tag{4}$$

where t^s, t^s_{brdc}, C_0, and c represent satellite precise clock, satellite clock calculated from the broadcast ephemeris, PPP-B2b clock corrections, and the speed of light in a vacuum, respectively.

It should be indicated that the BDS-3 precision clock provided by PPP-B2b is based on the B3I frequency, and all RTS clocks are based on the IF combination [26]. Therefore, all of these clocks for a GNSS system should be projected to one frequency. According to [26], the following method can be adopted to realize such projection,

$$t^s_j = t^s - b^s_j \tag{5}$$

where t^s_j is the satellite precise clock of signal j; b^s_j represents the DCB between signal j and B3I.

2.3. Orbit and Clock Evaluation Methods

Several methods can be utilized to evaluate the accuracy of precise orbit and clock. For example, the overlapping orbit comparison, satellite laser ranging inspection, and comparison with a reference product. The first two usually are used to evaluate the final products of IGS precise products. Thus, the third method is adopted in this paper by using the final products provided by WHU as references. The orbit differences can be expressed as,

$$\begin{bmatrix} \Delta X \\ \Delta Y \\ \Delta Z \end{bmatrix}_t = \begin{bmatrix} X_{RT} \\ Y_{RT} \\ Z_{RT} \end{bmatrix}_t - \begin{bmatrix} X_{WHU} \\ Y_{WHU} \\ Z_{WHU} \end{bmatrix}_t \tag{6}$$

where t is the epoch number; $[\Delta X \ \ \Delta Y \ \ \Delta Z]^T$, $[X_{RT} \ \ Y_{RT} \ \ Z_{RT}]^T$, and $[X_{WHU} \ \ Y_{WHU} \ \ Z_{WHU}]^T$ are the orbit differences, real-time orbit, and final orbit vector.

The clock products from different ACs contain a benchmark inconsistency error. In this paper, to eliminate the system inconsistency, the GPS and BDS-3 constellations use G01 and C19 as reference satellites for the single difference from other satellites [28]. The STD of the clock difference between real-time and the reference clock after a single difference is estimated. The degree of clock consistency greatly influenced the solution accuracy of PPP [29]. The PPP-B2b products are estimated using the real-time observations collected by the regional tracking stations in China. Thus, the satellites rise or fall more frequently compared with using global observations. Furthermore, it is challenging to select a stable reference BDS satellite to generate the inter-satellite differences. To assess the accuracy of PPP-B2b clock products, we re-edited the inter-satellite difference clock using the method proposed by [23]. The expression of this method can be written as,

$$\nabla\Delta t^s = (t^s_{RT} - t^s_{WHU}) - \frac{1}{M}\sum_{i=1}^{M}(t^i_{RT} - t^i_{WHU}) \tag{7}$$

where $\nabla\Delta t^s$ is the double-difference clock value; t^s_{RT} is PPP-B2b real-time clock; t^s_{WHU} is the final precise clock of WHU; M is the number of available satellites.

Due to the observation discontinuity of the BDS satellites, the clock values estimated in Equation (7) are segmented, which affects the clock STD value. Therefore, the discontinuity is compensated by,

$$\nabla\Delta t^s_t = \nabla\Delta t^s_t - \Delta D_{t,t-1} \tag{8}$$

With

$$\Delta D_{t,t-1} = \begin{cases} \frac{1}{M}\sum\limits_{i=0}^{M} \Delta\nabla\Delta t^{i}_{t,t-1}, \left|\frac{1}{M}\sum\limits_{i=0}^{M} \Delta\nabla\Delta t^{i}_{t,t-1}\right| \geq 0.1 \text{ ns} \\ 0, \left|\frac{1}{M}\sum\limits_{i=0}^{M} \Delta\nabla\Delta t^{i}_{t,t-1}\right| < 0.1 \text{ ns} \end{cases} \tag{9}$$

where $\Delta D_{t,t-1}$ is the compensation term to compensate the discontinuity in the double-difference clock; $\Delta\nabla\Delta t^{i}_{t,t-1}$ is the inter-epoch difference clock based on double difference clock.

2.4. Mathematical Model of Real-Time Single-/Dual-Frequency PPP

The measurements used for RT-PPP are pseudo-range (P) and carrier-phase (L), which can be expressed as [29,30],

$$P_j = \rho_r^s + c(t_r - t^s) + I_{r,j}^s + T_r^s + c(b_{r,j} - b_j^s) + \varepsilon_{P,j} \tag{10}$$

$$L_j = \rho_r^s + c(t_r - t^s) - I_{r,j}^s + T_r^s - \lambda_j(N_j + B_{r,j} - B_j^s) + \varepsilon_{L,j} \tag{11}$$

where ρ_r^s indicates the geometric distance from the satellite s to the receiver r; t_r and t^s are the receiver clock and the satellite clock offset, respectively; $I_{r,j}^s$ and T_r^s are the ionospheric delay and tropospheric delay, respectively; $b_{r,j}$ and b_j^s are the DCB of the pseudo-range on receiver and satellite, respectively; λ_j and N_j are wavelength and ambiguity of carrier-phase; $B_{r,j}$ and B_j^s denote the Un-calibrated Phase Delay (UPD) on the satellite and receiver, respectively; $\varepsilon_{P,j}$ and $\varepsilon_{L,j}$ are observing noise of pseudo-range and carrier-phase, respectively.

2.4.1. Dual-Frequency PPP

The dual-frequency ionosphere-free combination pseudo-range and carrier-phase are used to form the observations of dual-frequency PPP [31],

$$P_{IF} = \frac{f_i^2}{f_i^2 - f_j^2}P_i - \frac{f_j^2}{f_i^2 - f_j^2}P_j = \rho_r^s + c(t_r - t^s) + T_r^s + \varepsilon_{P,IF} \tag{12}$$

$$L_{IF} = \frac{f_i^2}{f_i^2 - f_j^2}L_i - \frac{f_j^2}{f_i^2 - f_j^2}L_j = \rho_r^s + c(t_r - t^s) - \lambda_{IF}N_{IF} + T_r^s + \varepsilon_{P,IF} \tag{13}$$

where f_i is the frequency number; the other symbols have the same means as above.

In addition, if the observed frequencies in Equations (12) and (13) are different from that of precise real-time products, a DCB correction should be applied [32]. In this study, both B1I+B3I and B1I+B2b IF combinations are employed for static and kinematic PPP solutions, and the following equations should be employed,

$$\begin{cases} t^s_{B1I,B3I} = t^s - \frac{f^2_{B1I}}{f^2_{B1I} - f^2_{B3I}} b^s_{B1I} \\ t^s_{B1I,B2b} = t^s - \left(\frac{f^2_{B1I}}{f^2_{B1I} - f^2_{B2b}} b^s_{B1I} - \frac{f^2_{B2b}}{f^2_{B1I} - f^2_{B2b}} b^s_{B2b}\right) \end{cases} \tag{14}$$

where $t^s_{B1I,B3I}$ and $t^s_{B1I,B2b}$ represent satellite precise clocks using the B1I+B3I and B1I+B2b IF combinations; f_{B1I}, f_{B2b} and f_{B3I} represent the frequencies corresponding to the B1I, B2b, and B3I observations, respectively.

In this case, the corresponding parameters that should be estimated are position vector, receiver clock offset, residual of troposphere delay, and IF ambiguity, which can be written as,

$$\boldsymbol{x} = \begin{bmatrix} \delta x_r & \delta y_r & \delta z_r & c \cdot \delta t_r & d_{wet} & \boldsymbol{N}_{IF} \end{bmatrix}^T \tag{15}$$

where δx_r, δy_r and δz_r indicate the corrected value of the receiver coordinates in three directions; δt_r and d_{wet} indicate the receiver clock offset and the zenith wet tropospheric delay; $\boldsymbol{N}_{IF}$ denotes the IF ambiguity vector.

2.4.2. Single-Frequency Model

For single-frequency PPP, Equations (10) and (11) form the observation equations. Here, the B1C and B2b measurements will be used in this paper. Thus, the satellite DCB corrections should be considered,

$$\begin{cases} t^s_{B1C} = t^s - b^s_{B1C} \\ t^s_{B2b} = t^s - b^s_{B2b} \end{cases} \tag{16}$$

Compared with the IF model, the ionospheric delay of each satellite and the receiver DCB cannot be eliminated. Therefore, they are estimated as parameters [33–35]. The corresponding state vector can be expressed as,

$$x = \begin{bmatrix} \delta x_r & \delta y_r & \delta z_r & ct_r & d_{wet} & \boldsymbol{N}_j & b_r & \boldsymbol{I}_r \end{bmatrix}^T \tag{17}$$

where b_r and $\boldsymbol{I}_r$ are the receiver DCB and the ionospheric delay vector.

Due to the correlations between the ionosphere and receiver DCB, we introduce the following pseudo-observations to improve the estimation accuracy of single-frequency PPP [33].

$$I^s_{r,j} = 40.28STEC/f_j^2 + \omega_{I^s_{r,j}}, \omega_{I^s_{r,j}} \sim (0, \sigma^2_{I^s_{r,j}}) \tag{18}$$

where $STEC$ denotes the vertical electronic content obtained from the GIM model; $\sigma^2_{I^s_{r,j}}$ is the variance of the priori ionospheric model errors ($\omega_{I^s_{r,j}}$).

2.4.3. Parameter Modeling and Estimation

According to the observation functions and state parameters mentioned above, the sequential least square algorithm is adopted. The corresponding expressions can be described by,

$$\boldsymbol{V}_k = \boldsymbol{A}_k \boldsymbol{X}_k - \boldsymbol{L}_k \tag{19}$$

where $\boldsymbol{V}_k$, $\boldsymbol{A}_k$, $\boldsymbol{X}_k$, and $\boldsymbol{L}_k$ denote the observation residual vector, the design coefficient matrix, the state vector, and the observation vector at time k, respectively.

Target function is,

$$\boldsymbol{V}_k^T \boldsymbol{P}_k \boldsymbol{V}_k + \boldsymbol{V}_{X,k}^T \boldsymbol{P}_{X,k-1} \boldsymbol{V}_{X,k} = \min \tag{20}$$

where $\boldsymbol{V}_{X,k}$ denotes the difference between the current state vector $\boldsymbol{X}_k$ and its predicted vector $\bar{\boldsymbol{X}}_k$; $\boldsymbol{P}_{X,k-1}$ denotes the weight matrix of the predicted state vector. The equation can be calculated by,

$$\boldsymbol{X}_k = \left(\boldsymbol{A}_k^T \boldsymbol{P}_k \boldsymbol{A}_k + \boldsymbol{P}_{X,k-1}\right)^{-1} \left(\boldsymbol{A}_k^T \boldsymbol{P}_k \boldsymbol{L}_k + \boldsymbol{P}_{X,k-1} \bar{\boldsymbol{X}}_k\right) \tag{21}$$

$$\boldsymbol{P}_{X,k}^{-1} = \left(\boldsymbol{A}_k^T \boldsymbol{P}_k \boldsymbol{A}_k + \boldsymbol{P}_{X,k-1}\right)^{-1} \tag{22}$$

The corresponding posterior covariance matrix of the state vector

$$\boldsymbol{D}_{X,k-1} = \boldsymbol{P}_{X,k-1}^{-1} \left(\boldsymbol{V}_k^T \boldsymbol{P}_k \boldsymbol{V}_k + \boldsymbol{V}_{X,k}^T \boldsymbol{P}_{X,k-1} \boldsymbol{V}_{X,k}\right) / (n - m) \tag{23}$$

where n and m represent the number of satellites obtained at k and $k-1$ epochs, respectively.

3. Experiments and Discussions

In this section, we collected four-day data to compare the accuracies of PPP-B2b orbit and clock to those of IGS RTS products. The positioning accuracies of PPP-B2b on dual-frequencies and single-frequency PPP are further studied with static and dynamic tests.

3.1. Static Data Collection

The real-time orbit and clock SSR products of PPP-B2b, CAS, DLR, GFZ, and WHU, the ultra-rapid products from WHU, and the broadcast ephemeris during the Days of Year (DOY) 354–357 in 2021 were collected. The distribution of the stations is illustrated in Figure 1. The IGS real-time SSR corrections were collected from the internet via the open-source software BNC Ver2.12.17 [36]. The static data from four MGEX stations (URUM, ULAB, WHU2, and JFNG) during DOYs of 354–357 of 2021 were collected.

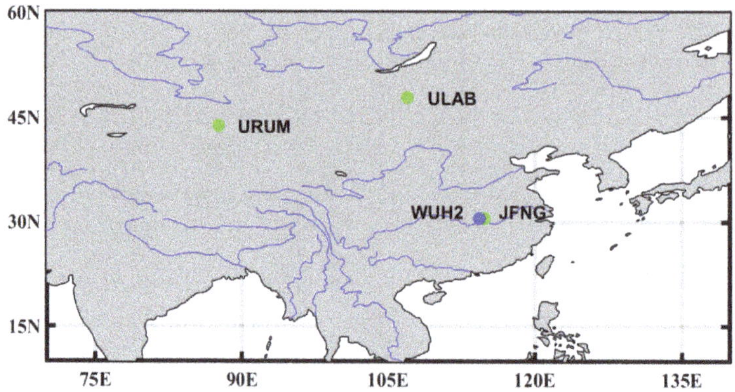

Figure 1. The location of 4 selected MGEX stations.

In the data analysis phase, (1) all of these orbit and clock products were projected to the same coordinate system and time system; (2) static test data were processed daily with dual-frequency IF PPP (B1I+B3I and L1+L2) and single-frequency PPP (B1C and L1); (3) the corresponding GPS data were also processed as a comparison; (4) the reference coordinates obtained from the MGEX weekly SINEX file. The details of the static data processing strategies are listed in Table 3. It is worth noting that the B1C observation is BDS-3 new signal, which is not used for orbit determination and clock calculation. Thus, the corresponding DCB correction should be applied to the observations.

Table 3. PPP processing strategies of static test.

Item	Processing Strategies
GNSS	BDS-3 and GPS
Signal selection of IF PPP	BDS-3: B1I+B3I GPS: L1+L2
Signal selection of SF PPP	BDS-3: B1C GPS: L1
Interval	30 s
Cutoff angle	10°
Weight method	Elevation angle dependent
Troposphere	Estimate the wet component
Ionospheric	IF PPP: IF combination SF PPP: Estimated
PCO/PCV	IGS14.atx
Ambiguity	Estimated
Adjustment	Sequential least square
Satellite DCB	Corrected by DCB products
Receiver DCB	IF PPP: IF combination SF PPP: Estimated

The available satellites and PDOP are the two critical indexes for positioning accuracy analysis; therefore, the number of visible satellites and the PDOP values for both BDS-3 and GPS are presented in Figure 2. The number of available BDS-3 satellites on average are 8.4, 8.1, 7.8, and 8.1, with the corresponding PDOP of 2.1, 2.2, 2.9, and 2.3 for JFNG, ULAB, URUM, and WUH2 stations. For GPS, the average number of available satellites are 6.8, 7.1, 6.8, and 6.6, with the corresponding PDOP of 2.5, 2.6, 2.9, and 2.7 for JFNG, ULAB, URUM, and WUH2 stations.

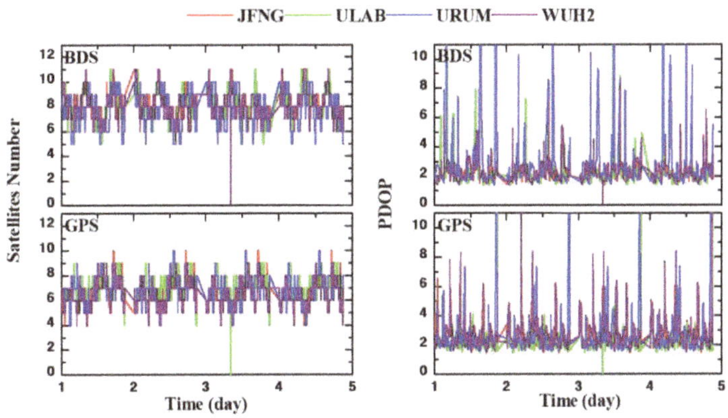

Figure 2. Number of visible satellites and PDOP values of the four MEGX stations.

3.2. Assessments of BDS-3 Real-Time Orbits and Clock Products

To evaluate the accuracies of PPP-B2b BDS-3 real-time orbits and clocks from CAS, DLR, GFZ, and WHU, as well as the broadcast ephemeris objectively, this study uses the final precision product from WHU as the reference. The B2b corrected GPS orbit and clock are also studied as a comparison. The time series of the orbit errors in radial, along-track, and cross-track directions, as well as the clock errors, are depicted in Figure 3. The corresponding statistics in terms of RMSE and STD are shown in Figures 4 and 5. In Figure 3, the orbit errors of different real-time products of BDS are within 20 cm, 60 cm, and 60 cm in radial, along-track, and cross-track directions. The BDS B2b corrected and broadcast orbit error in radial and along-track directions are similar to those of other real-time products; in the cross-track direction, however, the fluctuations of B2b-corrected and BDS broadcast orbit errors are more significant than those of other real-time products. The B2b corrected GPS orbit error is similar to that of the B2b corrected BDS orbit. As plotted in Figure 4, the statistic orbit accuracies of the IGSO satellites, C38~C40, are much lower than those of MEO satellites, especially while using the SSR corrections from GFZ. Generally, the orbit RMSE values of the real-time products from CAS, DLR, GFZ, WHU, and WHU_U (ultra-precise products) are within 10 cm in the radial, along-track, and cross-track components for MEO satellites. The RMSE of BRDC, PPP-B2b for BDS-3, are about 20 cm in the radial component and 50 cm in the along-track and cross-track components; the B2b corrected orbit for GPS has a similar performance as those for BDS in three directions. According to earlier studies, the higher accuracy in the radial direction is related to the high-quality satellite hydrogen and rubidium clocks [37]. In Figure 3, it is noted that systematic biases existed among different real-time clock products, especially the B2b corrected BDS and GPS clocks. This systematic clock bias is caused by pseudorange observation and can be absorbed by ambiguities [24,38]. In Figure 5, The clock STD of the real-time products from CAS, DLR, GFZ, WHU, and WHU_U are within 0.4 ns for BDS MEO satellites. In contrast, the STD of PPP-B2b is within 0.2 ns, and the STD of BRDC is within 1.0 ns. The STD of the B2b corrected GPS clock is 0.3 ns.

The average accuracies of real-time orbit and clock products are shown in Table 4. DLR did not provide SSR corrections for the three IGSO satellites, C38, C39, and C40, during this test; thus, the average accuracies of these three satellites are not considered. According to Table 4, the BDS-3 orbit accuracy in the radial component is higher than those in the along-track and cross-track components. The RMSE of the BDS-3 PPP-B2b orbits in the radial, along-track, and cross-track components are 9.42 cm, 21.26 cm, and 28.65 cm, respectively. The accuracy of BDS-3 PPP-B2b orbits in terms of RMSE is lower than that of IGS centers products due to the distribution and number of the monitoring stations. Compared to the broadcast orbit, the IGS real-time products provide about 30.62~54.95% improvements in the radial component, and the improvement provided by BDS PPP-B2b is 2.89%. The STD of PPP-B2b clocks is 0.18 ns, which has the highest accuracy among these real-time products of BDS-3. The improvements of the BDS-3 real-time clock provided by PPP-B2b is 79.87% compared to broadcast ephemeris. For the GPS correction, the B2b corrected orbit RMSE in radial, along-track, and cross-track directions are 13.73 cm, 23.83 cm, and 17.96 cm, respectively, and the clock STD is 0.25 ns. For the user side, the positioning accuracy is mainly affected by the orbit in radial and the clock. Therefore, the PPP-B2b SSR products theoretically can provide precise positioning solutions.

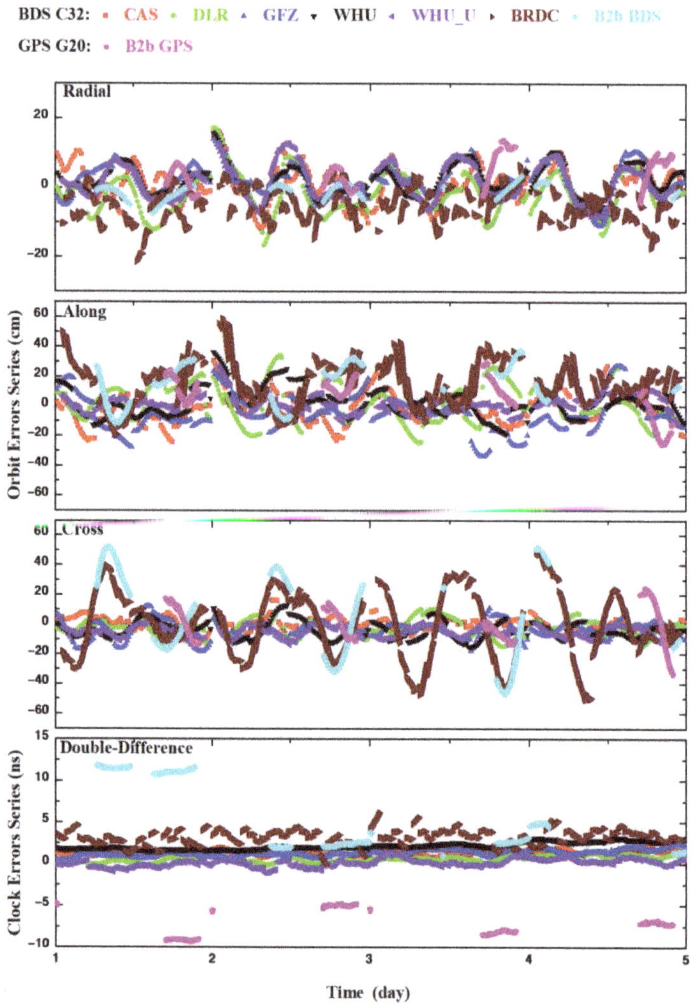

Figure 3. Time series of the orbit radial and clock errors of C32 and G20.

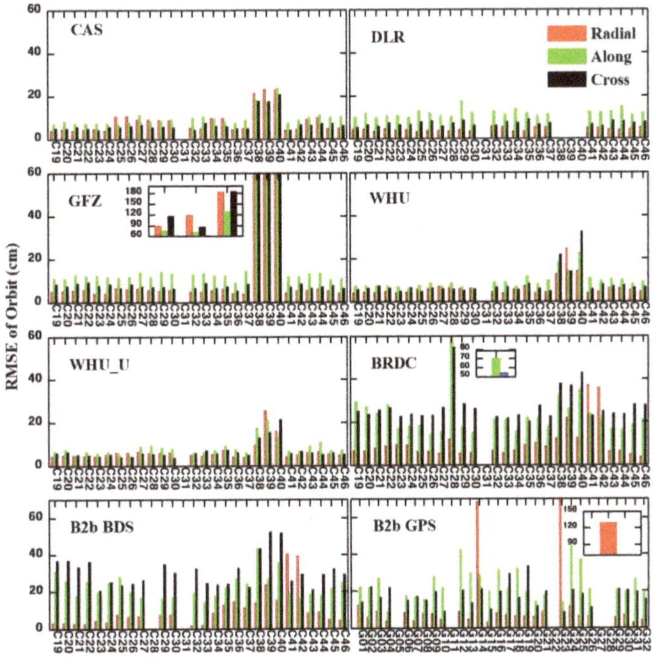

Figure 4. RMS of real-time orbit errors.

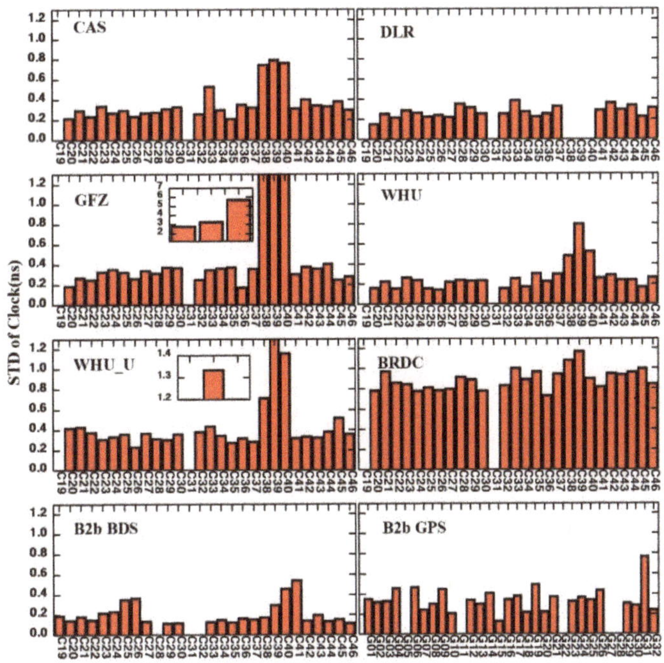

Figure 5. STD of real-time orbit clock errors.

Table 4. The average accuracies of real-time orbit and clock products.

Products	Orbit (cm)			Clock (ns)
	Radial	Along-Track	Cross-Track	
CAS	6.73	8.82	5.79	0.31
DLR	4.37	11.95	7.08	0.27
GFZ	5.41	12.40	7.63	0.31
WHU	5.38	8.57	6.41	0.22
WHU_U	5.39	7.25	5.54	0.35
BRDC	9.70	22.21	25.74	0.87
B2b BDS	9.42	21.26	28.65	0.18
B2b GPS	13.73	23.83	17.96	0.25

3.3. Accuracy and Convergence of RT-PPP in MGEX Stations

3.3.1. Dual-Frequency PPP

The time series of the position differences between the B1I+B3I PPP solutions of the four MEGX stations and the SINEX solutions on the E, N, and U components are presented in Figure 6. The average RMSE of the static PPP at JFNG, ULAB, URAM, and WUH2 stations is shown in Figure 7. Accordingly, the positioning accuracy of BDS-3 dual-frequency PPP based on the PPP-B2b real-time products is within 4.6 cm in the three directions for the four stations. In contrast, the GPS dual-frequency PPP with B2b correction is about 5.3 cm in three directions. The lower accuracy of GPS PPP with B2b corrections is caused by the product accuracy of GPS orbits and clocks, which has been studied in the above section. Additionally, the positioning accuracy of PPP-B2b-based positioning in the N component is higher than in the E component, which is related to the satellite's observation geometry strength of the carrier phase ambiguity. Most BDS-3 satellites have a north-south ground track and, thus, provide strong observation geometry strength in N directions and better positioning accuracy. In addition, the fixing ambiguity may improve the positioning accuracy in the E components [39], and similar results can be obtained in [22,40]. In general, the positioning accuracy of static RT-PPP with PPP-B2b is slightly lower than that of IGS real-time products, but higher than that of WHU ultra-rapid products and broadcast ephemeris.

The statistics accuracy in terms of RMSE of static B1I+B3I PPP using different real-time products is listed in Table 5. The positioning accuracies of BDS-3 PPP with PPP-B2b service are 4.8 cm and 5.4 cm in horizontal and 3D components, respectively. While using GPS, the positioning accuracy is slightly lower than that of BDS-3, with 5.4 cm and 5.9 cm in horizontal and 3D components, respectively. This demonstrates that the PPP-B2b service has the capability of providing centimeter-level positioning accuracy for users in China and neighboring countries while using the dual-frequency static PPP mode. However, the accuracy of PPP using PPP-B2B services is still lower than those solutions based on IGS real-time products. Among these solutions, the solution based on the products of WHU has the highest accuracy in horizontal components with 1.1 cm. In comparison, the solution based on the CAS products has the highest accuracy in 3D with 1.9 cm. The solutions based on DLR real-time products present the worst accuracy with 3.5 cm and 3.9 cm in horizontal and 3D components, respectively. Additionally, the horizontal and 3D positioning accuracies using WHU ultra-rapid products are 10.5 cm and 12.1 cm, respectively, while those based on broadcast ephemeris are 23.9 cm and 27.7 cm. Generally, the B1I+B3I BDS-3 PPP based on PPP-B2b SSR corrections can provide users with centimeter-level static positioning solutions, which is close to these based on IGS RTS SSR corrections, especially in horizontal components.

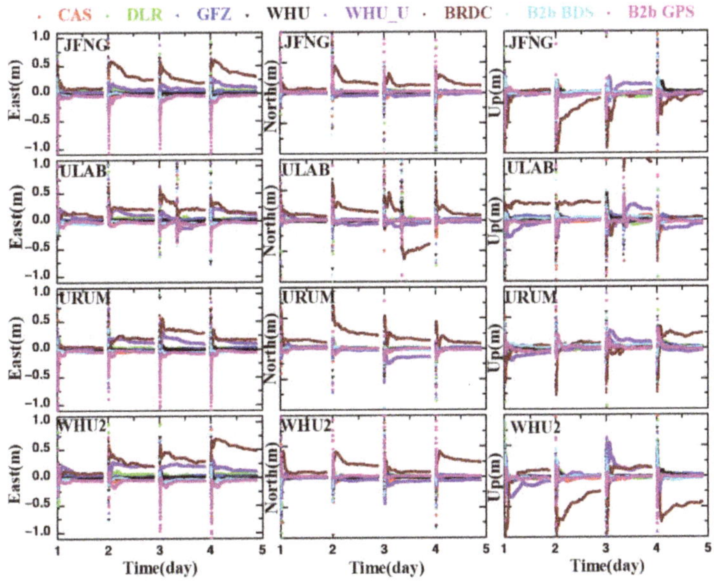

Figure 6. Error series of the four-day IF static PPP solution.

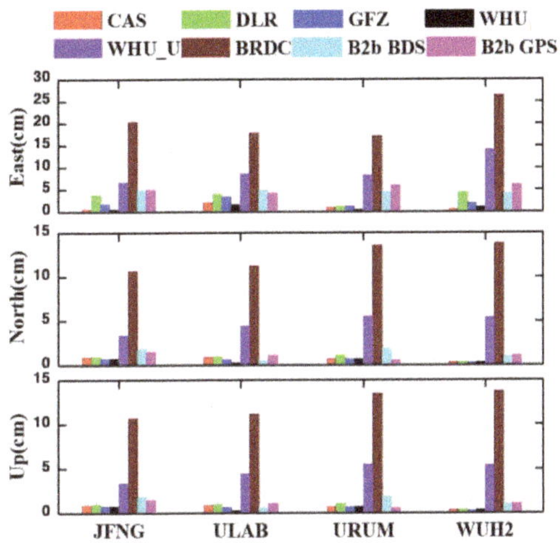

Figure 7. The average accuracy of the four-day IF static PPP solution.

Table 5. Average dual-frequency PPP positioning accuracy of the four MGEX stations (unit: cm).

Direction	CAS	DLR	GFZ	WHU	WHU_U	BRDC	B2b BDS	B2b GPS
E	1.1	3.3	2.1	0.9	9.4	20.4	4.6	5.3
N	0.7	0.8	0.6	0.5	4.7	12.3	1.2	1.0
U	1.4	1.8	1.4	1.8	6.0	13.2	2.4	2.2
2D	1.3	3.5	2.2	1.1	10.5	23.9	4.8	5.4
3D	1.9	3.9	2.7	2.1	12.1	27.7	5.4	5.9

Convergence time is another crucial index for PPP evaluation. In this paper, we define the convergence criteria of static PPP to be that the position accuracies in horizontal and vertical are continuously better than 10 cm and 20 cm for at least 10 min. The convergence times of static PPP for each MGEX station are presented in Figure 8. Since it is difficult to converge while using WHU ultra-rapid product and broadcast ephemeris, we only provide the convergence times of PPP using IGS real-time products and PPP-B2b products. As shown in Figure 8, the convergence times of BDS-3 PPP based on PPP-B2b products are within 15 min. The PPP-B2b-based GPS PPP presents the slowest convergence time, with an average convergence time of 52 min and the longest time of 109 min. The reason for the longer convergence time of JFNG may be caused by the inconsistency of the satellites' orbit errors, as shown in Figure 4, in which the fluctuations of B2b corrected orbit is more significant than that of B2b corrected BDS.

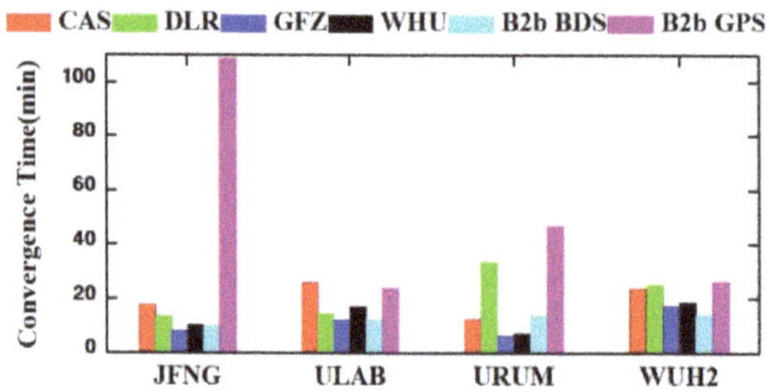

Figure 8. Convergence time of static dual-frequency PPP using different real-time orbit and clock products.

3.3.2. Single-Frequency PPP

In this section, a single-frequency PPP model with ionospheric and receiver DCB constraints is used. The time series of position differences between the static single-frequency PPP and the reference results in the E, N, and U for each station are shown in Figure 9, and the corresponding statistics are presented in Figure 10. The positioning accuracy in terms of RMSE of BDS B1C PPP with PPP-B2b SSR products is better than 5.5 cm in the E, N, and U for each station. While using GPS L1, the PPP accuracy is better than 5.6 cm in three components. The positioning accuracy of single-frequency PPP with PPP-B2b is comparable to that of IGS real-time RTS products, which have similar performance to the dual-frequency PPP. The positioning accuracy of static single-frequency PPP using WHU ultra-rapid product and broadcast ephemeris is much lower than that with PPP-B2b. The maximum positioning error for broadcast ephemeris-based solution exceeds 39.5 cm in the three components.

According to the statistics of average positioning error in Table 6, single-frequency RT-PPP based on PPP-B2b SSR products and IGS RTS SSR products (except DLR) can satisfy the demand for centimeter-level high-precision real-time positioning accuracy. While using PPP-B2b SSR products, the position accuracies in horizontal and 3D of B1C PPP are 5.4 cm and 7.7 cm, respectively. While using the IGS RTS products, the B1C PPP based on WHU products provides the highest accuracy within 3.5 cm in horizontal, and the PPP based on GFZ products provides the highest accuracy in 3D with 7.0 cm among evaluated the IGS RTS products. In contrast, the DLR real-time products provide the worst static B1C PPP positioning accuracy with 5.5 cm and 14.3 cm in the horizontal and 3D. The horizontal/3D positioning accuracies using WHU ultra-rapid products and broadcast ephemeris were about 7.6/13.1 cm and 18.3/43.5 cm, respectively, much lower than those of PPP-B2b PPP. As a comparison, the positioning accuracies of GPS L1 PPP based on PPP B2b SSR products are 5.4 cm and 7.8 cm, similar to those of BDS-3 B1C PPP.

To present the convergence speed of single-frequency B1C PPP, we redefine the convergence criteria of static PPP to be that the position accuracies in horizontal and vertical are continuously better than 20 cm and 40 cm for at least 10 min. The convergence times of static single-frequency PPP based on different SSR products for each MGEX station are shown in Figure 11, in which the average convergence times of BDS-3 B1C PPP and GPS L1 PPP using PPP-B2b SSR products are 80 min and 73 min, respectively. For solutions based on the four IGS real-time products, the converge times with CAS's products are the shortest among other products, with an average time of 65 min. The DLR products-based solutions converge the slowest with an average time of 95 min.

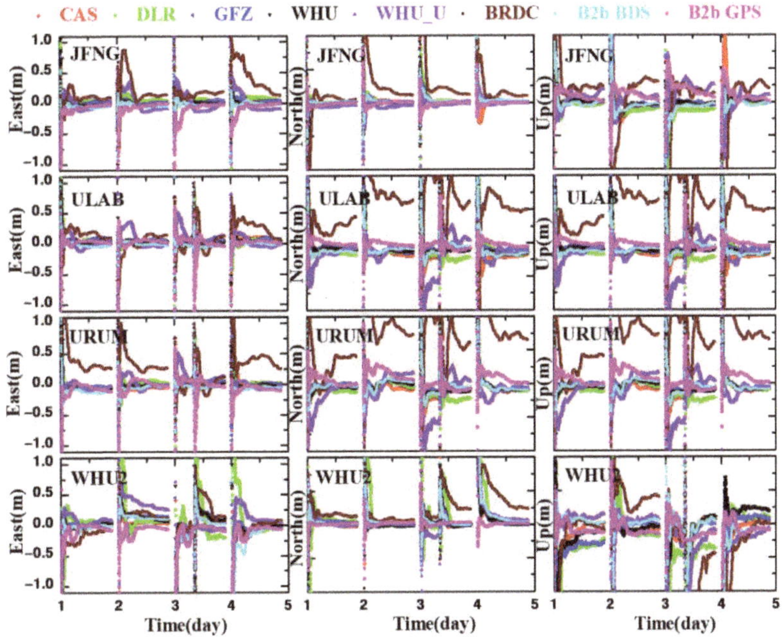

Figure 9. Time series of position errors of the four-day single-frequency static PPP.

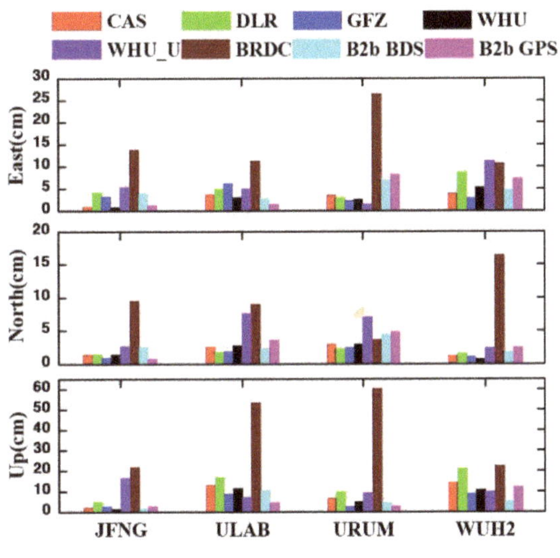

Figure 10. Average accuracy of the four-day single-frequency static PPP.

Table 6. The average positioning accuracy at the four stations (unit: cm).

Direction	CAS	DLR	GFZ	WHU	WHU_U	BRDC	B2b BDS	B2b GPS
E	3.0	5.2	3.7	2.9	5.8	15.5	4.6	4.6
N	2.0	1.8	1.6	2.0	4.9	9.7	2.8	2.9
U	9.1	13.2	5.7	7.0	10.7	39.5	5.5	5.6
2D	3.6	5.5	4.0	3.5	7.6	18.3	5.4	5.4
3D	9.8	14.3	7.0	7.8	13.1	43.5	7.7	7.8

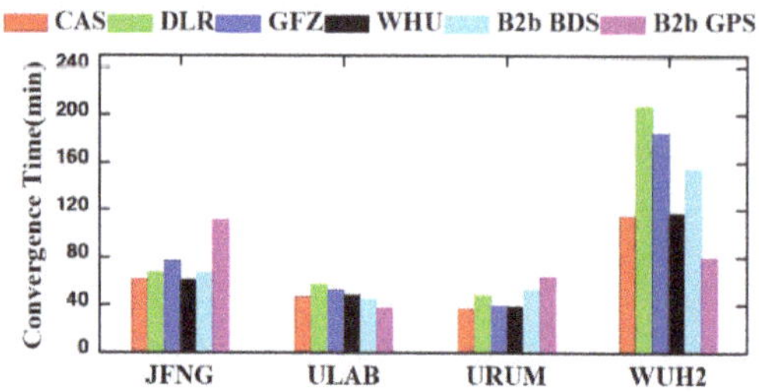

Figure 11. Convergence time of single-frequency PPP using different SSR products.

3.4. Accuracy of Real-Time Vehicle-Borne PPP

The dynamic test is carried out on DOY 357 in 2021 from 06:00 to 08:30 on the fifth ring road with Inertial Navigation System (INS) (NovAtel SPAN EPSON G370) and NovAtel GNSS receiver, which receives the B1I and B2b signal from BDS-3, and L1 and L2 signal from GPS. To study the PPP-B2b performance in complex environments, especially the performance of new signal B2b, we equipped a GNSS jammer to interfere with the received GNSS signals. The jammer reduces the number of visible satellites rather than affecting the observed distances. In the dynamic test, the vehicle drove about 40 km in the north–south direction and 30 km in the east–west direction. The tight integration of GPS+BDS RTK and INS provided by Inertial Explorer software were used as reference values for the dynamic test, and the corresponding trajectory is presented in Figure 12.

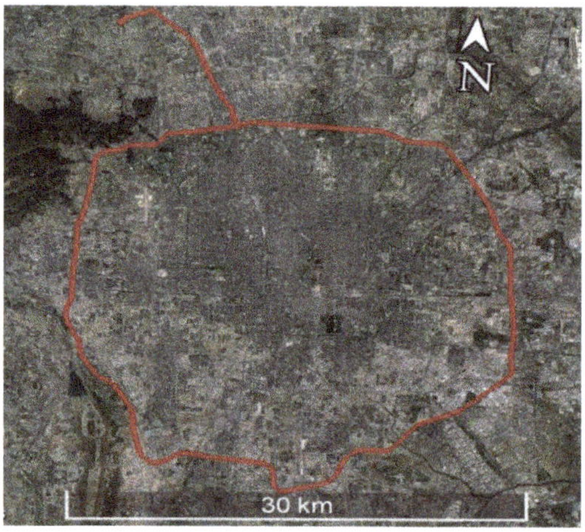

Figure 12. The trajectory of vehicle dynamic test.

The positioning strategies are listed in Table 7. It should be noted that the satellite DCB products from both real-time SSR and IGS final products do not contain the satellite

DCB value of C7D (B2b). Therefore, the DCB value of C7Z from CAS' was used to weaken the influence of satellite DCB on B2b frequency with ignoring the intra-frequency error. The number of visible satellites and corresponding PDOP values are presented in Figure 13. The average number of visible BDS-3 and GPS satellites during the 4-h dynamic test in Beijing are 8.9 and 2.0, respectively, and the average corresponding PDOP value of BDS-3 and GPS are 6.3 and 2.4, respectively. The sharp changes in the number of visible satellites are affected by the equipped GNSS jammer and further affect the PDOP values.

Table 7. PPP processing strategies of dynamic test.

Item	Processing Strategies
GNSS	BDS-3 and GPS
Signal selection of IF PPP	B1I+B2b GPS: L1+L2
Signal selection of SF PPP	BDS-3: B2b GPS: L1
Interval	1 s
Cutoff angle	10°
Weight method	Elevation angle dependent
Troposphere	Estimate the wet component
Ionospheric	IF PPP: IF combination SF PPP: Estimated
PCO/PCV	IGS14.atx
Ambiguity	Estimated
Adjustment	Sequential least square
Satellite DCB	Corrected by DCB products
Receiver DCB	IF PPP: IF combination SF PPP: Estimated

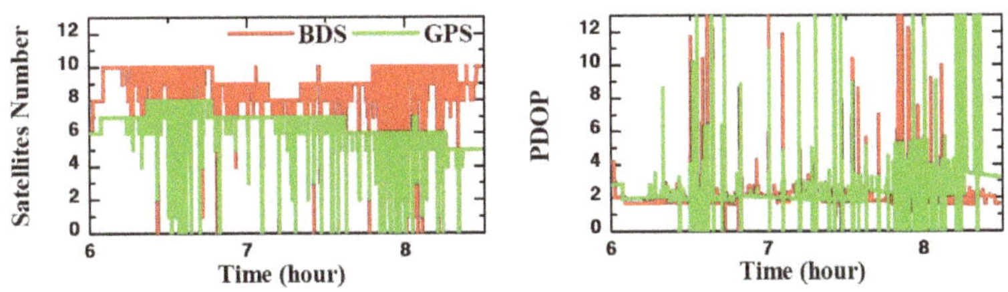

Figure 13. Number of visible satellites and PDOP values during the dynamic test.

Figure 14 presents the time series of position errors of vehicle-borne dual-frequencies (B1i+B2b and L1+L2) PPP, single frequency PPP (L1 and B2b) by comparing with the RTK/INS tight integration solutions in E, N, and U components. It is noted that the large fluctuations in the three components occur in dual-frequencies and single frequencies PPP, especially in the U components. Comparing to the visible satellite number presented in Figure 13, the presence of significant error components coincided with epochs of poor observations, which results in positioning discontinuity and further leads to PPP re-convergence. We chose dual-frequency PPP during epochs 3587 to 3588, red box in Figure 14, to further analyze the effect of GNSS signal blockage. At the period of epochs 3587 and 3588, the visible satellite is 0, and the BDS cannot provide continuous positioning results. At epoch 3589, the number of visible satellites is 8, and the positioning error at E, N, and U components are 4 m, −2 m, and −4 m, respectively. The average visible satellite number during epochs 3589 and 3900 is 8.9, and the B1I+B2b combination PPP with PPP-B2b converged to 0.5 and 1.0 m positioning accuracy in both horizontal and vertical components. The convergence time is 85 s in this period, and the average PDOP value is 2.2. The error components of single frequency PPP, L1 and B2b, present similar statuses.

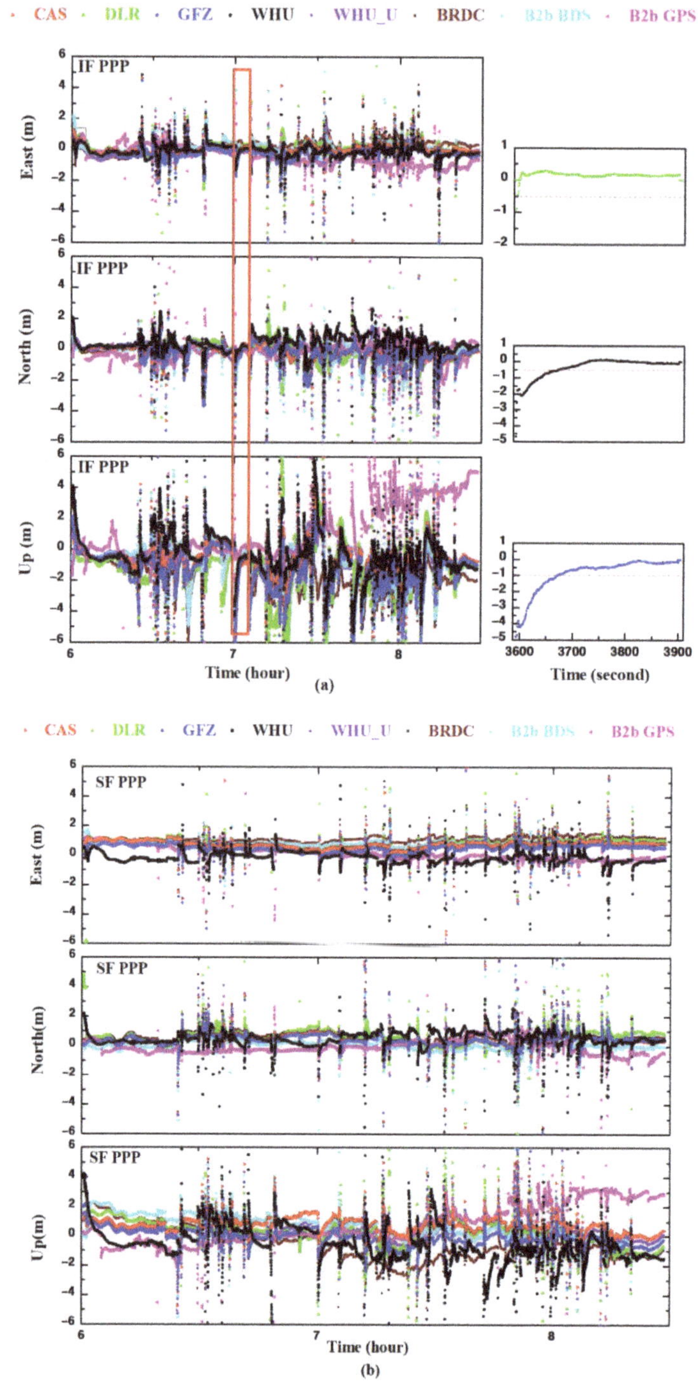

Figure 14. Time series of position errors of the dual-(**a**)/single (**b**)-frequency vehicle-borne RT-PPP using different SSR products.

The accuracies in terms of RMSE of vehicle-borne dual-frequency PPP and single-frequency PPP based on different SSR products are shown in Tables 8 and 9. For the dual-frequency PPP, the positioning accuracy of the BDS-3 B1I+B2b PPP using PPP-B2b SSR products is 35.8 cm, 55.0 cm, and 121.3 cm in the E, N, and U components, respectively. In contrast, the B1I+B2b PPP positioning accuracy using CAS products is the highest in horizontal with 58.9 cm, and WHU products are the highest in 3D with 129.7 cm. The solutions based on broadcast ephemeris (67.5 cm and 182.4 cm in horizontal and 3D components) have the worst positioning accuracy among the seven products. The GPS L1+L2 PPP is 57.0 cm, 47.9 cm, and 101.4 cm in the E, N, and U components. The GPS L1+L2 PPP solution has the highest 3D accuracy (126.0 cm) and the worst horizontal accuracy (74.5 cm) compared to the BDS-3 B1I+B2b PPP using the SSR product of both PPP-B2b and IGS RTS.

Table 8. Position RMSE of vehicle-borne real-time dual-frequency PPP using different SSR products (unit: cm).

Direction	CAS	DLR	GFZ	WHU	WHU_U	BRDC	B2b BDS	B2b GPS
E	27.8	33.0	35.4	34.9	35.5	49.2	35.8	57.0
N	61.9	51.8	53.9	62.1	49.7	46.1	55.0	47.9
U	115.7	139.4	138.2	108.4	125.5	169.4	121.3	101.4
2D	58.9	61.4	64.5	71.2	61.1	67.5	65.7	74.5
3D	129.9	152.3	152.5	129.7	139.5	182.4	137.9	126.0

Table 9. Position RMSE of vehicle-borne real-time single-frequency PPP using different SSR products (unit: cm).

Direction	CAS	DLR	GFZ	WHU	WHU_U	BRDC	B2b BDS	B2b GPS
E	80.0	88.6	54.3	79.4	73.9	111.8	87.1	42.3
N	53.9	72.0	53.1	156.7	57.3	72.7	32.9	47.8
U	88.6	87.8	60.0	178.6	57.3	126.3	103.5	129.2
2D	96.5	114.2	75.9	175.7	93.5	133.4	93.1	63.9
3D	131.0	144.0	96.8	250.5	109.6	183.7	139.2	144.1

For the single-frequency PPP, the positioning accuracy of B2b frequency PPP based on PPP-B2b products in E, N, and U components are 87.1 cm, 32.9 cm, and 103.5 cm, respectively; those positioning accuracies of GPS L1 PPP are 42.3 cm, 47.8 cm, and 63.9 cm, respectively. While using GPS, the positioning accuracy is slightly higher in horizontal components (42.3 cm and 47.8 cm) but much worse in the U direction (129.2 cm). For the solutions based on IGS RTS real-time products, the positioning accuracy using GFZ real-time products is the best, with 75.9 cm and 96.8 cm in horizontal and 3D components, respectively. The positioning accuracy with WHU real-time products is 175.7 cm and 250.5 cm in horizontal and 3D components, respectively, which are the worst among other real-time products. In Figure 14, a noticeable systematic error exists in the N and U components, which are caused by the unabsorbed satellite clock errors. Compared to BDS-3 B1I+B2b PPP, the BDS-3 B2b PPP provides smoother position solutions, especially after satellites signal re-tracking, which is caused by the greater noise of the dual-frequency IF and high-accuracy of prior ionosphere data used in single-frequency PPP.

Figure 15 shows the radial, along-track, and cross-track RMS values of the orbit's errors and clock STD values for each type of real-time product, as well as the positioning accuracy of static and dynamic with single/dual-frequency PPP. In static PPP, the IGS real-time product positioning accuracy is better than that of PPP-B2b, and BRDC positioning accuracy is the worst among the seven products. This is consistent with the orbit and clock accuracies of each real-time product. In dynamic PPP, affected by the complex positioning environment, the positioning accuracy of various products is similar.

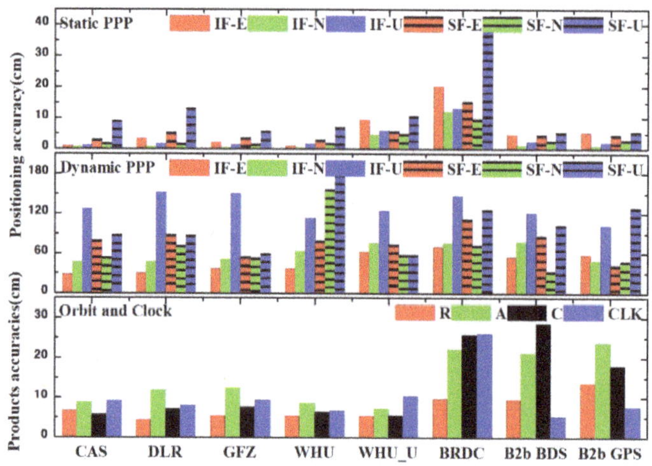

Figure 15. The average accuracy of positioning and real-time products.

It also can be found from Figure 15 that the accuracies of the real-time WHU orbits and clock products are similar to other ACs, but the positioning of single frequency PPP with WHU products is lower than that of other ACs. This is because the WHU BDS real-time clock products contain systematic errors. The RMSE of different BDS real-time clock products with respect to the WHU final clock product is presented in Figure 16. It can be clearly found that the RMSE value of WHU BDS real-time products is more significant than that of other products.

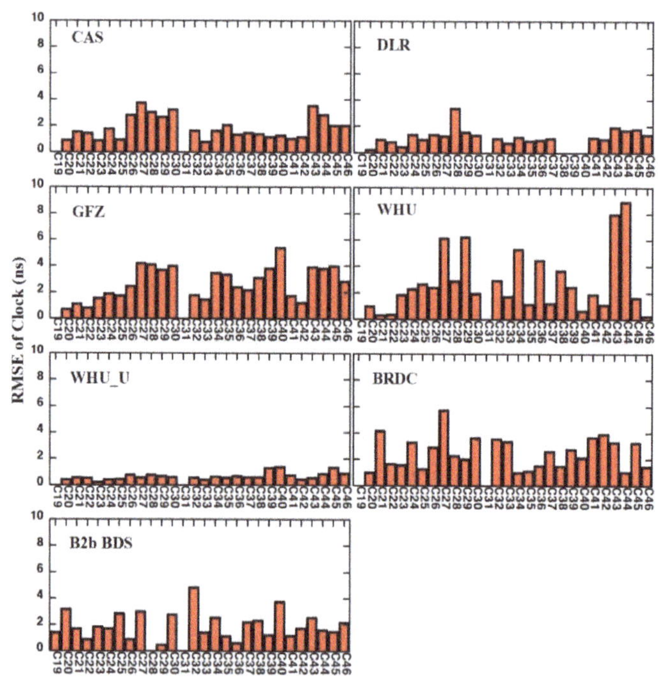

Figure 16. The RMSE of real-time orbit clock errors.

4. Conclusions

Real-time orbits and clocks are crucial factors for RT-PPP applications. To evaluate the impacts of existing real-time orbit and clock products (BDS-3 PPP-Bb and IGS RTS), this study provides comprehensive assessments of the accuracy of these real-time products and their influences on the positioning accuracy and convergence time of RT-PPP in both static and dynamic tests. After the descriptions of the evaluation methods for real-time orbits and clocks as well as the real-time single-/dual-frequency PPP models, BDS-3 new signals observations of B1C and B2b from the four MEGX stations and a set of the vehicle-borne test were used to find the conclusions of the assessment. The main conclusions can be summarized as follows.

For these real-time SSR products, the orbit accuracy of PPP-B2b SSR products is lower than these of IGS real-time products (CAS, DLR, GFZ, and WHU). However, the clock accuracy of PPP-B2b is better than these of IGS real-time products. Additionally, the accuracy of the real-time orbit and clock of BDS-3 provided by PPP-B2b is higher than that of GPS.

For real-time dual-frequency PPP, the position RMSEs of the four MEGX stations based on PPP-B2b SSR products are slightly lower than those based on the IGS RTS SSR products. Meanwhile, the BDS-3 B1I+B3I PPP positioning accuracy using the PPP-B2b service is centimeter-level which is better than that of GPS L1+L2 PPP. The dual-frequency B1I+B2b PPP positioning accuracy in the vehicle-borne test achieved submeter-level and meter-level in horizontal and vertical components in our test environment.

For the single-frequency RT-PPP based on PPP-B2b SSR products, the BDS-3 B1C PPP is lower than that of B1I+B3I PPP, providing about 10–20 cm-level positioning accuracy with MEGX data. While using the vehicle-borne B2b frequency data, the accuracy of BDS-3 PPP is very close to that of B1I+B2b PPP with submeter-level and meter-level in horizontal and vertical components.

In general, our work shows that both the BDS-3 PPP-B2b service and the IGS RTS can satisfy the demand for high-accuracy positioning in real time. However, these real-time products provided by IGS RTS are calculated from the global-distributed stations, which are more accurate than BDS-3 PPP-B2b, especially for GPS. However, IGS RTS are limited by the connection of the internet. Additionally, BDS-3 PPP-B2b introduces inter-satellite link technology and broadcasts SSR information by navigation signal, which gives it more significant potential in real-time precise positioning applications in the future.

Author Contributions: Conceptualization, C.Y., Y.Z. and Z.G.; Data curation, Q.X.; Funding acquisition, C.Y. and Z.G.; Investigation, R.L., C.Y. and Z.G.; Software, R.L.; Supervision, C.Y. and Z.G.; Visualization, R.L., Q.X. and J.L.; Writing—original draft, R.L.; Writing—review and editing, R.L., C.Y., Y.Z. and Z.G. All authors have read and agreed to the published version of the manuscript.

Funding: This research was partly supported by the National Key Research and Development Program of China (Grant No. 2020YFB0505802), the National Natural Science Foundation of China (NSFC) (Grants No. 42274022, No. 42274024).

Data Availability Statement: The datasets analyzed in this study are managed by the School of Land Science and Geomatics, China University of Geosciences, Beijing, and can be available on request from the corresponding author.

Acknowledgments: Thanks to the original observations provided by MGEX and the multi-system SSR products provided by the analysis centers CAS, DLR, GFZ, and WHU.

Conflicts of Interest: The authors declare no conflict of interest.

References

1. Yang, Y.; Gao, W.; Guo, S.; Mao, Y.; Yang, Y. Introduction to BeiDou-3 navigation satellite system. *Navigation* **2019**, *66*, 7–18. [CrossRef]
2. Yang, Y. Resilient PNT concept frame. *Acta Geod. Cartogr. Sin.* **2018**, *47*, 893.
3. Yang, Y.; Mao, Y.; Sun, B. Basic performance and future developments of BeiDou global navigation satellite system. *Satell. Navig.* **2020**, *1*, 1. [CrossRef]

4. Zumberge, J.; Heflin, M.; Jefferson, D.; Watkins, M.; Webb, F. Precise point positioning for the efficient and robust analysis of GPS data from large networks. *J. Geophys. Res. Solid Earth* **1997**, *102*, 5005–5017. [CrossRef]
5. Kouba, J.; Héroux, P. Precise point positioning using IGS orbit and clock products. *GPS Solut.* **2001**, *5*, 12–28. [CrossRef]
6. Gao, Y.; Shen, X. A New Method for Carrier-Phase-Based Precise Point Positioning. *Navigation* **2002**, *49*, 109–116. [CrossRef]
7. Abdel-Salam, M.A.-T. *Precise Point Positioning using Un-Differenced Code and Carrier Phase Observations*; University of Calgary: Calgary, AB, Canada, 2005; Volume 69. Available online: https://www.collectionscanada.gc.ca/obj/thesescanada/vol2/002/NR37676.PDF?is_thesis=1&oclc_number=302053384 (accessed on 17 October 2022).
8. Dow, J.M.; Neilan, R.E.; Rizos, C. The international GNSS service in a changing landscape of global navigation satellite systems. *J. Geod.* **2009**, *83*, 191–198. [CrossRef]
9. Ogutcu, S. Performance assessment of IGS combined/JPL individual rapid and ultra-rapid products: Consideration of Precise Point Positioning technique. *Int. J. Eng. Geosci.* **2020**, *5*, 1–14. [CrossRef]
10. Chu, C.-W.; Chiang, Y.-T.; Chang, F.-R.; Wang, H.-S. Toward real-time precise point positioning: Differential GPS based on IGS ultra rapid product. In Proceedings of the SICE Annual Conference, Taipei, Taiwan, 18–21 August 2010; pp. 355–358.
11. Chen, J.; Li, H.; Wu, B.; Zhang, Y.; Wang, J.; Hu, C. Performance of real-time precise point positioning. *Mar. Geod.* **2013**, *36*, 98–108. [CrossRef]
12. Duan, B.; Hugentobler, U.; Chen, J.; Selmke, I.; Wang, J. Prediction versus real-time orbit determination for GNSS satellites. *GPS Solut.* **2019**, *23*, 39. [CrossRef]
13. Pan, L.; Gao, X.; Hu, J.; Ma, F.; Zhang, Z.; Wu, W. Performance assessment of real-time multi-GNSS integrated PPP with uncombined and ionospheric-free combined observables. *Adv. Space Res.* **2021**, *67*, 234–252. [CrossRef]
14. Caissy, M.; Agrotis, L.; Weber, G.; Fisher, S. The IGS real-time service. In Proceedings of the EGU General Assembly Conference Abstracts, Vienna, Austria, 7–12 April 2013; p. EGU2013-11168.
15. Weber, G.; Mervart, L.; Lukes, Z.; Rocken, C.; Dousa, J. Real-time clock and orbit corrections for improved point positioning via NTRIP. In Proceedings of the 20th international technical meeting of the satellite division of the institute of navigation (ION GNSS 2007), Fort Worth, TX, USA, 25–28 September 2007; pp. 1992–1998.
16. Elsobeiey, M.; Al-Harbi, S. Performance of real-time Precise Point Positioning using IGS real-time service. *GPS Solut.* **2016**, *20*, 565–571. [CrossRef]
17. Li, Z.; Zhang, J.; Li, T.; He, X.; Wu, M. Analysis of static and dynamic real-time precise point positioning and precision based on SSR correction. In Proceedings of the 2016 IEEE International Conference on Information and Automation (ICIA), Ningbo, China, 1–3 August 2016; pp. 2022–2027.
18. Wang, Z.; Li, Z.; Wang, L.; Wang, X.; Yuan, H. Assessment of multiple GNSS real-time SSR products from different analysis centers. *ISPRS Int. J. Geo-Inf.* **2018**, *7*, 85. [CrossRef]
19. Ouyang, C.; Shi, J.; Huang, Y.; Guo, J.; Xu, C. Evaluation of BDS-2 real-time orbit and clock corrections from four IGS analysis centers. *Measurement* **2021**, *168*, 108441. [CrossRef]
20. Ge, Y.; Chen, S.; Wu, T.; Fan, C.; Qin, W.; Zhou, F.; Yang, X. An analysis of BDS-3 real-time PPP: Time transfer, positioning, and tropospheric delay retrieval. *Measurement* **2021**, *172*, 108871. [CrossRef]
21. Lu, X.; Chen, L.; Shen, N.; Wang, L.; Jiao, Z.; Chen, R. Decoding PPP corrections from BDS B2b signals using a software-defined receiver: An initial performance evaluation. *IEEE Sens. J.* **2020**, *21*, 7871–7883. [CrossRef]
22. Nie, Z.; Xu, X.; Wang, Z.; Du, J. Initial assessment of BDS PPP-B2b service: Precision of orbit and clock corrections, and PPP performance. *Remote Sens.* **2021**, *13*, 2050. [CrossRef]
23. Tao, J.; Liu, J.; Hu, Z.; Zhao, Q.; Chen, G.; Ju, B. Initial Assessment of the BDS-3 PPP-B2b RTS compared with the CNES RTS. *GPS Solut.* **2021**, *25*, 131. [CrossRef]
24. Xu, Y.; Yang, Y.; Li, J. Performance evaluation of BDS-3 PPP-B2b precise point positioning service. *GPS Solut.* **2021**, *25*, 142. [CrossRef]
25. Ren, Z.; Gong, H.; Peng, J.; Tang, C.; Huang, X.; Sun, G. Performance assessment of real-time precise point positioning using BDS PPP-B2b service signal. *Adv. Space Res.* **2021**, *68*, 3242–3254. [CrossRef]
26. BeiDou Navigation Satellite System Signal in Space Interface Control Document: Precise Point Positioning Service Signal PPP-B2b (Version 1.0). 2020. Available online: http://en.beidou.gov.cn/SYSTEMS/ICD/202008/P020200803538771492778.pdf (accessed on 20 October 2021).
27. Hadas, T.; Bosy, J. IGS RTS precise orbits and clocks verification and quality degradation over time. *GPS Solut.* **2015**, *19*, 93–105. [CrossRef]
28. Ge, M.; Chen, J.; Douša, J.; Gendt, G.; Wickert, J. A computationally efficient approach for estimating high-rate satellite clock corrections in realtime. *GPS Solut.* **2012**, *16*, 9–17. [CrossRef]
29. Li, X.; Ge, M.; Dai, X.; Ren, X.; Fritsche, M.; Wickert, J.; Schuh, H. Accuracy and reliability of multi-GNSS real-time precise positioning: GPS, GLONASS, BeiDou, and Galileo. *J. Geod.* **2015**, *89*, 607–635. [CrossRef]
30. Gao, Z.; Zhang, H.; Ge, M.; Niu, X.; Shen, W.; Wickert, J.; Schuh, H. Tightly coupled integration of multi-GNSS PPP and MEMS inertial measurement unit data. *GPS Solut.* **2017**, *21*, 377–391. [CrossRef]
31. An, X.; Meng, X.; Jiang, W. Multi-constellation GNSS precise point positioning with multi-frequency raw observations and dual-frequency observations of ionospheric-free linear combination. *Satell. Navig.* **2020**, *1*, 7. [CrossRef]

32. Guo, F.; Zhang, X.; Wang, J. Timing group delay and differential code bias corrections for BeiDou positioning. *J. Geod.* **2015**, *89*, 427–445. [CrossRef]
33. Gao, Z.; Ge, M.; Shen, W.; Zhang, H.; Niu, X. Ionospheric and receiver DCB-constrained multi-GNSS single-frequency PPP integrated with MEMS inertial measurements. *J. Geod.* **2017**, *91*, 1351–1366. [CrossRef]
34. Jin, S.; Su, K. PPP models and performances from single-to quad-frequency BDS observations. *Satell. Navig.* **2020**, *1*, 16. [CrossRef]
35. Zhang, B.; Zhao, C.; Odolinski, R.; Liu, T. Functional model modification of precise point positioning considering the time-varying code biases of a receiver. *Satell. Navig.* **2021**, *2*, 11. [CrossRef]
36. Weber, G.; Mervart, L.; Sturze, A.; Rülke, A.; Stöcker, D. *BKG Ntrip Client (BNC), Version 2.12*; Mitteilungen des Bundesamtes für Kartographie und Geodäsie: Frankfurt am Main, Germany, 2016.
37. Kazmierski, K.; Zajdel, R.; Sośnica, K. Evolution of orbit and clock quality for real-time multi-GNSS solutions. *GPS Solut.* **2020**, *24*, 111. [CrossRef]
38. Lv, Y.; Geng, T.; Zhao, Q.; Xie, X.; Zhou, R. Initial assessment of BDS-3 preliminary system signal-in-space range error. *GPS Solut.* **2019**, *24*, 16. [CrossRef]
39. Blewitt, G. Carrier phase ambiguity resolution for the Global Positioning System applied to geodetic baselines up to 2000 km. *J. Geophys. Res. Solid Earth* **1989**, *94*, 10187–10203. [CrossRef]
40. Liu, Y.; Yang, C.; Zhang, M. Comprehensive Analyses of PPP-B2b Performance in China and Surrounding Areas. *Remote Sens.* **2022**, *14*, 643. [CrossRef]

remote sensing

MDPI

Article

An Empirical Grid Model for Precipitable Water Vapor

Xinzhi Wang [1,2], Fayuan Chen [1,*], Fuyang Ke [1,2] and Chang Xu [3]

1 School of Remote Sensing and Geomatics Engineering, Nanjing University of Information Science and Technology, Nanjing 210044, China
2 Wuxi Institute, Nanjing University of Information Science and Technology, Wuxi 214100, China
3 Zhejiang Institute of Communications, Hangzhou 311112, China
* Correspondence: 20201248037@nuist.edu.cn

Abstract: Atmospheric precipitable water vapor (PWV) is a key variable for weather forecast and climate research. Various techniques (e.g., radiosondes, global navigation satellite system, satellite remote sensing and reanalysis products by data assimilation) can be used to measure (or retrieve) PWV. However, gathering PWV data with high spatial and temporal resolutions remains a challenge. In this study, we propose a new empirical PWV grid model (called ASV-PWV) using the zenith wet delay from the Askne model and improved by the spherical harmonic function and vertical correction. Our method is convenient and enables the user to gain PWV data with only four input parameters (e.g., the longitude and latitude, time, and atmospheric pressure of the desired position). Profiles of 20 radiosonde stations in Qinghai Tibet Plateau, China, along with the latest publicly available C-PWVC2 model are used to validate the local performance. The PWV data from ASV-PWV and C-PWVC2 is generally consistent with radiosonde (the average annual bias is −0.44 mm for ASV-PWV and −1.36 mm for C-PWVC2, the root mean square error (RMSE) is 3.44 mm for ASV-PWV and 2.51 mm for C-PWVC2, respectively). Our ASV-PWV performs better than C-PWVC2 in terms of seasonal characteristics. In general, a sound consistency exists between PWV values of ASV-PWV and the fifth generation of European Centre for Medium-Range Weather Forecasts Atmospheric Reanalysis (ERA5) (total 7381 grid points in 2020). The average annual bias and RMSE are −0.73 mm and 4.28 mm, respectively. ASV-PWV has a similar performance as ERA5 reanalysis products, indicating that ASV-PWV is a potentially alternative option for rapidly gaining PWV.

Keywords: ASV-PWV model; radiosonde; ERA5; Qinghai Tibet Plateau

Citation: Wang, X.; Chen, F.; Ke, F.; Xu, C. An Empirical Grid Model for Precipitable Water Vapor. *Remote Sens.* **2022**, *14*, 6174. https://doi.org/10.3390/rs14236174

Academic Editor: Chung-yen Kuo

Received: 3 November 2022
Accepted: 2 December 2022
Published: 6 December 2022

1. Introduction

Atmospheric water vapor is mostly stored in the troposphere and accounts for a relatively small proportion, but it has a profound influence on the hydrological cycle, the atmospheric circulation, and the evolution of synoptic systems [1–3]. The precipitable water vapor (PWV) is often used to characterize the water vapor content. Therefore, monitoring PWV is crucial for climate research and extreme weather warning [4–6]. Plenty of techniques (e.g., ground-based photometers, radiosondes, satellite remote sensing, global navigation satellite system (GNSS), and data assimilation) can be used to measure (or retrieve) the PWV. Using ground-based photometers to obtain PWV is simple and inexpensive. Such an approach can offer a higher spatial and temporal resolution than radiosonde data, but less coverage than from satellite measurements, the downside being that it is vulnerable to cloud cover when in use [7,8]. In-situ balloon borne radiosonde is one of the most accurate methods to map the PWV but for local to regional scale with a low spatial and temporal resolution [9,10]. Moreover, radiosonde has a high operating cost and its observations are often disrupted by the artificial shifts caused by equipment changes and the operating procedure [11–13]. Satellite remote sensing using IR emission channels [14] provides an alternative option for retrieving the high temporal and spatial resolution PWV data, but it suffers strongly from the effects from clouds, precipitation, and surface reflection spectrum uncertainty [15], and PWV values from IR emission channels are available only over the oceans. In many regions, numerous scholars have compared PWV obtained by IR emission channels and GNSS. Kumar Sanjay et al. [16] presented an analysis of GPS data on the campus of Banaras Hindu University, Varanasi and compared the variability of water vapor from Kanpur GPS, AERONET, and MODIS water vapor data. The results indicate that the observation of PWV from GPS is well-correlated with satellite observation (MODIS), and the PWV estimated from GPS data is found to be very sensitive with surface temperature, monsoon onset and rainfall. Shi et al. [17] evaluate the accuracies of the water vapor data MODIS, MERSI, and the visible infrared radiometer (VIRR) using ground-based data (CARE-China and AERONET) as a reference and characterize the local water vapor variations in China. The results indicate that PWV values in China exhibited large spatial and temporal variations (especially in the Qinghai-Tibet Plateau to the South China Sea). The PWV values were obtained from MODIS, VIRR, and MERSI data, and the accuracies varied widely with relative errors from 10% to 899%. Khaniani et al. [18] used one year of ground-based observation from a network of 38 GPS stations to evaluate the performance of MODIS Near Infrared (Near-IR) PWV over Iran. The results indicate that times series between MODIS and GPS both methods agree very well, and the bias values and the height of the stations have a linear relation. Due to the unique superiority (e.g., all-weather, low cost and high accuracy), the GNSS-based PWV detection has achieved dramatic developments in recent years. Unfortunately, the GNSS-derived PWV are often inhomogeneous and have relatively low temporal resolutions [19], resulting from the sparse distribution of GNSS stations. Another powerful method to gather PWV data is the reanalysis data set, which is produced by data assimilation using different atmospheric observations and circulation models [20]. With respect to the sensor-based methods, reanalysis products have advantages of global coverage and spatial integrity, but large uncertainties still exist in the areas where no or limited observations are available for data assimilation [21,22].

Since the distribution in the troposphere is heterogeneous and anisotropic [10], gathering PWV data with high spatial and temporal resolutions confronts the climatology community with a challenge. In recent years, many attempts have been made for this issue. Cortés et al. [23] validate the use of submillimeter tippers to be used at other sites and use the PWV database to detect a potential signature for Chajnantor area climate change over 20 years. Shikhovtsev et al. [24] present the results of a comparison of the PWV estimated from GNSS measurements and the ERA5 reanalysis values corrected for the relief. It is shown that the best correlation between these quantities is observed under clear sky conditions. Ma et al. [25] proposed a dual-scale method for retrieving PWV maps, based on heterogeneous earth data, which can obtain sub-kilometer resolution PWV, while maintaining accuracy. The academic team led by Yao Yibin in Wuhan University, China, focused on the data fusion methods (e.g., the Gaussian Processes method, spherical cap harmonic analysis, Helmert variance component estimation and generalized regression neural network as well) and established some large-scale PWV fusion models for North America by combining different PWV data (e.g., GNSS, the moderate resolution imaging spectroradiometer (MODIS) and the European Centre for Medium-range Weather Forecasts (ECMWF)) (for details, see Yao et al. [26], Zhang et al. [27], Zhang and Yao [28]). The academic team led by Huang Liangke in Guilin University of Technology, China, systematically evaluated the PWV products of MERRA-2 and ERA5 reanalysis in China based on GNSS site and in-situ ground meteorological observations [29], and proposed a ZTD vertical stratification model based on MERRA-2 data considering spatiotemporal factors [30]. In addition, they also analyzed the spatiotemporal characteristics of GNSS-derived PWV by establishing an atmospheric weighted mean temperature model [31,32]. As an alternative, PWV data can be converted by the zenith wet delay (ZWD) obtained from many empirical models. Classic empirical models by Leckner [33] and Kouba [34] have a certain potential for calculating ZWD, but the water vapor decrease factor is usually constant, yielding systematic errors in different seasons. This has given rise to a growing literature on PWV vertical correction modeling. Dousa and Elias [35] propose an improved

ZWD analytical model and its vertical approximation derived from the original profile of meteorological data can be superior to the conventional methods by a factor of 2–3. More recently, Huang et al. [36] proposed a new PWV vertical correction model, considering the time-varying lapse rate and geographical divisions, which can be applied for whole areas in mainland China. No doubt varying model parameters as well as input reanalysis data have great impacts on both of the data fusion methods and PWV vertical correction methods. This issue motivates us to perform further studies in this direction. We follow the concept of the Askne model [37] and establish a new empirical PWV grid model (called ASV-PWV), which is improved by the spherical harmonic function and vertical correction.

The plan of this paper is as follows. Section 2 describes the fifth generation of ECMWF reanalysis (ERA5) data and field Radiosonde observations included in this study. Particular attention is paid to the empirical ZWD deriving and its improvement by the spherical harmonic function and vertical correction. Section 3 discusses the model evaluation. Both the local and area performances are examined. Some concluding remarks are made in Section 4.

2. Materials and Methods

2.1. ERA5 Reanalysis Products

ERA5 is the latest ECMWF atmospheric reanalysis product with the highest temporal and spatial resolution, which can provide global atmospheric, land, and ocean parameters [38]. ERA5 replaces the ERA-Interim reanalysis, which began in 2006 [39]. ERA5 uses the latest integrated forecasting systems (IFS) to increase ERA5's horizontal resolution to 31 km (ERA-Interim's horizontal resolution is 80 km) and its time resolution from 6 h to 1 h. As a result, ERA5 captures atmospheric phenomena in greater detail than previous, lower-resolution global reanalyses. A large number of assimilated datasets have been reprocessed to improve ERA5 accuracy, especially in the troposphere. Compared with ERA-Interim, another advantage of ERA5 is that its delay time is 5 days, not 2–3 months [40–42]. The products used in this study include meteorological data (37 atmospheric pressure layers from 1000 hPa to 1 hPa). We focus on the 7381 grid points from 75°E to 105°E and 25°N to 40°N.

2.2. Radiosonde Profiles

The Integrated Global Radiosonde Archive (IGRA) is a collection of historical and near real time global radiosonde observations [43]. Sounding-derived parameters are recorded according to separated station files and continue to be updated daily. PWV is one of the derived parameters that can be calculated when the pressure, temperature, and dew point depression are available from the surface to a level of 500 hPa [44].

Meteorological profiles of 20 radiosonde stations (see Figure 1) in Qinghai Tibet Plateau, China, are used for model validation. The radiosonde data are obtained from the Integrated Global Radiosonde Archive Version 2 (IGRA2) [45], which is maintained by the National Climate Data Center (NCDC) of the United States (ftp://ftp.ncdc.noaa.gov/pub/data/igra (accessed on 1 June 2022)). In order to improve the data quality, Radiosonde profiles with less than ten pressure levels of valid data were excluded. The PWV data from ERA5 and radiosonde can be calculated from an integral of the meteorological data as follows [46,47]:

$$PWV = \frac{\sum\limits_{i=2}^{n} (P_i - P_{i-1})(q_i + q_{i-1})}{2g} \quad (1)$$

where g is gravitational acceleration(m/s^2), n is the number of atmospheric pressure layers; P_i and q_i are the pressure and the humidity at ith atmospheric layer, respectively.

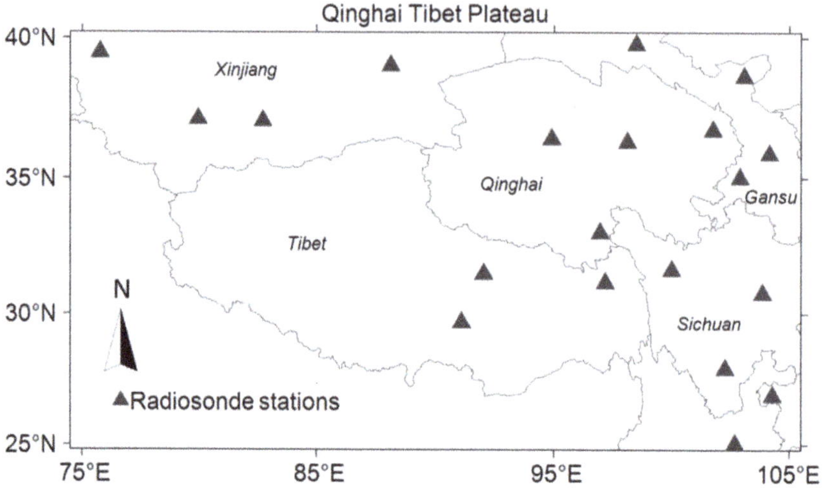

Figure 1. Distribution of the 20 radiosonde stations.

2.3. ASV-PWV Deriving and Correction

(1) Fundamentals of ASV-PWV model

Following the model of Askne and Nordius [37], we calculated the approximate ZWD by using the following formula:

$$ZWD = 10^{-6}(k_2' + k_3/T_m)\frac{R_d}{(\lambda+1)g_m}e_s \quad (2)$$

where k_2' and k_3 represent empirically determined refractivity constants, here k_2' = 22.1 ± 2.2 K/hPa, k_3 = 3.739 × 10^5 ± 0.012 × 10^5 K^2/hPa; T_m is the mean temperature weighted with water vapor pressure (K); λ is the water vapor decrease factor; g_m is the mean gravity which equals 9.80665 m/s^2; R_d is specific gas constant for dry air which equals 287.0464 J/K/kg; and e_s is the water vapor pressure (hPa) which can be calculated by

$$e_s = \frac{q}{0.378q + 0.622}P \quad (3)$$

where q is specific humidity(kg/kg) and P is atmospheric pressure(hPa).

Then, PWV can be converted from ZWD by

$$PWV = ZWD \times II \quad (4)$$

where II is the water vapor conversion coefficient, which can be calculated by

$$II = \frac{10^6}{(k_3 \times T_m^{-1} + k_2')R_v} \quad (5)$$

where R_v is the gas constant of water vapor(J/K/kg).

Substituting (4) and (5) into (2), we get

$$PWV = \frac{R_d}{(\lambda+1)\cdot g_m \cdot R_v} \cdot \frac{q}{0.378q + 0.622}P \quad (6)$$

We find that the preliminary ASV-PWV is built up on the three basic parameters p, q, and λ. There is evidence that some of these parameters exhibit seasonal characteristics (see Figure 2). In Equation (6), the humidity at a specific atmospheric pressure of 1000 hPa,

PWV is obtained by using 5 years (from 2015 to 2019, 1825 days in total) ERA5 atmospheric meteorological data according to Equation (1).

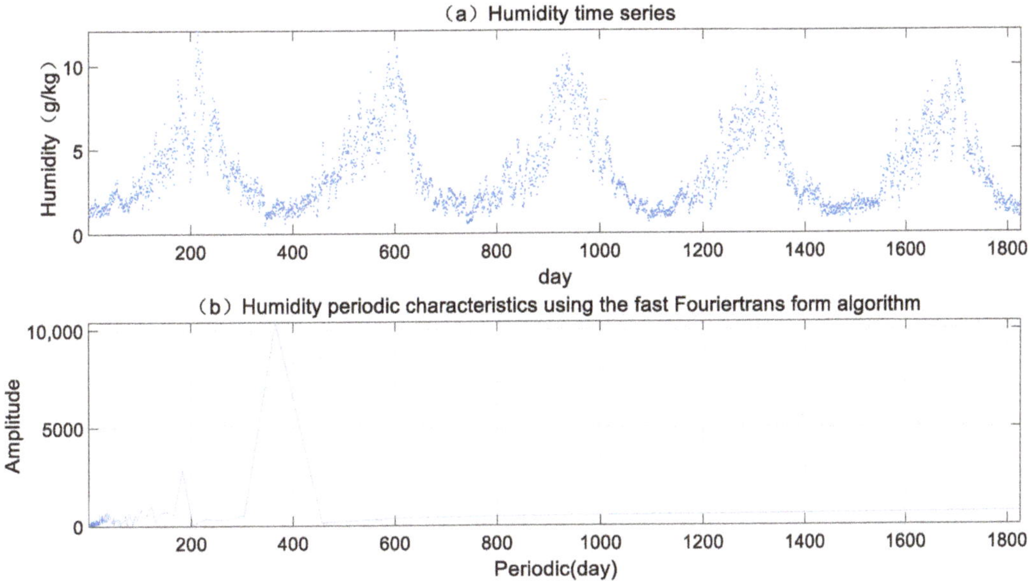

Figure 2. (**a**) humidity time series at a specific atmospheric pressure of 1000 hPa and (**b**) its periodic characteristics.

Using the PWV outputs from (1), we can obtain the water vapor decrease factor λ by (6). The results are shown in Figure 3. Given their significant seasonal nature, we adopt the following seasonal fit formula [48] to deduce empirical temporal information for the coefficients q and λ. For q, it appears as:

$$q = A_0 + A_1 \sin(2\pi \frac{JD}{365.25}) + A_2 \cos(2\pi \frac{JD}{365.25}) + A_3 \sin(4\pi \frac{JD}{365.25}) + A_4 \cos(4\pi \frac{JD}{365.25}) \tag{7}$$

where A_0 is intercept, A_1 and A_2 are annual amplitudes, A_3 and A_4 are semi-annual amplitudes of the coefficient, and JD is Julian day. Least-squares adjustments are used to fit these parameters to the 7381 grid points that we are interested.

(2) PWV correction using spherical harmonic function

From the foregoing analysis, we find that both Equations (1) and (6) can be used to determine the PWV data. Assuming d_{PWV} is the difference between the empirical PWV and reanalysis products based PWV, it can be fitted by the spherical harmonic function [49]

$$d_{PWV} = \sum_{n=0}^{N} \sum_{m=0}^{M} (A_{nm} \cdot a_{nm} + B_{nm} \cdot b_{nm}) \tag{8}$$

where N and M are the maximum degree and order of spherical harmonic function. Following Zhao et al. [50], we set the degree to $N = M = 12$ in this study. A_{nm} and B_{nm} are spherical harmonic coefficients, which can be calculated by

$$a_{nm} = P_{nm}(\sin(\varphi)) \cdot \cos(m\gamma) \tag{9}$$

$$b_{nm} = P_{nm}(\sin(\varphi)) \cdot \sin(m\gamma) \tag{10}$$

where γ and φ are longitude and latitude of the grid point; $P_{nm}(t)$ is the Legendre function with the form

$$P_{nm}(t) = \frac{1}{2^n}(1-t^2)^{\frac{m}{2}} \sum_{k=0}^{\frac{n-m}{2}} (-1)^k \frac{(2n-2k)!}{k!(n-k)!(n-m-2k)!} t^{n-m-k} \tag{11}$$

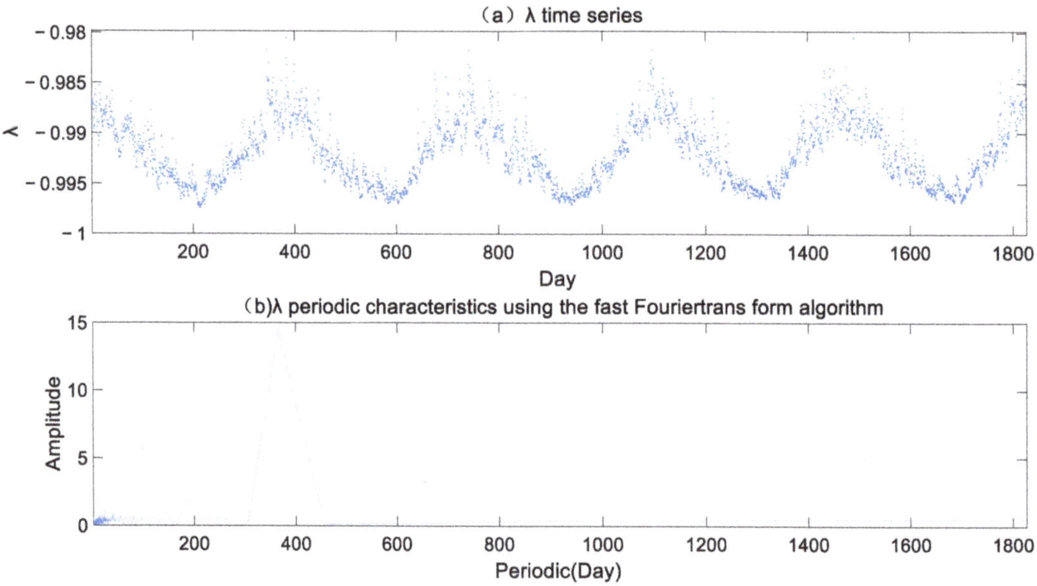

Figure 3. Same as Figure 2 but for the water vapor decrease factor.

(3) Vertical correction

A PWV difference occurs when the elevation difference between the reanalysis product point and the desired position is large [51]. In this study, we selected the ERA5 atmospheric layered meteorological data from 2015 to 2019 to calculate the PWV of different atmospheric pressure altitude and its corresponding atmospheric pressure according to Equation (1). Then, regression analysis is used to analyze the relationship between PWV and pressure. The results are shown in Figure 4.

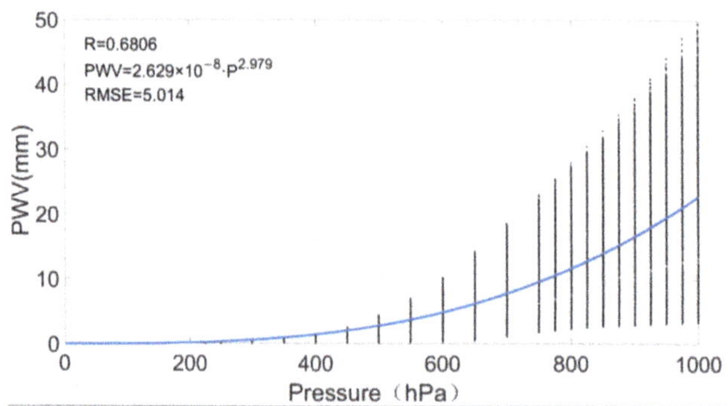

Figure 4. Variation relation between PWV and pressure.

Figure 4 shows that there is a significant nonlinear relationship between pressure and *PWV*, which can be expressed by

$$PWV = a \cdot P^b \tag{12}$$

where a is the coefficient and b is the increase coefficient of *PWV*.

Assuming there are two different atmospheric pressure altitudes P1 and P2 (P2 > P1), we obtain:

$$\frac{PWV_{P_2}}{PWV_{P_1}} = \frac{a \cdot P_2^b}{a \cdot P_1^b} = (\frac{P_2}{P_1})^b \tag{13}$$

$$PWV_{P_2} = PWV_{P_1} \cdot (\frac{P_2}{P_1})^b \tag{14}$$

where PWV_{P_1} and PWV_{P_2} are respectively the PWV of different atmospheric pressure altitudes P_1 and P_2.

Similar to the parameters q and λ, the parameter b changes with season (see Figure 5). As such, the empirical temporal information for the parameters b can again be deduced by using the seasonal fit Equation (7).

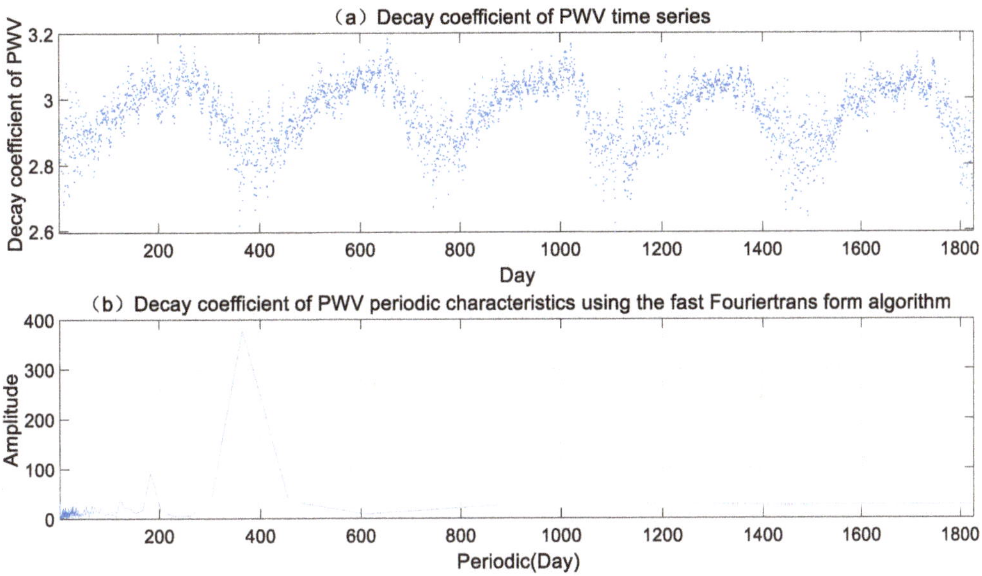

Figure 5. Same as Figures 2 and 3 but for the decay coefficient of PWV.

Finally, the main expressions of ASV-PWV are Equations (15) and (16):

$$PWV = \frac{R_d}{(\lambda+1) \cdot g_m \cdot R_v} \cdot \frac{1000q}{0.378q + 0.622} + dPWV \tag{15}$$

$$PWV_{P_2} = PWV_{P_1} \cdot (\frac{P_2}{P_1})^b \ (P_2 > P_1) \tag{16}$$

As mentioned already at an earlier stage, the coefficients of q, λ, b, and their amplitudes can be calculated by least-squares based seasonal fitting. For convenience, we store these results in a grid, from which the user could then spatially interpolate the desired position. As a result, ASV-PWV is an alternative option for rapidly gaining PWV with only four parameters of the desired position (e.g., the longitude and latitude, time and atmospheric pressure). Additionally, ASV-PWV has the same spatial resolution as the reanalysis products adopted. For example, if the ERA5 reanalysis data are adopted, ASV-PWV has a high spatial resolution of 0.25° × 0.25°. The next section follows the performance evaluation of our ASV-PWV model.

3. Model Validation of ASV-PWV

3.1. Statistical Indicators

Radiosonde profiles and ERA5 reanalysis PWV products are used to evaluate the PWV performance of ASV-PWV model. Statistical indicators to measure the performance of ASV-PWV include bias (perfect value = 0) and root mean square error (RMSE, perfect value = 0). Two calculations of indicators are presented as follows:

$$Bias = \sum_{i=1}^{N} \frac{(PWV_{mi} - PWV_{pi})}{N} \tag{17}$$

$$RMSE = \sqrt{\sum_{i=1}^{N} \frac{(PWV_{mi} - PWV_{pi})^2}{N}} \tag{18}$$

where N is the total number of matchups, PWV_{mi} is the PWV of model, and PWV_{pi} is the PWV of Radiosonde profiles or ERA5 reanalysis products.

3.2. Validation with Radiosonde Profiles

Profiles of 20 radiosonde stations in Qinghai Tibet Plateau, China, along with the latest publicly-available C-PWVC2 (Huang et al., 2021 [36]) are used to validate the local performance of our model. The results of C-PWVC2 are obtained using the vertical corrected ERA5 reanalysis PWV products (2012–2017), the expressions of C-PWVC2 are Equations (19) and (20):

$$PWV_{h_1} = PWV_{h_2} \cdot \exp(\beta(h_1 - h_2)) \tag{19}$$

$$\begin{aligned}\beta(\mathrm{DOY}) = &-0.453 - 0.037\sin(2\pi\tfrac{\mathrm{DOY}}{365.25}) - 0.087\cos(2\pi\tfrac{\mathrm{DOY}}{365.25}) + \\ &0.036\sin(4\pi\tfrac{\mathrm{DOY}}{365.25}) + 0.023\cos(4\pi\tfrac{\mathrm{DOY}}{365.25})\end{aligned} \tag{20}$$

where PWV_{h_1} and PWV_{h_2} are respectively the PWV of different height h_1 and h_2, β is decreasing coefficient, DOY is day of year. For more details, see Huang et al. (2021 [36]).

Table 1 and Figure 6 give the annual bias and root mean square error (RMSE) of the ASV-PWV and C-PWVC2 in 2020. With respect to radiosonde data, the annual bias of ASV-PWV is in the range from −2.30 mm to 0.79 mm with a mean of −0.44 mm and the annual bias of C-PWVC2 is in the range from −2.68 mm to −0.32 mm with a mean of −1.36 mm, respectively. The annual RMSE of ASV-PWV is in the range from 1.54 mm to 7.79 mm with a mean of 3.44 mm and the annual RMSE of C-PWVC2 is in the range from 1.42 mm to 4.58 mm with a mean of 2.51 mm, respectively.

Figure 6 shows that the bias of ASV-PWV is mainly distributed around 0 mm, with the range from −1 mm to 1 mm, whereas the bias of C-PWVC2 are all negative and mainly distributed around −1.5 mm. Particularly for the radiosonde in latitude greater than 35° or in longitude greater than 100°, ASV-PWV has larger RMSE than C-PWVC2 accompanied by some abnormally large value, whereas C-PWVC2 is relatively uniform in terms of RMSE and mainly smaller than 3 mm.

Table 1. Average annual bias and RMSE of the ASV-PWV and C-PWVC2 in 2020.

Model	Bias (mm)			RMSE (mm)		
	Avg	Max	Min	Avg	Max	Min
ASV-PWV	−0.44	0.79	−2.30	3.44	7.79	1.54
C-PWVC2	−1.36	−0.32	−2.68	2.51	4.58	1.42

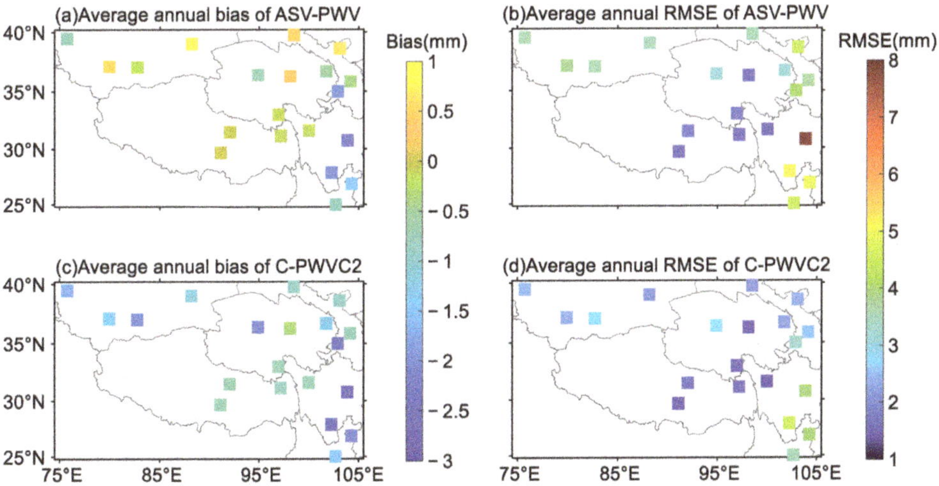

Figure 6. Average annual bias and RMSE of the ASV-PWV and C-PWVC2 in 2020.

Table 2, Figures 7 and 8 give the statistical bias and RMSE of ASV-PWV and C-PWVC2 models in four seasons of the year. We find that ASV-PWV has the minimum average bias (0.1 mm) and average RMSE (1.81 mm) in spring and has the maximum average bias (−1.34 mm) and average RMSE (4.83 mm) in autumn, respectively. C-PWVC2 has the minimum average bias (−0.99 mm) in autumn and has the maximum average bias (−1.81 mm) in spring. C-PWVC2 has the minimum average RMSE (1.39 mm) in spring and has the maximum average RMSE (3.31 mm) in summer.

Table 2. Performance statistics of the ASV-PWV and C-PWVC2 in 4 seasons of 2020.

Model	Indicator (mm)	Spring			Summer			Autumn			Winner		
		Avg	Max	Min	Avg	Max	Min	Avg	Max	Min	Avg	Max	Min
ASV-PWV	Bias	0.10	0.95	−2.86	0.14	2.71	−1.61	−1.34	1.34	−5.37	−0.65	0.50	−3.49
	RMSE	1.81	4.73	0.72	3.55	8.35	1.43	4.83	10.89	2.10	2.81	6.68	0.83
C-PWVC2	Bias	−1.01	−0.34	−1.88	−1.81	−0.42	−3.72	−0.99	0.67	−2.81	−1.76	−0.53	−5.40
	RMSE	1.39	2.77	0.48	2.64	4.69	1.20	3.13	4.66	1.68	2.51	7.58	0.70

For both ASV-PWV and C-PWVC2, their bias distribution in the four seasons is generally consistent with that of the entire year (see Figure 7). Some abnormal bias with large values are found in the individual radiosonde located in the southeast region and in autumn and winter. No significant difference of the bias distribution is found in C-PWVC2 in four seasons. Figure 8 shows that the RMSE of ASV-PWV is uniformly distributed in spring and winter but increases in summer and autumn with some larger RMSE values in the radiosonde with latitude greater than 35° or longitude greater than 100°. Meanwhile, the RMSE of C-PWVC2 is uniformly distributed in four seasons. Overall, the ASV-PWV performs better in terms of performance but with lower stability than C-PWVC2. For ASV-PWV, the performance in spring and winter is better than that in summer and autumn, whereas no obvious seasonal changes are found for C-PWVC2. These results reveal that our ASV-PWV performs better than C-PWVC2 in terms of seasonal characteristics.

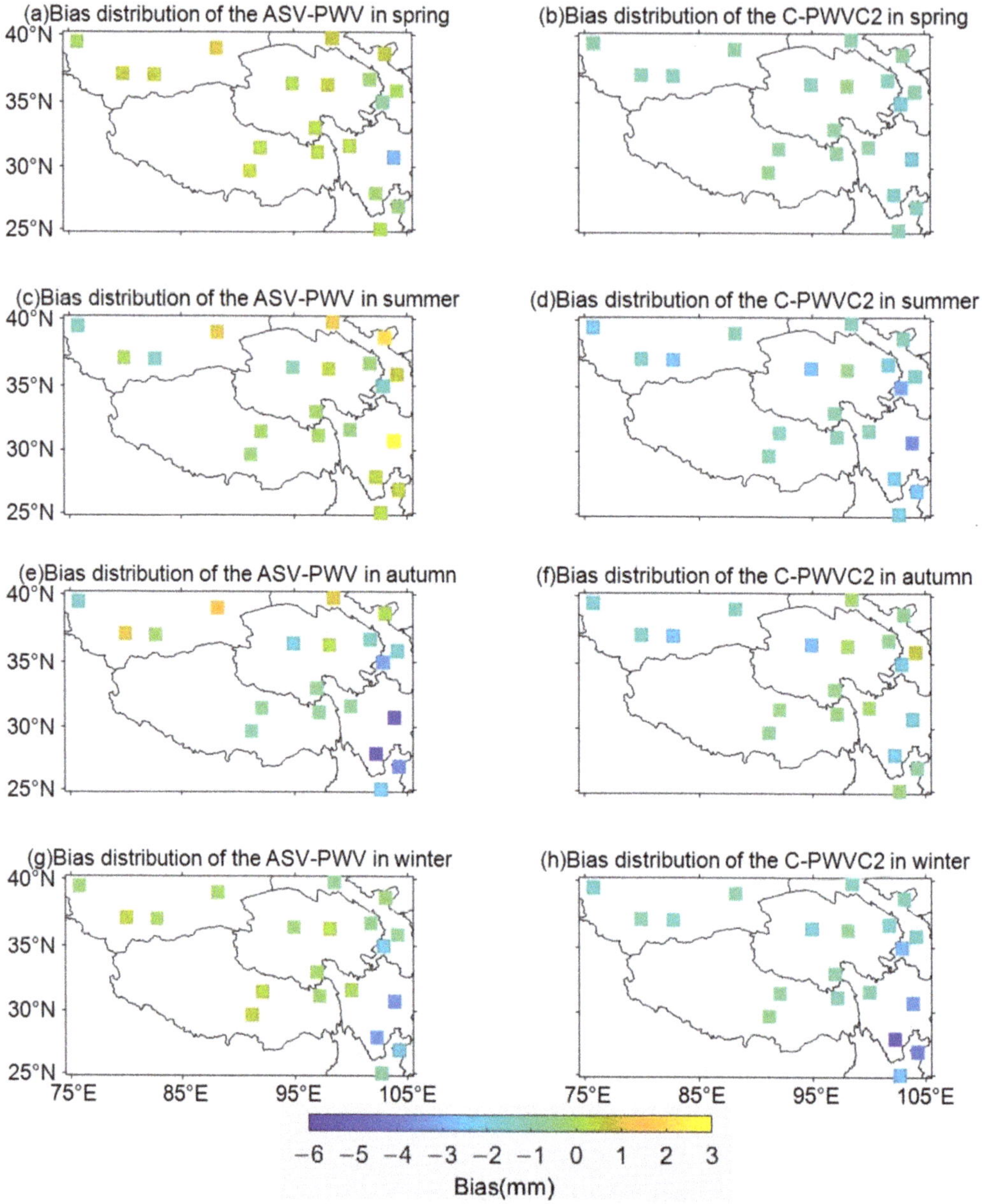

Figure 7. Bias distribution of the ASV-PWV and C-PWVC2 in 4 seasons of 2020.

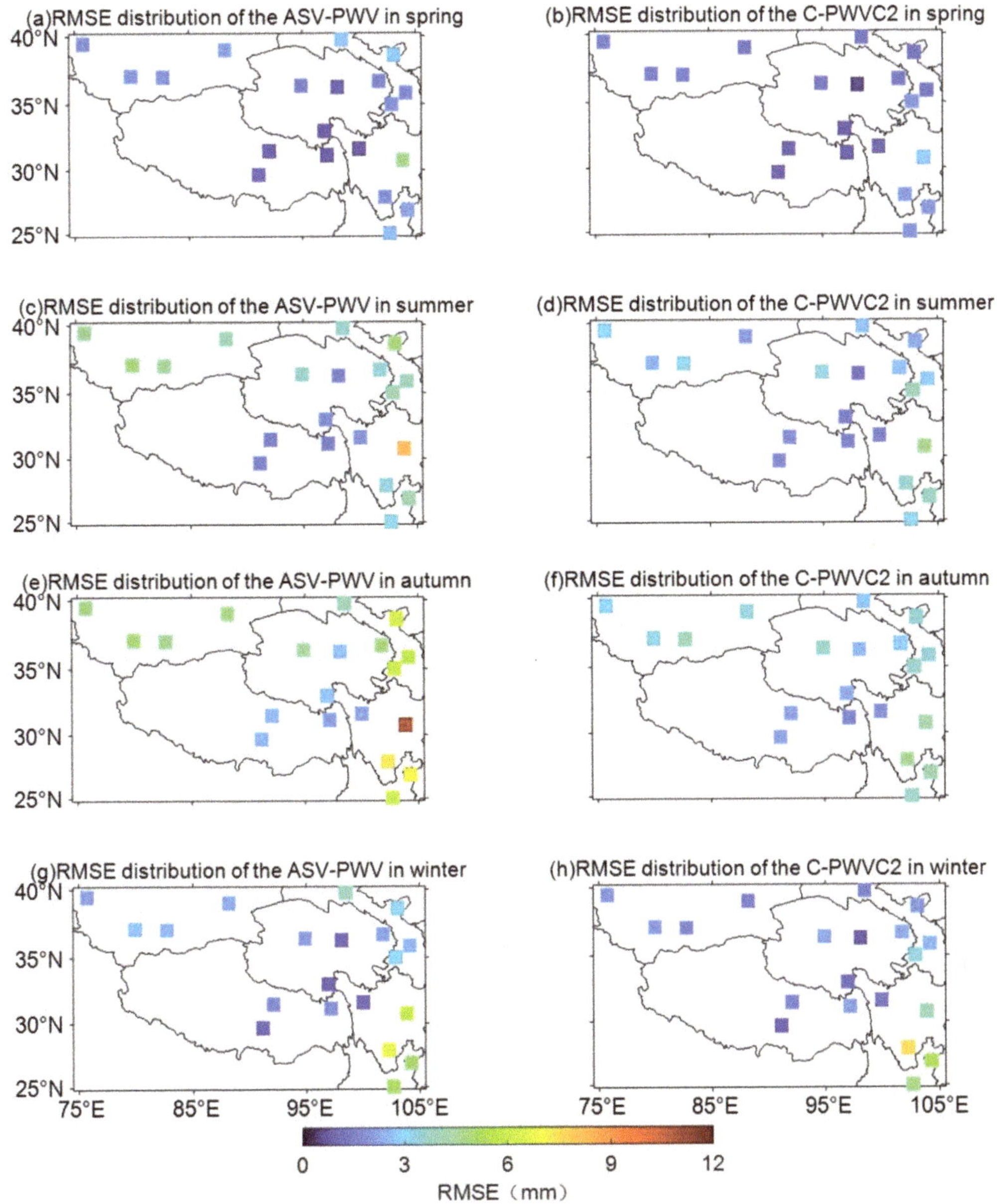

Figure 8. RMSE distribution of the ASV-PWV and C-PWVC2 in 4 seasons of 2020.

3.3. Validation with the ERA5 Reanalysis PWV Products

In order to further understand the spatial features of our ASV-PWV model in Qinghai Tibet Plateau, a comparison with PWV data from ERA5 is included in this study. Hence, 7381 grid points in 2020 of ERA5 reanalysis products are adopted. Figure 9 shows that the average annual bias and RMSE of ASV-PWV are −0.73 mm and 4.28 mm respectively. Obvious regional characteristics is also found in bias and RMSE distribution (see Figure 9). The bias in the northwest region is lower than that in the southeast region, where the bias is mainly negative, indicating that the PWV of ASV-PWV is smaller than the ERA5 reanalysis PWV products. Additionally, the RMSE in the central region is obviously better than that in other regions, and large RMSE values are found in southern region. Figures 10 and 11 provide the detail variation of bias and RMSE of ASV-PWV. We find that 90.59% regions have bias in the range of −2.5–0.5 mm, only 2.91% regions have bias less than −2.5 mm, and 6.50% regions have the bias in the range of 0.5–1.5 mm (see Figure 10). We also find that 78.30% regions have RMSE in the range of 1–6 mm, 13.14% regions have RMSE in the range of 6–9 mm, 6.03% regions have RMSE in the range of 9–12 mm, and only 2.22% regions have RMSE larger than 12 mm (see Figure 11).

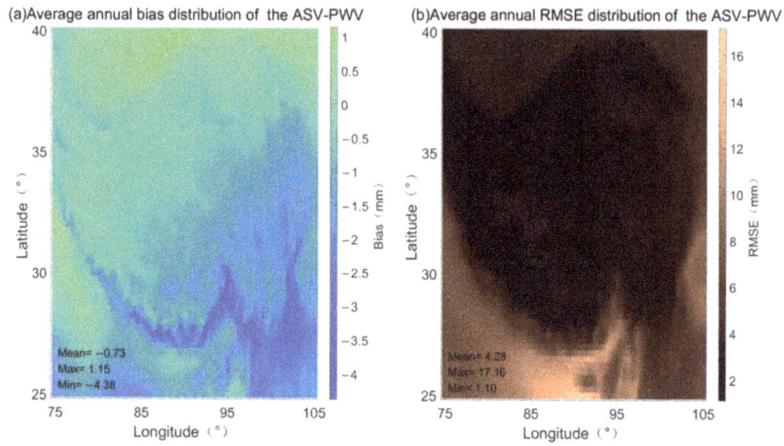

Figure 9. Average annual bias and RMSE distribution of the ASV-PWV (with respect to the ERA5 reanalysis PWV products).

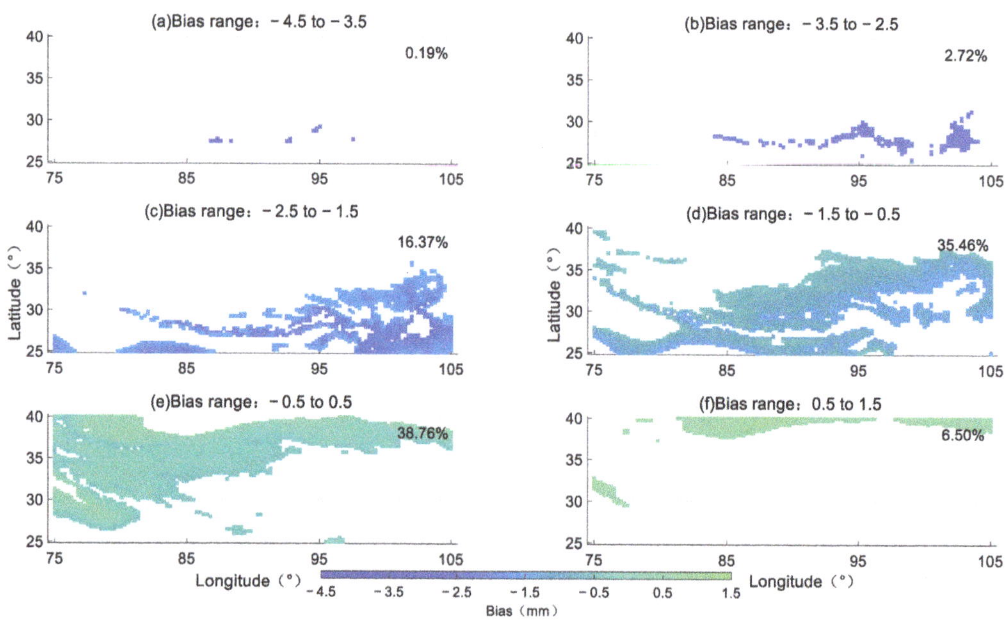

Figure 10. The bias distribution of ASV-PWV.

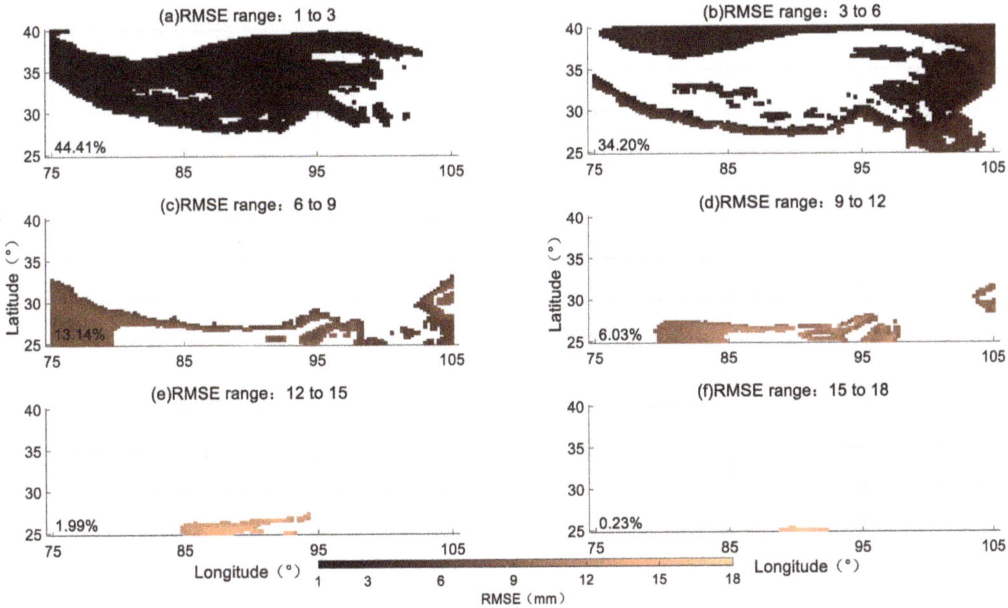

Figure 11. The RMSE distribution of ASV-PWV.

Since altitude has a profound influence on PWV as aforementioned, we divide the annual average atmospheric pressure of grid points into different altitudes, to better evaluate the performance of our method (see Figure 12). We find that the bias of ASV-PWV is mainly negative, indicating that the ASV-PWV derived PWV is generally lower than the ERA5 reanalysis PWV products. We also find that the bias generally increases first in the pressure range of 750~800 hPa and then decreases at 850~900 hPa with an average annual bias of −0.73 mm, while the RMSE increases with the atmospheric pressure with a mean of 4.28 mm. All the results show that the performance of our empirical model is basically equivalent to that of ERA5, indicating that ASV-PWV can be applied as a potentially alternative option for rapidly gaining PWV.

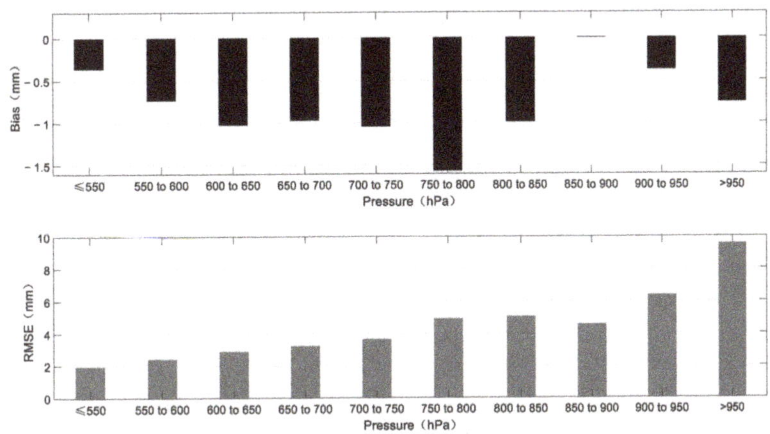

Figure 12. Annual average bias and RMSE of ASV-PWV for different atmospheric pressures for ten grid points (with No. of 911, 1774, 798, 500, 441, 403, 409, 731, 279 and 1135 respectively).

4. Conclusions

This paper introduces a new empirical PWV grid model (called ASV-PWV) based on the empirical zenith wet delay and improved by the spherical harmonic function and vertical correction. Based on the ERA5 atmospheric meteorological data, some fundamental parameters for empirical PWV model are calculated by least-squares based seasonal fitting and are stored in a grid, from which the user then could spatially interpolate the desired position. As such, ASV-PWV is convenient and enables the user to rapidly gain PWV data with only four input parameters, e.g., the longitude and latitude, time, and atmospheric pressure of the desired position.

Profiles of 20 radiosonde stations in Qinghai Tibet Plateau, China, along with the latest publicly available C-PWVC2 model and PWV data from ERA5 are used to validate the local performance of our model. The PWV value of ASV-PWV and C-PWVC2 models is generally consistent with radiosonde. The results for the whole year show that the annual bias of ASV-PWV is in the range from −2.30 mm to 0.79 mm with a mean of −0.44 mm and the annual bias of C-PWVC2 is in the range from −2.68 mm to −0.32 mm with a mean of −1.36 mm, respectively; the annual RMSE of ASV-PWV is in the range from 1.54 mm to 7.79 mm with a mean of 3.44 mm, and the annual RMSE of C-PWVC2 is in the range from 1.42 mm to 4.58 mm with a mean of 2.51 mm, respectively. The statistical results of the distribution of each radiosonde stations show that the bias of ASV-PWV is mainly distributed around 0 mm, with the range from −1 mm to 1 mm, whereas the bias of C-PWVC2 are all negative and mainly distributed around −1.5 mm. The performance statistics of ASV-PWV and C-PWVC2 models in different seasons over the Qinghai-Tibet Plateau show that ASV-PWV has the minimum average bias (0.1 mm) and average RMSE (1.81 mm) in spring and the maximum average bias (−1.34 mm) and average RMSE (4.83 mm) in autumn, respectively. C-PWVC2 has the minimum average bias (−0.99 mm) in autumn and has the maximum average bias (−1.81 mm) in spring, C-PWVC2 has the minimum average RMSE (1.39 mm) in spring and has the maximum average RMSE (3.31 mm) in summer. For both ASV-PWV and C-PWVC2, their bias distribution in the four seasons is generally consistent with that of the entire year. The RMSE of ASV-PWV is uniformly distributed in spring and winter but increases in summer and autumn with some larger RMSE values in the radiosonde with latitude greater than 35° or longitude greater than 100°, and the RMSE of C-PWVC2 is uniformly distributed in four seasons.

In general, a sound consistency exists between PWV values of ASV-PWV and ERA5 (total 7381 grid points in 2020); their average annual bias and RMSE are −0.73 mm and 4.28 mm, respectively. Obvious regional characteristics are found in bias and RMSE distribution (the bias in the northwest region is lower than that in the southeast region, where the bias is mainly negative, indicating that the PWV data from ASV-PWV is smaller than the PWV data from ERA5. Additionally, the RMSE in the central region is obviously better than that in other regions, and large RMSE values are found in southern region). The statistical results of range distribution of BIAS and RMSE show that 90.59% regions have bias in the range from −2.5 mm to 0.5 mm, only 2.91% regions have bias less than −2.5 mm, 6.50% regions have bias in the range from 0.5 mm to 1.5 mm, and 78.30% regions have RMSE in the range from 1 mm to 6 mm, 13.14% regions have RMSE in the range from 6 mm to 9 mm, 6.03% regions have RMSE in the range from 9 mm to 12 mm, and only 2.22% regions have RMSE larger than 12 mm. Meanwhile, we divide the annual average atmospheric pressure of grid points into different altitudes to evaluate the performance of our method. The results show that the PWV data from ASV-PWV is generally lower than the PWV data from ERA5. The bias generally increases first in the pressure range of 750~800 hPa and then decreases at 850~900 hPa with an average annual bias of −0.73 mm, while the RMSE increases with the atmospheric pressure with a mean of 4.28 mm. ASV-PWV achieves similar performance as the PWV data from ERA5 and radiosonde data, and performs better than C-PWVC2 in terms of seasonal characteristics, indicating that ASV-PWV is a potentially alternative option for rapidly gaining PWV.

Although our ASV-PWV model focuses on the Qinghai Tibet Plateau, it can also be applied to a larger region. Some critical issues (e.g., regional systematic model errors) are not completely taken into consideration. More quality assessments are desirable in future work to evaluate the possible benefits of the ASV-PWV model.

Author Contributions: Conceptualization, X.W. and F.C.; methodology, X.W. and F.C.; software, F.C.; validation, X.W., F.C., F.K. and C.X.; formal analysis, X.W., F.C., F.K. and C.X.; investigation, X.W. and F.C.; resources, X.W., F.C., F.K. and C.X.; data curation, X.W. and F.C.; writing—original draft preparation, X.W. and F.C.; writing—review and editing, F.K. and C.X.; supervision, F.K. and C.X.; project administration, X.W. and F.C. All authors have read and agreed to the published version of the manuscript.

Funding: This research was funded by the Key Research and Development Program of Jiangsu Province (Social Development Project) [BE2021622], the Natural Science Foundation of Jiangsu Province [BK20211037], the Higher Education Reform Educational Project of Jiangsu Province [2021JSJG219], the Postgraduate Research & Practice Innovation Program of Jiangsu Province [SJCX21_0373], the Science and Technology Project of Changzhou City [CE20225026], the Science and Technology Development Fund Project of Wuxi city [N20201011] and the APC was funded by the Key Research and Development Program of Jiangsu Province (Social Development Project).

Acknowledgments: We are grateful to ECMWF for providing ERA5 products and the NCDC for providing IGRA2 products.

Conflicts of Interest: The authors declare no conflict of interest.

References

1. Yao, Y.; Zhang, S.; Kong, J. Research Progress and Prospect of GNSS Space Environment Science. *Acta Geod. Cartogr. Sin.* **2017**, *46*, 1408–1420. [CrossRef]
2. Dessler, A.E.; Sherwood, S.C. A matter of humidity. *Science* **2009**, *323*, 1020–1021. [CrossRef] [PubMed]
3. Chung, E.; Soden, B.; Sohn, B.J.; Shi, L. Upper-tropospheric moistening in response to anthropogenic warming. *Proc. Natl. Acad. Sci. USA* **2014**, *111*, 11636–11641. [CrossRef] [PubMed]
4. Trenberth, K.E.; Fasullo, J.; Smith, L. Trends and variability in column-integrated atmospheric water vapor. *Clim. Dyn.* **2005**, *24*, 741–758. [CrossRef]
5. Allan, R.P.; Soden, B.J. Atmospheric Warming and the Amplification of Precipitation Extremes. *Science* **2008**, *321*, 1481–1484. [CrossRef]
6. King, M.D.; Kaufman, Y.J.; Menzel, W.P.; Tanre, D. Remote sensing of cloud, aerosol, and water vapor properties from the moderate resolution imaging spectrometer (MODIS). *IEEE Trans. Geosci. Remote Sens.* **1992**, *30*, 2–27. [CrossRef]
7. Berezin, I.A.; Timofeyev, Y.M.; Virolainen, Y.A.; Frantsuzova, I.S.; Volkova, K.A.; Poberovsky, A.V.; Holben, B.N.; Smirnov, A.; Slutsker, I. Error analysis of integrated water vapor measured by CIMEL photometer. *Izvestiya Atmos. Ocean. Phys.* **2017**, *53*, 58–64. [CrossRef]
8. Campmany, E.; Bech, J.; Rodríguez-Marcos, J.; Sola, Y.; Lorente, J. A comparison of total precipitable water measurements from radiosonde and sunphotometers. *Atmos. Res.* **2010**, *97*, 385–392. [CrossRef]
9. Ross, R.J.; Elliott, W.P. Radiosonde-Based Northern Hemisphere Tropospheric Water Vapor Trends. *J. Clim.* **2001**, *14*, 1602–1612. [CrossRef]
10. Gurbuz, G. On Variations of the Decadal Precipitable Water Vapor (PWV) over Turkey. *Adv. Space Res.* **2021**, *68*, 292–300. [CrossRef]
11. Easterling, D.; Peterson, T. A new method for detecting undocumented discontinuities in climatological time series. *Int. J. Climatol.* **1995**, *15*, 369–377. [CrossRef]
12. Rowe, P.M.; Miloshevich, L.M.; Turner, D.D.; Walden, V.P. Dry Bias in Vaisala RS90 Radiosonde Humidity Profiles over Antarctica. *J. Atmos. Ocean. Technol.* **2008**, *25*, 1529–1541. [CrossRef]
13. Zhao, T.; Dai, A.; Wang, J. Trends in Tropospheric Humidity from 1970 to 2008 over China from a Homogenized Radiosonde Dataset. *J. Clim.* **2012**, *25*, 4549–4567. [CrossRef]
14. Gao, B.; Kaufman, Y.J. Water vapor retrievals using Moderate Resolution Imaging Spectroradiometer (MODIS) near-infrared channels. *J. Geophys. Res. Atmos.* **2003**, *108*, 108:1–108:10. [CrossRef]
15. Liu, B.; Wang, Y.; Lou, Z.; Zhan, W. The MODIS PWV correction based on CMONOC in Chinese mainland. *Acta Geod. Cartogr. Sin.* **2019**, *48*, 1207–1215. [CrossRef]
16. Kumar, S.; Singh, A.K.; Prasad, A.K.; Singh, R.P. Variability of GPS derived water vapor and comparison with MODIS data over the Indo-Gangetic plains. *Phys. Chem. Earth* **2010**, *55–57*, 11–18. [CrossRef]
17. Shi, F.; Xin, J.; Yang, L.; Cong, Z.; Liu, R.; Ma, Y.; Wang, Y.; Lu, X.; Zhao, L. The first validation of the precipitable water vapor of multisensor satellites over the typical regions in China. *Remote Sens. Environ.* **2017**, *206*, 107–122. [CrossRef]

18. Khaniani, A.S.; Nikraftar, Z.; Zakeri, S. Evaluation of MODIS Near-IR water vapor product over Iran using ground-based GPS measurements. *Atmos. Res.* **2020**, *231*, 104657. [CrossRef]
19. Dee, D.P.; Uppala, S.M.; Simmons, A.J.; Berrisford, P.; Poli, P.; Kobayashi, S.; Andrae, U.; Balmaseda, M.A. The ERA-Interim reanalysis: Configuration and performance of the data assimilation system. *Q. J. R. Meteorol. Soc.* **2011**, *137*, 553–597. [CrossRef]
20. Zhang, L.; Wu, L.; Gan, B. Modes and Mechanisms of Global Water Vapor Variability over the Twentieth Century. *J. Clim.* **2013**, *26*, 5578–5593. [CrossRef]
21. Sherwood, S.C.; Roca, R.; Weckwerth, T.M.; Andronova, N.G. Tropospheric water vapor, convection, and climate. *Rev. Geophys.* **2010**, *48*, RG2001. [CrossRef]
22. Chen, B.; Liu, Z. Global water vapor variability and trend from the latest 36 year (1979 to 2014) data of ECMWF and NCEP reanalyses, radiosonde, GPS, and microwave satellite. *J. Geophys. Res. Atmos.* **2016**, *121*, 11442–11462. [CrossRef]
23. Cortés, F.; Cortés, K.; Reeves, R.; Bustos, R.; Radford, S. Twenty years of precipitable water vapor measurements in the Chajnantor area. *Astron. Astrophys.* **2020**, *640*, A126. [CrossRef]
24. Shikhovtsev, A.Y.; Khaikin, V.B.; Mironov, A.P.; Kovadlo, P.G. Statistical Analysis of the Water Vapor Content in North Caucasus and Crimea. *Atmos. Ocean. Opt.* **2022**, *35*, 168–175. [CrossRef]
25. Ma, X.; Yao, Y.; Zhang, B.; He, C. Retrieval of high spatial resolution precipitable water vapor maps using heterogeneous earth observation data. *Remote Sens. Environ.* **2022**, *278*, 113100. [CrossRef]
26. Yao, Y.; Xu, X.; Hu, Y. Establishment of a regional precipitable water vapor model based on the combination of GNSS and ECMWF data. *Atmos. Meas. Tech. Discuss.* **2018**, 1–21. [CrossRef]
27. Zhang, B.; Yao, Y.; Xin, L.; Xu, X. Precipitable water vapor fusion: An approach based on spherical cap harmonic analysis and Helmert variance component estimation. *J. Geod.* **2019**, *93*, 2605–2620. [CrossRef]
28. Zhang, B.; Yao, Y. Precipitable water vapor fusion based on a generalized regression neural network. *J. Geod.* **2021**, *95*, 36. [CrossRef]
29. Huang, L.; Wang, X.; Xiong, S.; Li, J.; Liu, L.; Mo, Z.; Fu, B.; He, H. High-precision GNSS PWV retrieval using dense GNSS sites and in-situ meteorological observations for the evaluation of MERRA-2 and ERA5 reanalysis products over China. *Atmos. Res.* **2022**, *276*, 106247. [CrossRef]
30. Huang, L.; Zhu, G.; Liu, L.; Chen, H.; Jiang, W. A global grid model for the correction of the vertical zenith total delay based on a sliding window algorithm. *GPS Solut.* **2021**, *25*, 98. [CrossRef]
31. Huang, L.; Mo, Z.; Xie, S.; Liu, L.; Chen, J.; Kang, C.; Wang, S. Spatiotemporal characteristics of GNSS-derived precipitable water vapor during heavy rainfall events in Guilin, China. *Satell. Navig.* **2021**, *2*, 13. [CrossRef]
32. Huang, L.; Liu, L.; Chen, H.; Jiang, W. An improved atmospheric weighted mean temperature model and its impact on GNSS precipitable water vapor estimates for China. *GPS Solut.* **2019**, *23*, 51. [CrossRef]
33. Leckner, B. The spectral distribution of solar radiation at the earth's surface–elements of a model. *Sol. Energy* **1978**, *20*, 143–150. [CrossRef]
34. Kouba, J. Implementation and testing of the gridded Vienna Mapping Function 1 (VMF1). *J. Geod. Vol.* **2018**, *82*, 193–205. [CrossRef]
35. Dousa, J.; Elias, M. An improved model for calculating tropospheric wet delay. *Geophys. Res. Lett.* **2014**, *41*, 4389–4397. [CrossRef]
36. Huang, L.; Mo, Z.; Liu, L.; Xie, S. An empirical model for the vertical correction of precipitable water vapor considering the time-varying lapse rate for Mainland China. *Acta Geod. Cartogr. Sin.* **2021**, *50*, 1320–1330. [CrossRef]
37. Askne, J.; Nordius, H. Estimation of tropospheric delay for microwaves from surface weather data. *Radio Sci.* **1987**, *22*, 379–386. [CrossRef]
38. Hersbach, H.; Dee, D. ERA5 Reanalysis is in Production. ECMWF Newsletter. 2016. Available online: https://www.ecmwf.int/en/newsletter/147/news/era145-reanalysis-production (accessed on 17 September 2022).
39. Hersbach, H.; Bell, B.; Berrisford, P.; Hirahara, S.; Horányi, A.; Muñoz-Sabater, J.; Nicolas, J.; Peubey, C.; Radu, R.; Schepers, R.; et al. The ERA5 global reanalysis. *Q. J. R. Meteorol. Soc.* **2020**, *146*, 1999–2049. [CrossRef]
40. Bill, B.; Hans, H.; Adrian, S.; Paul, B.; Per, D.; András, H.; Joaquín, M.; Julien, N.; Raluca, R.; Dinand, S.; et al. The ERA5 global reanalysis: Preliminary extension to 1950. *Q. J. R. Meteorol. Soc.* **2021**, *147*, 4186–4227. [CrossRef]
41. Zhao, Q.; Yang, P.; Yao, W.; Yao, Y. Hourly PWV Dataset Derived from GNSS Observations in China. *Sensors* **2019**, *20*, 231. [CrossRef]
42. Peng, Y.; Addisu, H.; Fadwa, A.; Joseph, A.; Anna, K.; Norman, T.F.; Hansjörg, K. Feasibility of ERA5 integrated water vapor trends for climate change analysis in continental Europe: An evaluation with GPS (1994–2019) by considering statistical significance. *Remote Sens. Environ.* **2021**, *260*, 112416. [CrossRef]
43. Zhang, W.; Wang, L.; Yu, Y.; Xu, G.; Hu, X.; Fu, Z.; Cui, C. Global evaluation of the precipitable-water-vapor product from MERSI-II (Medium Resolution Spectral Imager) on board the Fengyun-3D satellite. *Atmos. Meas. Tech.* **2021**, *14*, 7821–7834. [CrossRef]
44. Durre, I.; Yin, X.; Vose, R.S.; Applequist, S.; Arnfield, J. Enhancing the Data Coverage in the Integrated Global Radiosonde Archive. *J. Atmos. Ocean. Technol.* **2018**, *35*, 1753–1770. [CrossRef]
45. Durre, I.; Vose, R.S.; Wuertz, D.B. Overview of the Integrated Global Radiosonde Archive. *J. Clim.* **2006**, *19*, 53–68. [CrossRef]
46. Gui, K.; Che, H.; Chen, Q.; Zeng, Z.; Zheng, Y.; Long, Q.; Sun, T.; Liu, X.; Wang, Y.; Liao, T.; et al. Water vapor variation and the effect of aerosols in China. *Atmos. Environ.* **2017**, *165*, 322–335. [CrossRef]

47. He, Q. Water Vapor Retrieved from Ground-based GNSS and Its Applications in Extreme Weather Studies. Ph.D. Dissertation, China University of Mining and Technology, Beijing, China. Available online: https://kns.cnki.net/KCMS/detail/detail.aspx?dbname=CDFDLAST2022&filename=1021773689.nh (accessed on 24 September 2022).
48. Böhm, J.; Möller, G.; Schindelegger, M.; Pain, G.; Weber, R. Development of an improved blind model for slant delays in the troposphere (GPT2w). *GPS Solut.* **2015**, *19*, 433–441. [CrossRef]
49. Liu, J.; Chen, R.; Wang, Z.; Zhang, H. Spherical cap harmonic model for mapping and predicting regional TEC. *GPS Solut.* **2011**, *15*, 109–119. [CrossRef]
50. Zhao, Q.; Du, Z.; Wu, M.; Yao, Y.; Yao, W. Analysis of Influencing Factors and Accuracy Evaluation of PWV in Loess Plateau. Geomatics and Information Science of Wuhan University. 2021. Available online: http://kns.cnki.net/kcms/detail/42.1676.TN.20211028.20211233.20211008.html (accessed on 30 September 2022).
51. Wang, S.; Xu, T.; Nie, W.; Jiang, C.; Zhang, Z. Evaluation of Precipitable Water Vapor from Five Reanalysis Products with Ground-Based GNSS Observations. *Remote Sens.* **2020**, *12*, 1817. [CrossRef]

remote sensing

MDPI

Article

Low-Cost IMU and Odometer Tightly Augmented PPP-B2b-Based Inter-Satellite Differenced PPP in Urban Environments

Yu Min [1,2], Zhouzheng Gao [1,*], Jie Lv [1], Ruohua Lan [1], Qiaozhuang Xu [1] and Cheng Yang [1]

1 School of Land Science and Technology, China University of Geosciences Beijing, Beijing 100083, China
2 Aerospace Information Research Institute, Chinese Academy of Sciences, Beijing 100094, China
* Correspondence: zz.gao@cugb.edu.cn; Tel.: +86-186-2791-5172

Abstract: Since 23 June 2020, BDS-3 has been entirely operated and obtained the ability of global PNT (Positioning, Navigation, and Timing) services. Afterward, real-time Precise Point Positioning (PPP) service is available in China's surrounding areas via BDS-3 PPP-B2b signal. However, such a real-time PPP service cannot maintain the high accuracy and continuity of positioning solutions in challenging environments, such as urban environments. For that, we carried out a model by integrating between-satellite single-differenced (BSSD) PPP, a low-cost Inertial Navigation System (INS), and an odometer via an extended Kalman filter. The performance of this integration model was assessed with vehicle-borne data. Results demonstrated that (1) the position RMS (Root Mean Square) of BSSD PPP are 64.33 cm, 53.47 cm, and 154.11 cm. Compared with BSSD PPP, about 31.2%, 23.3%, and 27.3% position improvements can be achieved by using INS. Further enhancements of position RMS benefiting from the odometer are 1.34%, 1.41%, and 1.73% in the three directions. (2) Anyway, the accuracy of BSSD PPP/INS/Odometer tightly coupled integration is slightly higher than that of undifferenced PPP/INS/Odometer integration, with average improvement percentages of 7.71%, 3.09%, and 0.27%. Meanwhile, the performance of BSSD PPP/INS/Odometer integration during the periods with satellite outages is better than the undifferenced PPP-based solutions. (3) The improvements in attitudes from an odometer are more significant on heading angle than the other two attitudes, with percentages of 25.00%. (4) During frequent GNSS outage periods, the reduction in average maximum position drifts provided by INS are 18.01%, 8.95%, and 20.74%. After integrating with an odometer, the drifts can be furtherly decreased by 25.11%, 15.96%, and 20.69%. For attitude, about 41.67% reduction in average maximum drifts of heading angles is obtained.

Keywords: real-time Precise Point Positioning (RT-PPP); Inertial Navigation Pystem (INS); odometer; PPP-B2b service; tightly coupled integration

Citation: Min, Y.; Gao, Z.; Lv, J.; Lan, R.; Xu, Q.; Yang, C. Low-Cost IMU and Odometer Tightly Augmented PPP-B2b-Based Inter-Satellite Differenced PPP in Urban Environments. *Remote Sens.* **2023**, *15*, 199. https://doi.org/10.3390/rs15010199

Academic Editor: Xiaogong Hu

Received: 17 November 2022
Revised: 15 December 2022
Accepted: 26 December 2022
Published: 30 December 2022

1. Introduction

China began to build BeiDou Navigation Satellite System (BDS) at the end of the 20th century according to the three-step development strategies [1]. As planned, the first-generation BDS (BDS-1), the second-generation BDS (BDS-2), and the third-generation BDS (BDS-3) were completed successively in 2003, 2012, and 2020, with the corresponding satellite constellations of 3 Geostationary Orbit (GEO) satellites, 5 GEO satellites+5 IGSO (Inclined Geosynchronous Orbit) satellites+4 MEO (Medium Earth Orbit) satellites, and 3 GEO+3 IGSO+24 MEO satellites, respectively. Currently, the global Positioning, Navigation, and Timing (PNT) services of BDS are supported by signals on frequencies B1I (1561.098 MHz), B2I (1207.14 MHz), B3I (1268.52 MHz), B1C (1575.42 MHz), B2a (1176.45 MHz), and B2b (1207.14 MHz) [2–7]. Among these services, the Precise Point Positioning (PPP)-B2b enhancement service is of great significance, and also it is considered to be the core support for smart city development in China.

PPP, which was proposed by Zumberger et al. [8] in 1997, is the favored technology for high-accuracy positioning applications. The corresponding model was furtherly developed in the works [8–10]. PPP can provide an accurate positioning solution using only a single GNSS receiver by utilizing precise satellite products with about two weeks delay [11–13]. Consequently, PPP is mainly for applications in post-processing currently [14,15]. To satisfy the demands of real-time PPP, BDS-3 transmits the orbit/clock corrections of broadcast ephemeris by B2b signal [16–19]. Multi-GNSS Experiment (MGEX)/iGMAS stations were adopted in [17] to verify the real-time PPP positioning performance using PPP-B2b service in static and simulated-kinematic modes by comparing with the solutions based on Geodetic Benchmark (GBM) final products. It is shown that the positioning performance of real-time PPP is slightly worse than the post-processing PPP in general, according to the statistical results. However, the convergence time of real-time PPP is slightly shorter for the BDS-only in a static model. Tao et al. [16] compared PPP-B2b service with Real-Time Service (RTS) provided by Centre National d'Etudes Spatiales (CNES). Based on the analysis from six stations distributed in China, the positioning accuracy of BDS-3-only PPP with PPP-B2b service in kinematic mode can achieve decimeter-level positioning results, which is consistent with the accuracy of GPS PPP using products of CNES.

However, such a PPP-B2b service-based PPP cannot maintain positioning accuracy and continuity in urban environments [20,21], such as under bridges or trees, etc. In order to overcome the shortcomings of PPP in those circumstances, an Inertial Navigation System (INS) is integrated. INS is capable of providing position, velocity, and attitude results by using measurements from Inertial Measurement Units (IMU) without external observations. However, the position errors of INS will accumulate rapidly over time [22,23]. Meanwhile, integrating PPP and INS can estimate and compensate IMU errors to restrain the divergence [20,22,24]. According to previous works, more reliable position results can be obtained by PPP/INS [20,21,25–29].

Le et al. [25] investigated the Loosely Coupled Integration (LCI) of Single Frequency (SF) PPP/INS, which was validated by a flight experiment. Results showed that the SF-PPP-only positioning performance is visibly improved in the horizontal and vertical components. LCI mode cannot work when there are not enough GNSS observations. Martell in [26] further applied the Tightly Coupled Integration (TCI) of PPP and INS using different grade IMUs and different cut-off satellite angles. The results showed that reliable results could be obtained even if the number of satellites is less than 4. In [27], TCI was compared with LCI by using a tactical-grade IMU to illustrate the benefits of TCI. The position differences of TCI are within 1.0 m, and such errors of LCI are within 5.0 m. The studies above are mainly based on undifferenced GNSS observations. The Dual Frequency (DF) PPP/INS integration using Single-Difference Between-Satellites (BSSD) GPS observations was applied in [28]. During the simulated outages of 10 s~30 s, the position accuracy of BSSD PPP/INS TCI can be decimeter-level. Such accuracy is higher than those using undifferenced observations. Owing to the evolution of multi-constellation GNSS (multi-GNSS), more available observations can be adopted to enhance the integration performance. Gao et al. [21] developed the multi-constellation (GLONASS, BDS, and GPS) TCI of SF PPP/INS, and it was verified by a set of land-borne experiment data. Results showed that significant positum improvements in terms of accuracy, continuity, and reliability could be obtained by INS aiding. Anyway, the performance of conventional GPS SF-PPP can be improved by utilizing the multi-GNSS observation. The enhancement of multi-GNSS on PPP/INS is also illustrated in [29]. According to the results, the positioning and convergence performance of PPP is enhanced significantly by multi-GNSS and INS. However, such impacts in terms of velocity and attitude are invisible.

Based on the works in [20–29], continuous solutions with high accuracy can be provided by the PPP/INS integration during GNSS outages. However, the positioning errors of PPP/INS still accumulate rapidly, especially for a low-cost IMU when the GNSS signal seriously deteriorates, even interrupts completely around high buildings or under tunnels [30,31]. Such a circumstance can be facilitated by using the velocity information from an odometer. In [31], GPS + GLONASS DF-PPP was integrated with INS and odometer, and simulated GNSS outages were utilized to evaluate the performance in challenging circumstances. According to the results, the position accuracy was furtherly ameliorated by the odometer.

In this paper, we implied PPP/INS/odometer tightly coupled integration model. In comparison with previous works, the contribution of this paper is that such a tightly coupled integration is based on the BSSD model and the BDS-3 PPP-B2b orbit/clock corrections. In order to assess the performance of this algorithm, vehicle-borne data acquired in urban environments are processed and analyzed. The enhancements of a low-cost INS, BSSD BDS-3 PPP model, and an odometer on positioning and attitude determination are discussed in detail.

2. Methodology

In this section, the method to recover precise orbit and clock offsets by using the PPP-B2b service is discussed first. Then, the models of PPP/INS TCI and PPP/INS/odometer TCI based on the recovered PPP-B2b orbit/clock offsets and BSSD observation are presented in detail.

2.1. Recovery of Precise Satellite Orbit/Clock with PPP-B2b

The information provided by the PPP-B2b service includes the orbit and satellite clock offset corrections of broadcast products in the satellite-fixed frame (radial, along, and cross directions), which cannot be used directly in positioning. Thus, to apply the corrections in PPP, the precise orbit and clock offsets need to be recovered. The satellite positions computed by broadcast ephemeris are given in the Earth-Centered Earth-Fixed reference (ECEF) frame (*e*-frame). Therefore, the orbit corrections must be transformed into the ECEF frame by [32]

$$\begin{bmatrix} \delta O_x \\ \delta O_y \\ \delta O_z \end{bmatrix} = \begin{bmatrix} e_{radial} & e_{along} & e_{cross} \end{bmatrix} \begin{bmatrix} \delta O_{radial} \\ \delta O_{along} \\ \delta O_{cross} \end{bmatrix} \tag{1}$$

With

$$\begin{cases} e_{radial} = r/|r| \\ e_{cross} = (r \times \dot{r})/|r \times \dot{r}| \\ e_{along} = e_{radial} \times e_{cross} \end{cases} \tag{2}$$

where $\begin{bmatrix} \delta O_{radial} & \delta O_{cross} & \delta O_{along} \end{bmatrix}^T$ is the orbit correction vector in the satellite-fixed frame; $\begin{bmatrix} \delta O_x & \delta O_y & \delta O_z \end{bmatrix}^T$ represents the orbit correction vector in the *e*-frame; r is the satellite position vector and $\dot{r}$ represents the satellite velocity vector, which can be computed by broadcast ephemeris.

By applying the corrections in Equation (1) to broadcast ephemeris, precise satellite positions can be calculated by

$$\begin{bmatrix} X \\ Y \\ Z \end{bmatrix}_{prec,B2b} = \begin{bmatrix} X \\ Y \\ Z \end{bmatrix}_{brdc} - \begin{bmatrix} \delta O_x \\ \delta O_y \\ \delta O_z \end{bmatrix} \tag{3}$$

where $\begin{bmatrix} X & Y & Z \end{bmatrix}^T_{prec,B2b}$ denotes the vector of precise satellite positions after using the corrections provided by PPP-B2b service; $\begin{bmatrix} X & Y & Z \end{bmatrix}^T_{brdc}$ is the vector of satellite positions calculated by broadcast ephemeris.

PPP-B2b service also provides the correction of satellite clock offset. The precise clock offsets can be obtained by

$$dt^s_{prec,B2b} = dt^s_{brdc} - \frac{C_0}{c} \tag{4}$$

where $dt^s_{prec,B2b}$ is the precise clock offset calculated by using the clock offset derived from broadcast ephemeris (dt^s_{brdc}) and the PPP-B2b clock correction parameter (C_0); c denotes the velocity of light.

2.2. Single-Difference between-Satellites Observational Model

The linearized undifferenced Ionosphere-Free (IF) model of the satellite (k) can be written as

$$P_{IF}^{k} - \rho^{k} - \Delta\rho_{P,IF}^{k} = e_1\delta x_r + e_2\delta y_r + e_3\delta z_r + ct_r + m_{wet}\delta d_{wet} + \varepsilon_{P,IF} \quad (5)$$

$$L_{IF}^{k} - \rho^{k} - \Delta\rho_{L,IF}^{k} = e_1\delta x_r + e_2\delta y_r + e_3\delta z_r + ct_r + m_{wet}\delta d_{wet} - \lambda_{IF}N_{IF}^{k} + \varepsilon_{L,IF} \quad (6)$$

where $P_{IF}^{k} = \alpha P_1 - \beta P_2$ and $L_{IF}^{k} = \alpha L_1 - \beta L_2$ are IF pseudorange and carrier-phase, wherein $\alpha = f_1^2/(f_1^2 - f_2^2)$ and $\beta = f_2^2/(f_1^2 - f_2^2)$ are IF combination coefficients based on frequencies f_1 and f_2; r and k represent receiver and satellite; ρ^k is the geometric distance between receiver and satellite; t_r represents the receiver clock offset; m_{wet} is the tropospheric wet delay mapping function; δd_{wet} is the corrections of zenith wet delay; $E = \begin{bmatrix} e_1 & e_2 & e_3 \end{bmatrix}^T$ is the orientation cosine vector; $\delta p_{GNSS}^{e} = \begin{bmatrix} \delta x_r & \delta y_r & \delta z_r \end{bmatrix}$ represents receiver position correction in e-frame; λ_{IF} is the IF wavelength; N_{IF}^{k} is the IF float ambiguity [13]; $\Delta\rho_{L,IF}^{k}$ and $\Delta\rho_{P,IF}^{k}$ are the errors sum of carrier-phase and pseudorange; $\varepsilon_{L,IF}$ and $\varepsilon_{P,IF}$ are the measurement noise of carrier-phase and pseudorange [33].

In addition to the pseudorange and carrier-phase measurements, the Doppler measurements are essential for calculating receiver velocity. The linearized observation equation of Doppler can be expressed as

$$D_{IF}^{k} - \dot{\rho}^{k} - \Delta\dot{\rho}_{D,IF}^{k} = e_1\delta v_{x,r} + e_2\delta v_{y,r} + e_3\delta v_{z,r} + c\dot{t}_r + \varepsilon_{D,IF} \quad (7)$$

where $D_{IF}^{k} = \alpha D_1^{k} - \beta D_2^{k}$ is the Doppler measurement in meters; j represents frequency; the dot above the symbol indicates the variation. In general, except for the variation of receiver clocks ($\dot{t}_r$), satellite clocks, and geometric distance ($\dot{\rho}^{k}$), other errors' variations are close to zero and are ignored. $\delta v_{GNSS}^{e} = \begin{bmatrix} \delta v_{x,r} & \delta v_{y,r} & \delta v_{z,r} \end{bmatrix}$ are the vector of receiver velocity corrections; $\Delta\dot{\rho}_{D,IF}^{k}$ denotes the sum of Doppler errors; ε_D is Doppler noise.

Compared to the undifferenced model, the BSSD model has the advantage that receiver-related errors such as the receiver clock, receiver hardware delay, and unmodelled errors can be eliminated [28,34]. The BSSD IF model can be obtained by

$$\Delta P_{IF}^{km} - \Delta\rho^{km} - \Delta\rho_{P,IF}^{km} = e_1^{km}\delta x_r + e_2^{km}\delta y_r + e_3^{km}\delta z_r + m_{wet}^{km}\delta d_{wet} + \varepsilon_{P,IF}^{km} \quad (8)$$

$$\Delta L_{IF}^{km} - \Delta\rho^{km} - \Delta\rho_{L,IF}^{km} = e_1^{km}\delta x_r + e_2^{km}\delta y_r + e_3^{km}\delta z_r + m_{wet}^{km}\delta d_{wet} - \boldsymbol{\lambda}_{IF}\delta\mathbf{N}_{IF}^{km} + \varepsilon_{L,IF}^{km} \quad (9)$$

$$\Delta D_{IF}^{km} - \Delta\dot{\rho}^{km} - \Delta\dot{\rho}_{D,IF}^{km} = e_1^{km}\delta v_{x,r} + e_2^{km}\delta v_{y,r} + e_3^{km}\delta v_{z,r} + \varepsilon_{D,IF}^{km} \quad (10)$$

where $\Delta L_{IF}^{km} = L_{IF}^{m} - L_{IF}^{k}$; $\Delta P_{IF}^{km} = P_{IF}^{m} - P_{IF}^{k}$; $\Delta D_{IF}^{km} = D_{IF}^{m} - D_{IF}^{k}$; $\Delta\rho^{km} = \rho^{m} - \rho^{k}$; $\Delta\dot{\rho}^{km} = \dot{\rho}^{m} - \dot{\rho}^{k}$; $\delta\mathbf{N}_{IF}^{km} = \delta\mathbf{N}_{IF}^{m} - \delta\mathbf{N}_{IF}^{k}$; superscripts k and m represent satellite and the reference satellite. In this paper, GPS and BDS separately choose the reference satellite.

2.3. BSSD PPP/INS Tightly Coupled Integration

The innovation vector of TCI is based on the difference between GNSS observations (pseudorange, carrier-phase, and doppler) and the corresponding values predicted by INS [20,21,28]. The state equation and observation equation can be expressed as

$$X_{TCI,k} = \phi_{TCI,k,k-1}X_{TCI,k-1} + \mu_{TCI,k-1}, \mu_{TCI,k-1} \sim (0, Q_{TCI,k}) \quad (11)$$

$$Z_{TCI,k} = H_{TCI,k}X_{TCI,k} + \eta_{TCI,k}, \eta_{TCI,k} \sim N(0, R_{TCI}) \quad (12)$$

$$Z_{TCI,k} = \begin{bmatrix} P_{GNSS,IF} - P_{INS,IF} \\ L_{GNSS,IF} - L_{INS,IF} \\ D_{GNSS,IF} - D_{INS,IF} \end{bmatrix} = \begin{bmatrix} Z_{P_{IF}} \\ Z_{L_{IF}} \\ Z_{D_{IF}} \end{bmatrix} \quad (13)$$

where $\phi_{TCI,k,k-1}$ is the system transform matrix from epoch $k-1$ to epoch k; $\mu_{TCI,k-1}$ represent the state noise with the covariance of $Q_{TCI,k}$; $Z_{P_{IF}}$, $Z_{L_{IF}}$, and $Z_{D_{IF}}$ represent the innovation vector of pseudorange, carrier-phase, and doppler, respectively; $P_{INS,IF}$, $L_{INS,IF}$, and $D_{INS,IF}$ are the INS-predicted values; $P_{GNSS,IF}$, $L_{GNSS,IF}$, $D_{GNSS,IF}$ are the GNSS measurement; $\eta_{TCI,k}$ represents observation noise with the prior covariance of R_{TCI}.

In order to obtain the INS-predicted values corresponding to GNSS measurements, the position and velocity of the receiver updated by INS mechanization are required. However, the reference centers of the INS and GNSS antenna are different from each other, which results in a lever-arm system offset. Therefore, the linearization functions $Z_{TCI,k}$ after considering the lever arm can be written as [22,29]

$$\delta Z_{TCI,k} = \begin{bmatrix} \delta Z_{P_{IF}} \\ \delta Z_{L_{IF}} \\ \delta Z_{D_{IF}} \end{bmatrix} = \begin{bmatrix} C_1\left(\delta p^n_{INS} + \left(C^n_b l^b \times\right)\delta\psi\right) + c\delta t_r - m_{wet}d_{wet} \\ C_1\left(\delta p^n_{INS} + \left(C^n_b l^b \times\right)\delta\psi\right) + c\delta t_r - m_{wet}d_{wet} + \delta N_{IF}\lambda_{IF} \\ C_2 D^{-1}\delta p^n_{INS} + C^e_n H_\psi \delta\psi + C^e_n \delta v^n_{INS} + C^e_n C^n_b\left(l^b \times\right)\delta\omega^b_{ib} + c\delta\dot{t}_r \end{bmatrix} \tag{14}$$

$$D^{-1} = \begin{bmatrix} 1/(R_M + h) & 0 & 0 \\ 0 & 1/(R_N + h)\cos(B) & 0 \\ 0 & 0 & -1 \end{bmatrix} \tag{15}$$

where $\delta Z_{P_{IF}}$, $\delta Z_{L_{IF}}$, and $\delta Z_{D_{IF}}$ represent the differential form of innovation vector; n, b, and i are the navigation frame (n), the body frame (b), and the inertial frame (i); C^j_k ($k = n$ and b, $j = e$ and n) represent the rotation matrix from the k-frame to the j-frame; l^b is the lever-arm; δp^n_{INS}, δv^n_{INS}, and $\delta\psi$ are the corrections of position, velocity, and attitude at IMU center in n-frame; δt_r and $\delta\dot{t}_r$ represent the corrections of receiver clock offset and drift; C_1 is the rotation matrix of position corrections from e-frame to n-frame; C_2 is the differential form of C^e_n; $\delta\omega^b_{ib}$ means gyroscope errors [22,29].

Based on Equation (12), the satellite single-difference matrix (SSDM) can be expressed as [28]

$$SSDM = \begin{bmatrix} SSDM_G & 0 \\ 0 & SSDM_B \end{bmatrix}_{3h \times 2m} \tag{16}$$

$$SSDM_B = \begin{bmatrix} 1 & 0 & \cdots & -1 & 0 & 0 & \cdots & 0 \\ 0 & \ddots & 0 & -1 & \vdots & \vdots & 0 & \vdots \\ \vdots & 0 & 1 & -1 & \vdots & \vdots & \vdots & \vdots \\ \vdots & \vdots & \vdots & \vdots & \vdots & \vdots & \vdots & \vdots \\ \vdots & \vdots & 0 & -1 & 1 & 0 & \cdots & \vdots \\ \vdots & \vdots & \vdots & -1 & \vdots & \ddots & 0 & \vdots \\ \vdots & 0 & \vdots & -1 & 0 & \cdots & 1 & 0 \\ 0 & \cdots & 0 & -1 & 0 & 0 & \cdots & 1 \end{bmatrix}_{h_2 \times m} \tag{17}$$

where $SSDM_G$ is obtained referring to $SSDM_B$; h_1 and h_2 ($h = h_1 + h_2$) are the total number of GPS and BDS available satellites; m represents the number of estimated parameters for the BSSD model. BSSD coefficient matrix $H_{BSSD,TCI,k}$ and innovation vector $Z_{BSSD,TCI,k}$ can be calculated by

$$H_{BSSD,TCI,k} = SSDM \times H_{TCI,k} \times SSDM^T \tag{18}$$

$$Z_{BSSD,TCI,k} = SSDM \times Z_{TCI,k} \times SSDM^T \tag{19}$$

$$R_{BSSD,TCI} = SSDM \times R_{TCI} \times SSDM^T \tag{20}$$

In addition, there is Inter-System Bias (ISB) between GPS and BDS [35,36]. To solve this problem, we consider the receiver clock offset of different GNSS as independent parameters and estimate them separately. This method is presented in detail in the works of [37,38]. Anyway, there are two other methods to process ISB [6,36,39]. It is worth mentioning that the receiver clock offset, and drift can be removed in the BSSD PPP/INS TCI model, with the state vector of

$$X_{BSSD,TCI,k} = \begin{bmatrix} \delta p^n_{INS} & \delta v^n_{INS} & \delta \psi & \delta B_a & \delta B_g & \delta S_a & \delta S_g & \delta d_{wet} & \delta N^{km}_{IF} \end{bmatrix} \quad (21)$$

where δp^n_{INS}, δv^n_{INS}, and $\delta\psi$ represent the corrections of position, velocity, and attitude; δS_g and δB_g represent the scale factor and bias of gyroscope; δS_a and δB_a represent the scale factor and bias of accelerometers [40]; δd_{wet} is a wet component of tropospheric zenith delay, and δN_{IF} represents ambiguity.

2.4. Odometer-Aided BSSD PPP/INS Tightly-Coupled Integration

In motion scenarios, the vehicle does not slip sideways and upward, which means that the velocity on lateral and vertical will be close to zero and only the forward speed exists [31,32]. Meanwhile, the forward velocity obtained from an odometer can be used as a pseudo-measurement. However, the odometer-measured velocity will be influenced by the scale factor error

$$v_o \approx \left[v^v_o/(1+S_o) \ \ 0 \ \ 0 \right]^T \quad (22)$$

where v^v_o is the forward vehicle velocity measured by the odometer in the vehicle frame (v); S_o represent the scale factor which can be modeled as a random walk process.

The innovation vector of the odometer can be calculated by the difference between the velocity measured by the odometer and predicted by INS predicted one. However, the v-frame and the b-frame are not usually aligned theoretically. Therefore, the innovation vector (Z_o) can be described by

$$Z_o = v^v_o - C^v_b v^b_{INS} \approx \begin{bmatrix} v^v_o/(1+S_o) \\ 0 \\ 0 \end{bmatrix} - \begin{bmatrix} v^F_{INS} \\ v^R_{INS} \\ v^D_{INS} \end{bmatrix} + \eta, \eta \sim N(0, \sigma^2_o) \quad (23)$$

where C^v_b denotes the rotation matrix from b-frame to v-frame; η is the vector of odometer innovation noise with the prior variance of σ^2_o.

According to Equation (11), $Z_{BSSD,TCI,k}$ can be furtherly expressed as

$$Z_{BSSD,TCI,k} = \begin{bmatrix} SSDM \times Z_{TCI,k} \times SSDM^T \\ v^v_o - C^v_b v^b_{INS} \end{bmatrix} \quad (24)$$

with the linearized form of

$$\delta Z_o = v^v_o \delta S_o - C^v_b \left(C^b_n \delta v^n_{INS} - C^b_n (v^n_{INS} \times) \delta\psi - \left(l^b_o \times \right) \omega^b_{ib} \delta S_g \right) \quad (25)$$

where δS_o is the correction of the odometer scale factor; l^b_o denotes the lever-arm between the reference center of the odometer and IMU in the b-frame. The corresponding BSSD state vector can be expressed as

$$X_{BSSD,TCI,k} = \begin{bmatrix} \delta p^n_{INS} & \delta v^n_{INS} & \delta \psi & \delta B_a & \delta B_g & \delta S_a & \delta S_g & \delta S_o & \delta d_{wet} & \delta N^{km}_{IF} \end{bmatrix} \quad (26)$$

the state vector can be estimated by the EKF [41,42]

$$X_{BSSD,TCI,k} = X_{BSSD,TCI,k,k-1} + K_k (Z_{BSSD,TCI,k} - H_{BSSD,TCI,k} X_{BSSD,TCI,k,k-1}) \quad (27)$$

$$P_{BSSD,TCI,k} = (I - K_k H_{BSSD,TCI,k}) P_{BSSD,TCI,k,k-1} (I - K_k H_{BSSD,TCI,k})^T + K_k R_{BSSD,TCI} K_k^T \quad (28)$$

The algorithm structure of PPP/INS/ODO TCI is shown in Figure 1.

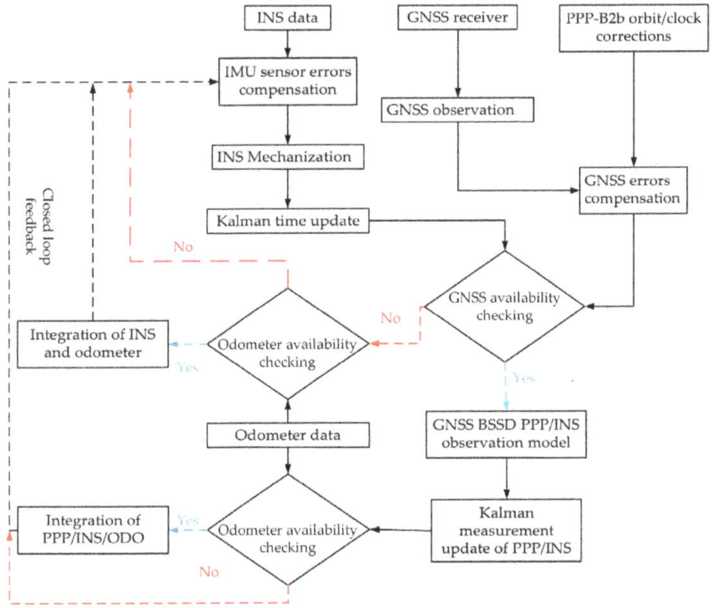

Figure 1. Algorithm of BSSD PPP/INS/ODO TCI structure.

3. Tests, Results, and Discussion

In order to evaluate the performance of the proposed model, a set of land-borne data was processed and analyzed. This first subsection demonstrates the experiment details, including the experimental equipment, satellite availability, and data processing schemes. The second subsection assesses the accuracy of the PPP-B2b corrections and the positioning performance of BSSD PPP. The third subsection validates the impacts of INS, the odometer on PPP positioning, and the effect of the BSSD model on the positioning performance of BSSD PPP/INS/ODO TCI. The last subsection analyzes the influences of the odometer and BSSD model on attitude determination.

3.1. Data Collection

The test vehicle was equipped with a NovAtel GNSS receiver, a low-cost INS616 IMU, and an odometer in Beijing on 23 December 2021. The sampling rates of GNSS, IMU, and odometer measurements were set to 1 Hz, 125 Hz, and 100 Hz, respectively. The designed test route, available satellite number, and PDOP, along with the trajectory, are shown in Figures 2 and 3. The trajectory is mainly on urban environments with many buildings on both sides of the road. The average number of satellites of GPS, BDS-3, and BDS-3/GPS are 6.1, 8.5, and 14.5. The corresponding PDOP are 2.3, 2.1, and 1.4, respectively. As is shown, the continuity of this test is poor, especially for GPS-only and BDS-only. The combination of GPS and BDS-3 can improve it, but there are still many periods with GNSS outages. For example, the 1500 s to 2500 s and the 6500 s to 7500 s are the most typical scenes. Based on the precise satellite orbit/clock recovered by the corrections obtained by BDS PPP-B2b service, six data processing strategies, namely undifferenced PPP, BSSD PPP, BSSD PPP/INS TCI, BSSD PPP/INS TCI, BSSD PPP/INS/ODO TCI, and undifferenced PPP/INS/ODO TCI will be implied. The results will be compared to the solutions of RTK/INS TCI calculated by the Inertial Explorer (IE) software.

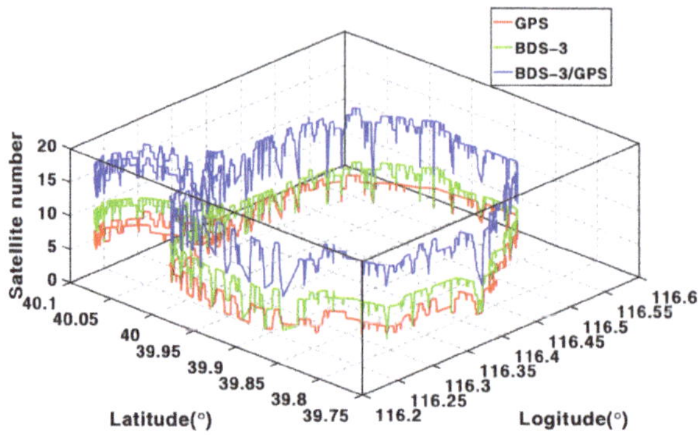

Figure 2. The available satellites number of different GNSS along with the test trajectory.

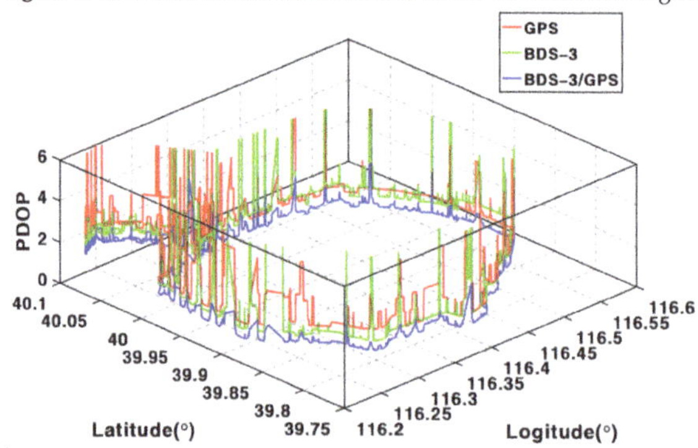

Figure 3. PDOP of different GNSS along with the test trajectory.

3.2. Accuracy of PPP-B2b Corrections and BSSD PPP

Precise orbit and clock offset products recovered by the corrections from the PPP-B2b service are utilized to process the satellite's orbit/clock errors. In order to assess the accuracy of real-time orbit/clock products obtained by the PPP-B2b service, the final products provided by WHU are adopted as references. Figures 4 and 5 depict the RMS of GPS and BDS-3 clock offset and orbit errors. The average orbit RMS of GPS and BDS-3 are listed in Table 1. As shown, the average RMS of GPS orbits are 13.95 cm, 20.51 cm, and 19.57 cm in the radial, along, and cross directions, and that of BDS-3 (MEO+IGSO) orbits are 10.33 cm, 20.31 cm, and 27.00 cm. The accuracy on the radial component for both GPS and BDS-3 is higher than the other two components For satellite clock offsets, the RMS of the GPS clock offset is 3.27 ns, and that of BDS-3 (MEO+IGSO) are 1.95 ns. Usually, orbit accuracy in radial components and clock accuracy are the major factors affecting positioning accuracy. Thus, the position solution with high accuracy can be obtained by using the orbit/clock corrections from the PPP-B2b service, which can also be illustrated in the works [17–19].

Table 1. Mean RMS of orbit errors and clock offset of real-time products recovered by PPP-B2b service.

	Orbit (cm)			Clock (ns)
	RMS-R	**RMS-A**	**RMS-C**	**RMS**
GPS	13.95	20.51	19.57	3.27
BDS-3 (MEO+IGSO)	10.33	20.31	27.00	1.95

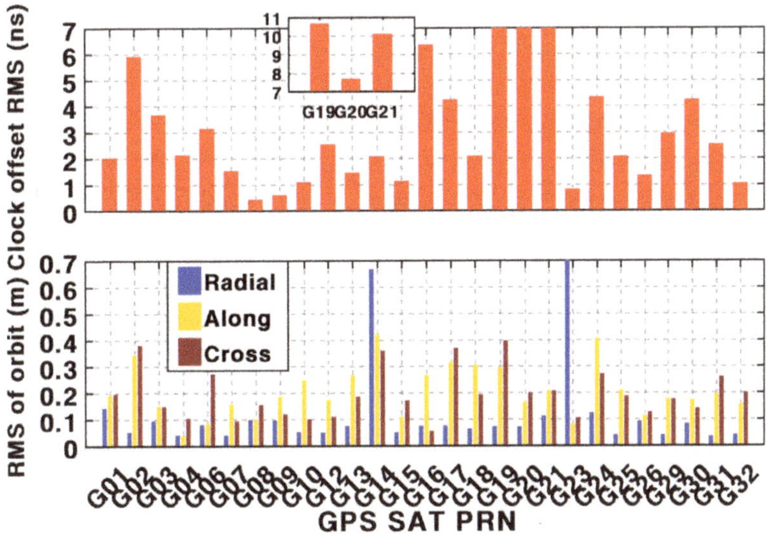

Figure 4. RMS of GPS clock offset, and orbit errors of real-time products recovered by PPP-B2b service.

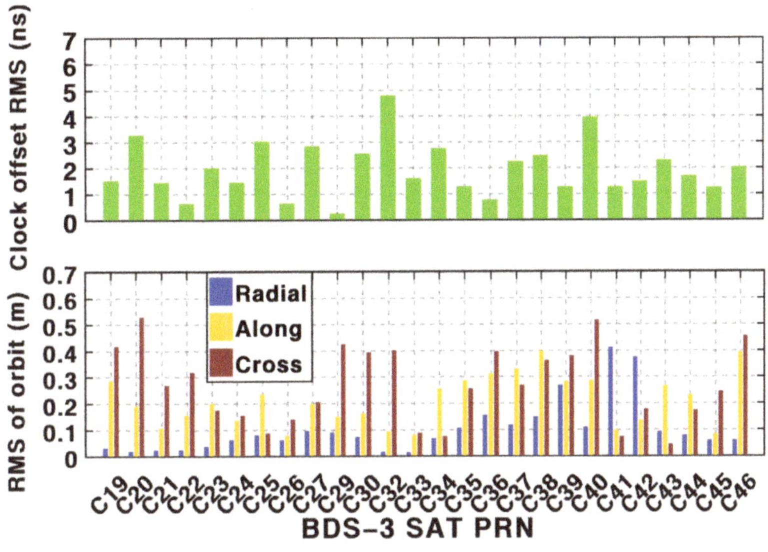

Figure 5. RMS of BDS-3 clock offset and orbit errors of real-time products recovered by PPP-B2b service.

Based on the PPP-B2b service, the position differences of BSSD PPP solutions in the north, east, and vertical with different GNSS systems are shown in Figure 6, and the corresponding RMS are listed in Table 2. After integrating BDS-3 and GPS, the improvements of position RMS of BSSD BDS-3 PPP are 3.24%, 22.25%, and 49.50% in the north, east, and down directions. Such improvements are 12.97%, 50.50%, and 62.45% in three components for BSSD GPS PPP. The improvements are due to the improved satellite spatial distribution, which is also verified in works [43–46]. Moreover, compared with undifferenced PPP, the average position RMS improved by 13.17%, 2.64%, and 11.21% in three components, which due to the receiver-related errors, can be eliminated by the BSSD model.

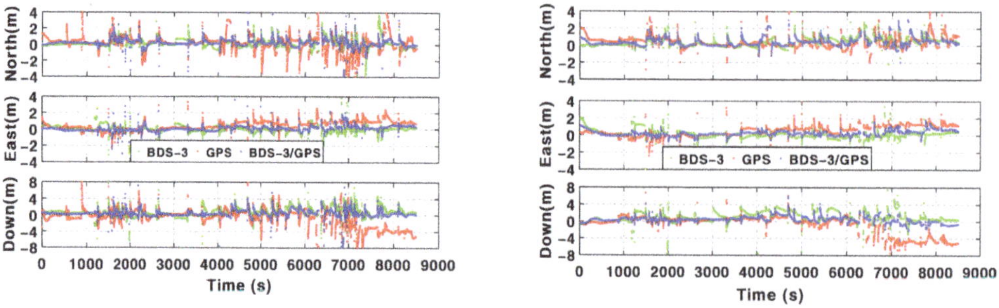

Figure 6. Positioning errors of undifferenced PPP (**left**) and BSSD PPP (**right**) with different GNSS.

Table 2. Positioning errors of undifferenced PPP and BSSD PPP with different GNSS.

	BSSD PPP			Undifferenced PPP		
	North (cm)	East (cm)	Down (cm)	North (cm)	East (cm)	Down (cm)
GPS	69.68	75.25	218.18	95.65	76.73	218.77
BDS-3	62.67	47.91	162.23	63.89	49.97	186.07
BDS-3/GPS	60.64	37.25	81.93	62.73	38.05	103.12

3.3. *Performance of BSSD PPP/INS TCI*

In comparison with BSSD PPP, position errors can be reduced visibly by the addition of INS in all three components (shown in Figure 7). The average improvements provided by BSSD PPP/INS TCI are 31.24%, 23.35%, and 27.38% in the three directions (Table 3).

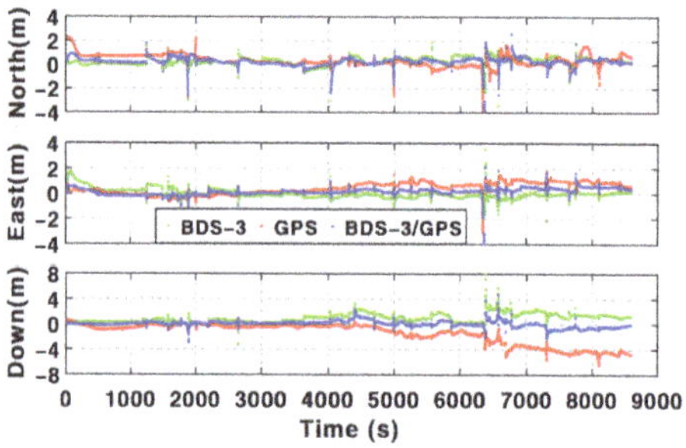

Figure 7. Positioning errors of BSSD PPP/INS TCI with different GNSS.

Table 3. Position RMS of BSSD PPP and BSSD PPP/INS TCI with different GNSS.

	BSSD PPP/INS TCI		
	North (cm)	East (cm)	Down (cm)
GPS	55.06	60.50	203.54
BDS-3	42.69	34.48	119.96
BDS-3/GPS	35.87	28.90	41.49

Figure 8 shows the position differences between BSSD PPP and BSSD PPP/INS TCI. Significantly, the accuracy of PPP is similar to that of PPP/INS TCI when there are sufficient satellites. That is because the absolute position accuracy mainly depends on PPP. However, such differences are significant during the periods with frequent GNSS outages, primarily from the 1500 s to 2500 s and 6500 s to 7500 s. That is because PPP/INS TCI can still provide high-accuracy position results when the number of satellites is less than 4, even if there is no available satellite in short-term time. The frequent GNSS outages between the 1500 s and 2500 s are displayed in Figure 9, from which we can see about ten partial and complete outages with the time last 1 s to 32 s happened. The details about the outage time and the average number of available satellites during these periods are listed in Table 4. As is shown, the maximum outage time for BDS-3, GPS, and BDS-3+GPS are 28 s, 32 s, and 26 s, respectively. During these periods, although a few satellites are still available, the number does not meet the minimum requirement of positioning both for single and dual systems PPP calculation. However, these available satellites can be used in BSSD PPP/INS TCI mode. The corresponding position differences of PPP and PPP/INS TCI during the 1500 s to 2500 s are shown in Figure 10. Visibly, BSSD PPP cannot provide position results, but BSSD PPP/INS TCI can work in partial outage periods. The divergence of position error can also be restrained in short-term outages. The average maximum drifts of position decreased from 107.96 cm, 59.90 cm, and 78.22 cm to 88.52 cm, 54.54 cm, and 62.00 cm after integrating with INS. Therefore, PPP/INS TCI can provide the position result with better continuity and accuracy. Nevertheless, in cases where the duration of complete outages is too long, such as outages 7 and 8, a divergence of the position results still can be found because INS would drift rapidly along with time.

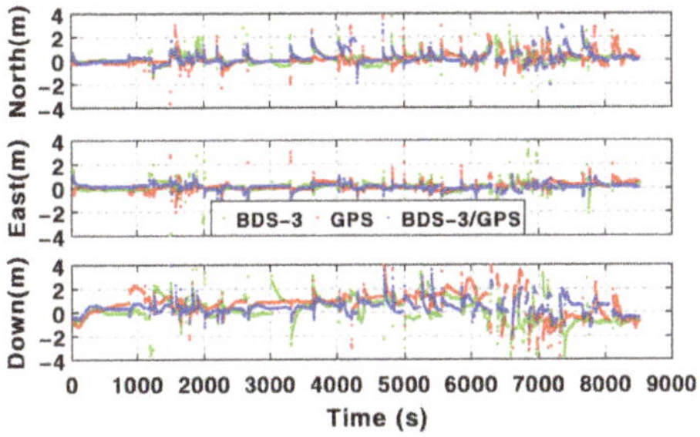

Figure 8. Positioning differences between BSSD PPP and BSSD PPP/INS TCI with different GNSS.

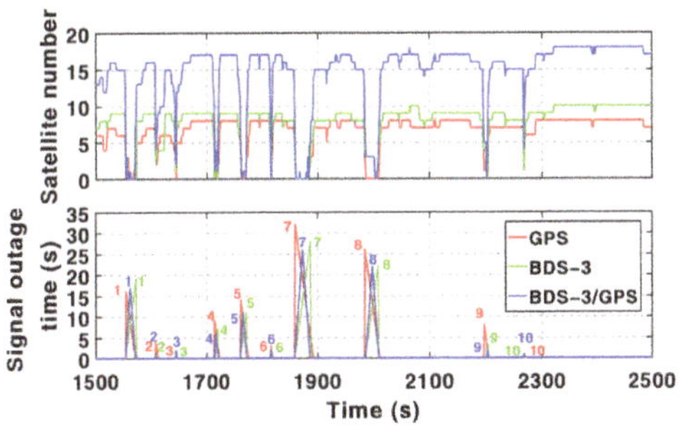

Figure 9. Satellite numbers and outages time between the 1500 s and 2500 s.

Table 4. The number of available satellites and the GNSS outages time.

Outages Scenes		1	2	3	4	5	6	7	8	9	10
Outages time(s)	GPS	16	4	2	9	14	3	32	26	8	1
	BDS-3	19	1	2	7	11	2	28	22	3	1
	GPS/BDS-3	17	1	2	6	11	2	26	22	2	1
Satellite number	GPS	0.6	2.3	0.5	1.4	0.9	1	0.4	0.3	1.3	3
	BDS-3	0.3	2	1.5	1.3	0.9	0.5	0.2	2.4	1.7	1
	GPS/BDS-3	0.5	4	2	2.5	1.2	0.5	0.5	2.5	1.5	4

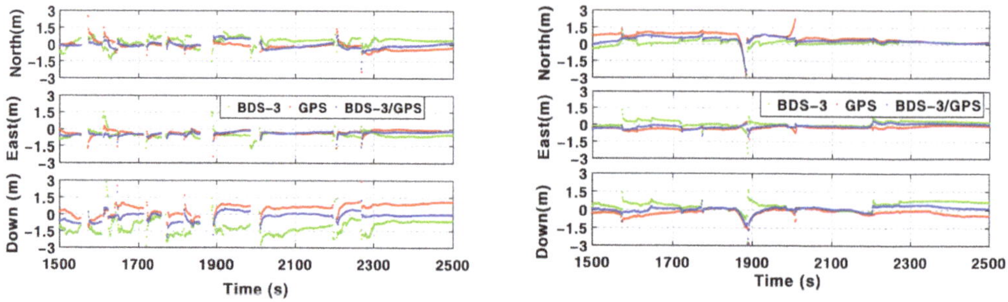

Figure 10. Positioning errors of BSSD PPP (**left**) and BSSD PPP/INS TCI (**right**) between the 1500 s and 2500 s.

3.4. BSSD PPP/INS/ODO TCI Positioning

In the case of a complete outage for a long time, position errors of PPP/INS TCI could accumulate rapidly. Therefore, constraint information is helpful. An odometer can measure the forward velocity of the vehicle. Based on such velocity and motion constraints, the problem above can be restrained by Equation (23). Figure 11 depicts the position errors of BSSD PPP/INS/ODO TCI with different GNSS, and the corresponding RMS for the common periods with BSSD PPP/INS TCI are calculated in Table 5. Accordingly, the position improvements caused by the addition of an odometer on average are 1.34%, 1.41%, and 1.73%. Such invisible enhancements are because the impact of the odometer on PPP/INS integration mainly affects these periods with poor or without GNSS observability. While there are enough satellites, the weight of the odometer observation function is much smaller than that of GNSS observations which can be obtained by making a comparison between Equations (14) and (23). The corresponding position differences between BSSD PPP/INS TCI and PPP/INS/ODO TCI are plotted in Figure 12. Significantly, the periods with sufficient satellites have small differences, and significant differences emerge along with long-term outages. Figure 13 depicts the position difference of BSSD PPP/INS/ODO TCI from 1500 s to 2500 s. Accordingly, the average position RMSs from 1500 s to 2500 s is reduced from 50.00 cm, 27.08 cm, and 37.42 cm to 44.40 cm, 26.85 cm, and 34.96 cm. The average of maximum positioning drifts with different GNSS during a satellite outage are plotted in Figure 14. The averages of maximum positioning drifts are reduced from 88.52 cm, 54.54 cm, and 62.00 cm to 66.29 cm, 45.84 cm, and 49.17 cm, with a maximum diminution of 74.94%, 33.0%, and 51.56%. Hence, the position performance of BSSD PPP/INS TCI can be further enhanced, especially in challenging environments, by using an odometer.

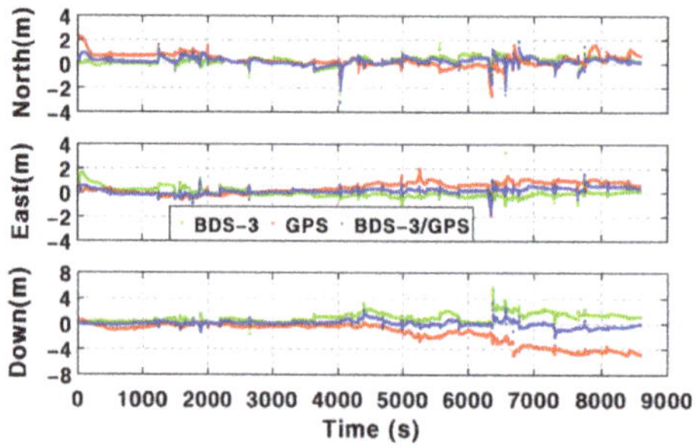

Figure 11. Positioning errors of BSSD PPP/INS/ODO TCI with different GNSS by using PPP-B2b corrections.

Table 5. Position RMS of BSSD PPP/INS/ODO TCI with different GNSS.

	BSSD PPP/INS/ODO TCI			Undifferenced PPP/INS/ODO		
	North (cm)	East (cm)	Down (cm)	North (cm)	East (cm)	Down (cm)
GPS	55.23	60.29	202.20	55.69	61.92	207.96
BDS-3	41.78	33.47	113.43	44.50	34.38	108.16
GPS/BDS-3	35.02	28.63	40.86	42.90	29.98	41.33

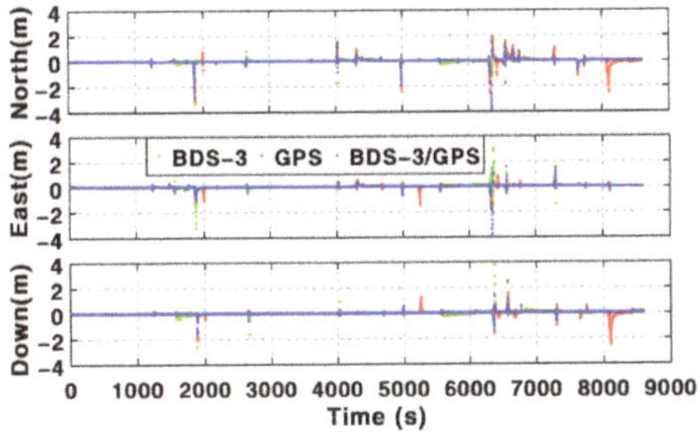

Figure 12. Positioning differences between BSSD PPP/INS TCI and BSSD PPP/INS/ODO TCI with different GNSS.

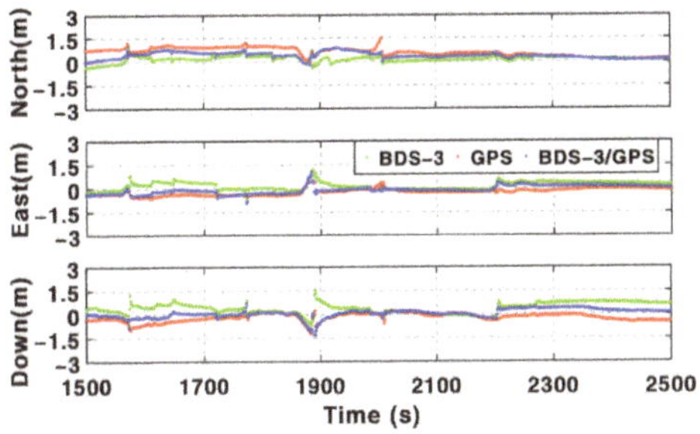

Figure 13. Positioning errors of BSSD PPP/INS/ODO TCI with frequent GNSS outages.

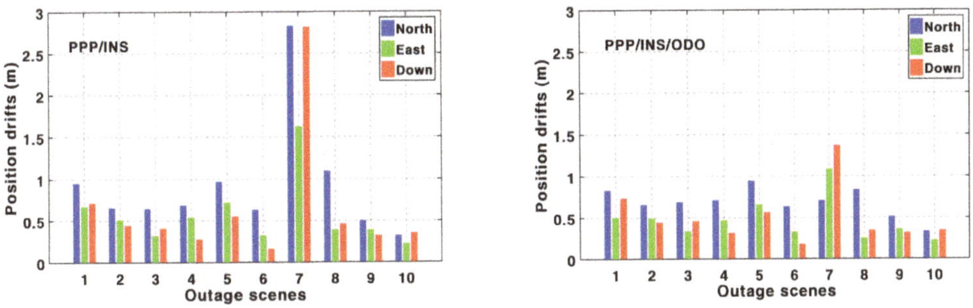

Figure 14. Average maximum position drifts of different GNSS for BSSD PPP/INS TCI (**left**) and PPP/INS/ODO TCI (**right**) on different GNSS outage scenes.

In addition, to analyze the impact of the BSSD model on PPP/INS integration, the time series of position errors of undifferenced PPP/INS/ODO TCI is shown in Figure 15. In contrast, the BSSD model provided about 7.71%, 3.09%, and 0.27% in the three directions on average, according to RMSs listed in Table 5. It can be seen that the position accuracy of the BSSD model-based integration mode is close to the undifferenced model-based integration in the time series in general but is slightly higher in position statistics. It is due to the reason that the BSSD model can remove the receiver-depended errors (i.e., receiver clock offset, receiver time delays on pseudorange and carrier-phase, unmodelled receiver errors, etc.) that impact initial convergence or re-convergence of PPP after satellite signal outages. Such a character is illustrated in Figure 16, which depicts the differences between the BSSD-based solutions and the undifferenced PPP-based solutions. The visible differences emerge during the periods of re-convergence caused by satellite outages, while the differences are invisible in periods with sufficient available satellites. Figure 17 shows the average maximum positioning drifts calculated by the odometer-aided PPP/INS TCI based on the BSSD model and the undifferenced PPP model in the satellite outage periods from 1500 s to 2500 s. The position drifts of the solutions based on the BSSD model are 66.29 cm, 45.84 cm, and 49.17 cm, which are smaller than those solutions based on the undifferenced model (82.38 cm, 50.66 cm, and 52.30 cm). It means that the BSSD model can provide visible enhancements in demanding user environments.

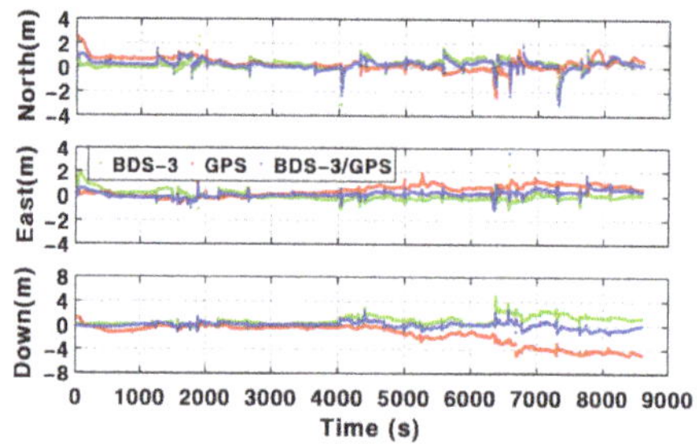

Figure 15. Positioning errors of undifferenced BSSD PPP/INS/ODO TCI with different GNSS using PPP-B2b correction.

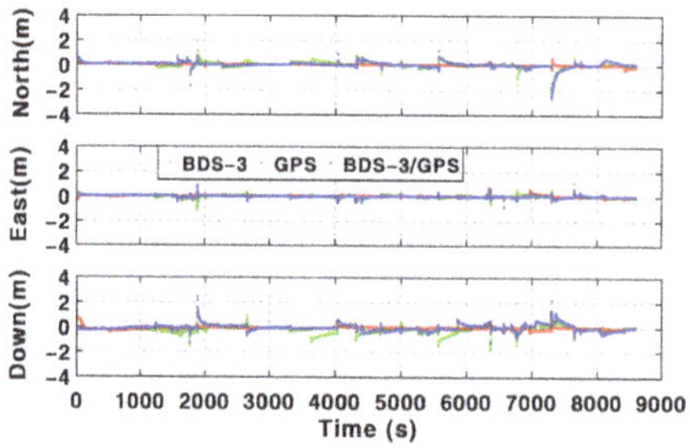

Figure 16. Positioning differences between undifferenced PPP/INS/ODO TCI and BSSD PPP/INS TCI/ODO.

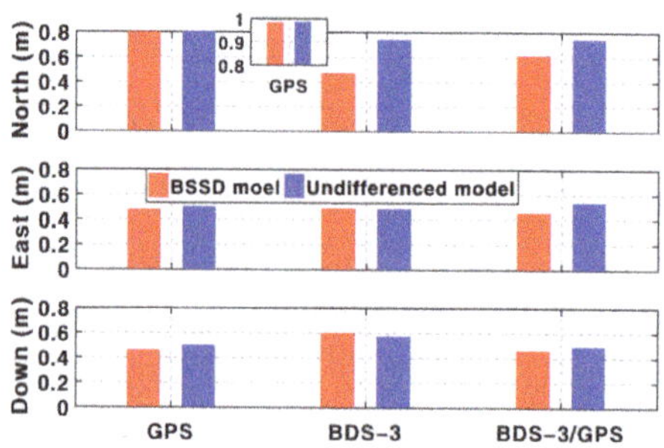

Figure 17. Mean maximum position drifts of different GNSS for BSSD PPP/INS/ODO TCI and undifferenced PPP/INS/ODO TCI.

3.5. BSSD PPP/INS/ODO TCI Attitude Determination

The attitude errors of BSSD PPP/INS TCI with and without odometer aid are shown in Figure 18. The results of the roll, pitch, and heading angles in the first 500 s are significantly different from the results in other periods. That is because the vehicle kept static in the first 500 s, which provided no observabilities for gyroscopes. Then, the accuracy of attitude determination during these times mainly depended on the accuracy of initial attitudes. While the vehicle moved ahead, the motions increased the observabilities of gyroscopes and upgraded the accuracy of attitudes. According to the statistics in Table 6, the average RMSs of BSSD PPP/INS TCI are 0.025°, 0.049°, and 0.184° in roll, pitch, and heading angles, respectively. Visibly, the accuracies of roll and pitch angles are more precise than heading angles because of the poor observability of heading angles measured by the gyroscope in the vertical direction [43]. Furthermore, attitude RMSs calculated using different GNSS are close to each other. That is because the attitudes are majorly determined by the gyroscope and are slightly affected by GNSS positioning accuracy while GNSS observations are available [28]. Therefore, the attitude solutions obtained by using different GNSS systems are somewhat different in accuracy.

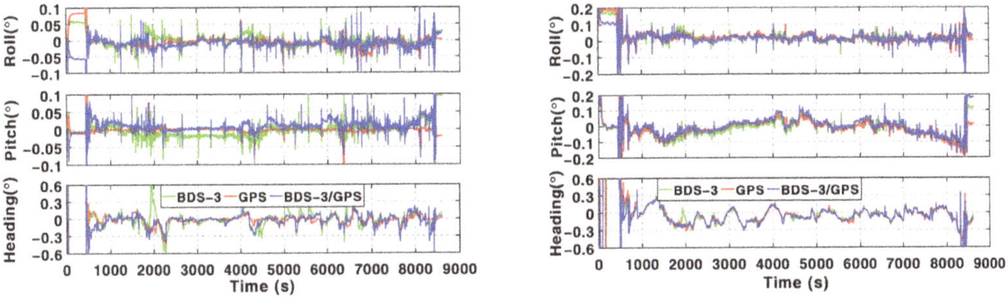

Figure 18. Attitude errors of BSSD PPP/INS TCI (**left**) and BSSD PPP/INS/ODO TCI (**right**) with different GNSS.

Table 6. Attitude RMS of BSSD PPP/INS TCI with and without odometer aid.

	BSSD PPP/INS TCI			BSSD PPP/INS/ODO TCI		
	Roll (°)	Pitch (°)	Heading (°)	Roll (°)	Pitch (°)	Heading (°)
GPS	0.025	0.050	0.178	0.025	0.049	0.137
BDS-3	0.027	0.049	0.194	0.028	0.048	0.140
GPS/BDS-3	0.024	0.048	0.181	0.025	0.048	0.138

According to the RMSs of BSSD PPP/INS/ODO TCI listed in Table 6, the average RMSs of roll, pitch, and heading angles are 0.026°, 0.048°, and 0.138°. In contrast, the odometer provides about 25.00% in heading angle and invisible enhancements in roll and pitch angles. It is due to the observability improvement on the vertical gyroscope by adding an odometer. To assess the impact of the odometer on attitude determination at each epoch, we also provided the differences between the solutions with and without the odometer in Figure 19. From it, there are visible differences in the three directions at every epoch. It means that the odometer affects the estimation of roll, pitch, and heading angles. Such effects on pitch and roll would become invisible after using the statistic index (i.e., RMS). Anyway, to furtherly illustrate the influence of the odometer on attitude determination under GNSS outages. The attitude differences of BSSD PPP/INS and BSSD PPP/INS/ODO during the 1500 s and 2500 s are plotted in Figure 20. It can be seen that the humps of heading angles appearing at the profile of attitudes errors of PPP/INS TCI can be restrained effectively by the addition of an odometer. According to the average attitude RMSs in Figure 21, the reduction in attitude drifts can be significantly constrained by using an odometer.

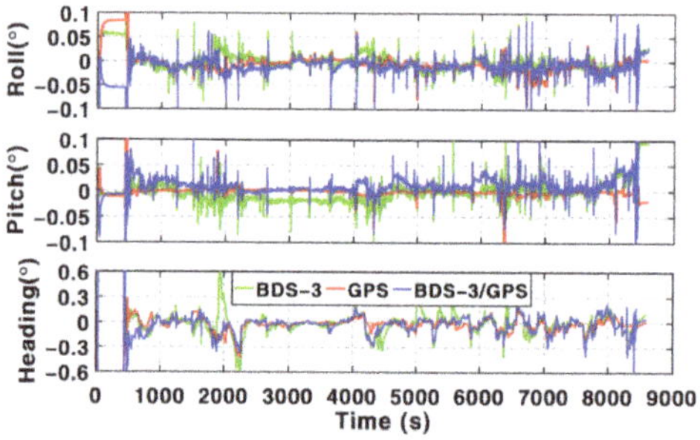

Figure 19. Attitude differences between BSSD PPP/INS TCI and BSSD PPP/INS/ODO with different GNSS.

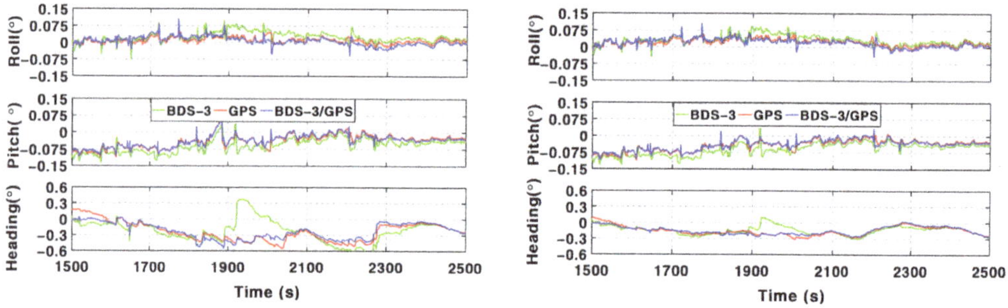

Figure 20. Attitude errors of BSSD PPP/INS TCI (**left**) and BSSD PPP/INS/ODO TCI (**right**) with frequent GNSS outages in the periods between the 1500 s and 2500 s.

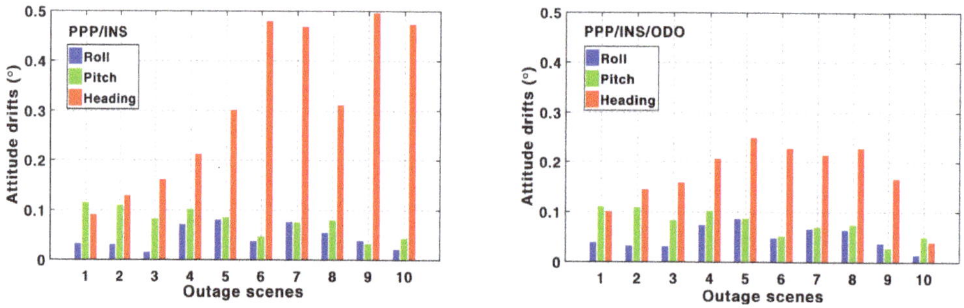

Figure 21. Average of Maximum attitude drifts of different GNSS for BSSD PPP/INS TCI (**left**) and PPP/INS/ODO TCI (**right**) on different GNSS outage scenes.

Similarly, the differences between the attitudes calculated by the undifferenced PPP-based integration (Figure 22) and the BSSD PPP-based integration are plotted in Figure 23. The attitudes RMSs based on undifferenced PPP are 0.031°, 0.048°, and 0.135° in three components, which are close to the solutions based on BSSD PPP. As the result shows, the two solutions are close to each other actually in terms of statistics index (RMS) but different in time series, which is due to the accuracies of attitudes being mainly determined by IMU ([28,46]). However, the GNSS data processing strategy would affect the estimation of attitudes at each epoch by Equation (14). A similar conclusion can also be obtained from the average maximum attitude drifts of the two methods from 1500 s to 2500 s with GNSS outages in Figure 24.

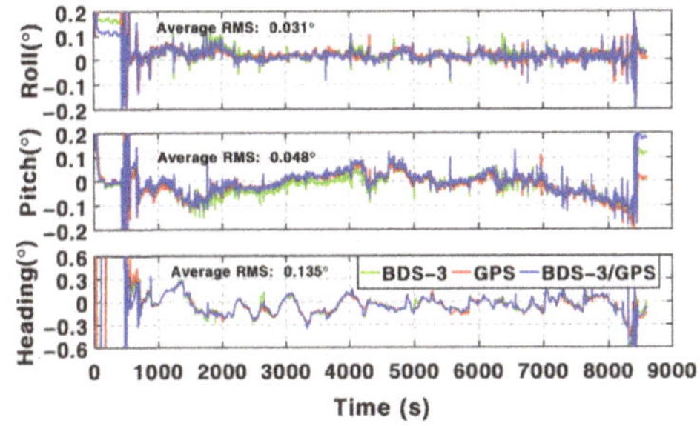

Figure 22. Attitude errors of undifferenced PPP/INS/ODO TCI with different GNSS.

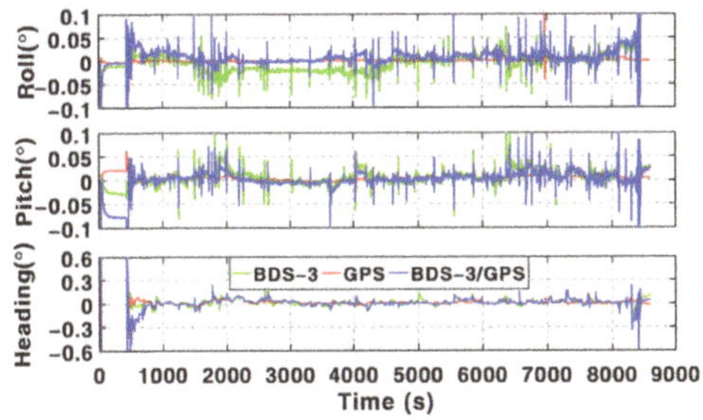

Figure 23. Attitudes differences between undifferenced PPP/INS/ODO TCI and BSSD PPP/INS/ODO TCI.

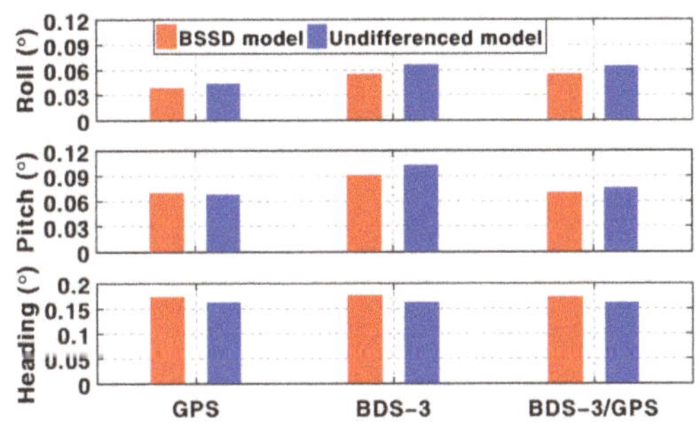

Figure 24. Mean maximum attitude drifts of different GNSS for BSSD PPP/INS/ODO TCI and undifferenced PPP/INS/ODO TCI.

4. Discussion

Based on the BDS-3 PPP-B2b service, real-time PPP can be used via B2b signals. However, it is still challenging in an urban environment. According to the assessments above and the results summarized in Table 7, the positioning performance of BSSD PPP can be enhanced visibly by the addition of INS and odometer, especially in periods with frequent GNSS outages. The mean position RMS of BSSD PPP is 64.33 cm, 53.47 cm, and 154.11 cm in three components based on PPP-B2b service. By integrating INS, the mean position RMS can be improved by 31.2%, 23.3%, and 27.3%. Such percentages can be furtherly increased by 1.34%, 1.41%, and 1.73% after using an odometer. The test data from 1500 s to 2500 s are adopted to validate the performance in the periods with frequent GNSS outages. The mean position maximum drifts during these periods decreased from 107.96 cm, 59.90 cm, and 78.22 cm of BSSD PPP to 88.52 cm, 54.54 cm, and 62.00 cm of BSSD PPP/INS TCI. After adding an odometer, such values are 66.29 cm, 45.84 cm, and 49.17 cm.

Table 7. Statistic of position and attitude errors.

Positioning	Mean RMS (cm)			Mean Maximum Drifts (cm)		
	North	East	Down	North	East	Down
BSSD PPP	64.33	53.47	154.11	107.96	59.90	78.22
BSSD PPP/INS TCI	44.54	41.29	121.66	88.52	54.54	62.00
BSSD PPP/INS/ODO TCI	44.01	40.79	118.83	66.29	45.84	49.17
Undifferenced PPP/INS/ODO TCI	47.69	42.09	119.15	82.38	50.66	52.30
Attitude	**Mean RMS (°)**			**Mean Maximum Drifts (°)**		
	Roll	**Pitch**	**Heading**	**Roll**	**Pitch**	**Heading**
BSSD PPP/INS TCI	0.025	0.049	0.184	0.047	0.077	0.313
BSSD PPP/INS/ODO TCI	0.026	0.048	0.138	0.049	0.076	0.174
Undifferenced PPP/INS/ODO TCI	0.031	0.048	0.135	0.058	0.082	0.163

Anyway, the position accuracy of PPP/INS/ODO TCI based on the BSSD model is slightly higher than the solutions based on the undifferenced model. Compared with undifferenced PPP/INS/ODO TCI, the mean position RMS of BSSD PPP/INS/ODO TCI is improved by 7.71%, 3.09%, and 0.27%. The mean maximum drifts can be reduced from 82.38 cm, 50.66 cm, and 52.30 cm to 66.29 cm, 45.84 cm, and 49.17 cm by utilizing the BSSD model.

For attitude determination, the mean attitude RMS of PPP/INS TCI is 0.025°, 0.049°, and 0.184° in three components. The addition of an odometer brings a 25% improvement to the RMS of heading angles and reduces the mean maximum drifts from 0.313° to 0.174°. The results of the other two components are comparable. Moreover, the accuracy of PPP/INS/ODO TCI with and without the BSSD model is similar to each other.

5. Conclusions

In this contribution, we implied the tightly coupled integration of BDS-3/GPS, low-cost IMU, and odometer based on the inter-satellite differenced PPP model and the orbit/clock corrections of PPP-B2b. A vehicle experiment in urban circumstances was implemented to validate the performance of positioning and attitude determination of the developed model. The following conclusions can be obtained. (1) With the addition of INS, the improvements of BSSD PPP position accuracy on average are more than 31.2%, 23.3%, and 27.3% in the north, east, and down directions. Further enhancements in position accuracy are achievable with the aid of an odometer, especially while suffering GNSS outages. (2) By using the odometer, the accuracies of pitch and heading angles are improved by about 2.04% and 25%. (3) In comparison with the PPP/INS/ODO TCI based on the undifferenced PPP model, the developed BSSD model can provide results with higher accuracy, especially in the re-convergence periods. For attitude determination, comparable results can be obtained by both the BSSD model and the undifferenced model.

Author Contributions: Conceptualization, Y.M. and Z.G.; data curation, Y.M., Q.X. and R.L.; funding acquisition, C.Y. and Z.G.; investigation, Y.M. and Z.G.; software, Y.M. and Z.G.; visualization, Y.M. and J.L.; writing—original draft preparation, Y.M.; writing—review and editing, Y.M., Z.G. and C.Y. All authors have read and agreed to the published version of the manuscript.

Funding: This research was partly supported by the National Key Research and Development Program of China (Grant No. 2021YFB3901301) and the National Natural Science Foundation of China (NSFC) (Grant No. 42274022).

Data Availability Statement: The datasets adopted in this paper are managed by the School of Land Science and Technology, China University of Geosciences Beijing and can be available on request from the corresponding author.

Acknowledgments: The authors would like to thank anonymous reviewers who gave valuable suggestions that helped to improve the quality of the manuscripts.

Conflicts of Interest: The authors declare no conflict of interest.

References

1. Yang, Y.; Xu, Y.; Li, J.; Yang, C. Progress and Performance Evaluation of BeiDou Global Navigation Satellite System: Data Analysis Based on BDS-3 Demonstration System. *Sci. China Earth Sci.* **2018**, *61*, 614–624. [CrossRef]
2. Yang, Y.; Li, J.; Wang, A.; Xu, J.; He, H.; Guo, H.; Shen, J.; Dai, X. Preliminary assessment of the navigation and positioning performance of BeiDou regional navigation satellite system. *Sci. China Earth Sci.* **2014**, *57*, 144–152. [CrossRef]
3. Yang, Y.; Tang, J.; Montenbruck, O. Chinese navigation satellite systems. In *Handbook of Global Navigation Satellite Systems*; Teunissen, M., Ed.; Springer: Cham, Switzerland, 2017; Chapter 10; pp. 273–304.
4. Lv, J.; Gao, Z.; Kan, J.; Lan, R.; Li, Y.; Lou, Y.; Yang, H.; Peng, J. Modeling and Assessment of Multi-Frequency GPS/BDS-2/BDS-3 Kinematic Precise Point Positioning Based on Vehicle-Borne Data. *Measurement* **2022**, *189*, 110453. [CrossRef]
5. Shi, C.; Wu, X.; Zheng, F.; Wang, X.; Wang, J. Modeling of BDS-2/BDS-3 Single-Frequency PPP with B1I and B1C Signals and Positioning Performance Analysis. *Meas. J. Int. Meas. Confed.* **2021**, *178*, 109355. [CrossRef]
6. Jiao, G.; Song, S.; Jiao, W. Improving BDS-2 and BDS-3 Joint Precise Point Positioning with Time Delay Bias Estimation. *Meas. Sci. Technol.* **2019**, *31*, 25001. [CrossRef]
7. Lu, M.; Li, W. Overview of BDS III New Signals. *Navigation* **2019**, *66*, 19–35. [CrossRef]
8. Zumberge, J.F.; Heflin, M.B.; Jefferson, D.C.; Watkins, M.M.; Webb, F.H. Precise Point Positioning for the Efficient and Robust Analysis of GPS Data from Large Networks. *J. Geophys. Res. Solid Earth* **1997**, *102*, 5005–5017. [CrossRef]
9. Gao, Y.; Shen, X. A New Method for Carrier-Phase-Based Precise Point Positioning. *Navigation* **2002**, *49*, 109–116. [CrossRef]
10. Li, X.; Ge, M.; Zhang, H.; Wickert, J. A Method for Improving Uncalibrated Phase Delay Estimation and Ambiguity-Fixing in Real-Time Precise Point Positioning. *J. Geod.* **2013**, *87*, 405–416. [CrossRef]
11. Elsobeiey, M.; El-Rabbany, A. Efficient Between-Satellite Single-Difference Precise Point Positioning Model. *J. Surv. Eng.* **2014**, *140*, 04014007. [CrossRef]
12. Kouba, J. A Guide to Using International GNSS Service (IGS) Products. IGS. 2015. Available online: http://igscb.jpl.nasa.gov/igscb/resource/pubs/UsingIGSProductsVer21.pdf (accessed on 9 September 2022).
13. Kouba, J.; Héroux, P. Precise Point Positioning Using IGS Orbit and Clock Products. *GPS Solut.* **2001**, *5*, 12–28. [CrossRef]
14. Zhang, W.; Lou, Y.; Song, W.; Sun, W.; Zou, X.; Gong, X. Initial Assessment of BDS-3 Precise Point Positioning Service on GEO B2b Signal. *Adv. Sp. Res.* **2022**, *69*, 690–700. [CrossRef]
15. Nie, Z.; Xu, X.; Wang, Z.; Du, J. Initial Assessment of BDS PPP-B2b Service: Precision of Orbit and Clock Corrections, and PPP Performance. *Remote Sens.* **2021**, *13*, 2050. [CrossRef]
16. Tao, J.; Liu, J.; Hu, Z.; Zhao, Q.; Chen, G.; Ju, B. Initial Assessment of the BDS-3 PPP-B2b RTS Compared with the CNES RTS. *GPS Solut.* **2021**, *25*, 131. [CrossRef]
17. Ren, Z.; Gong, H.; Peng, J.; Tang, C.; Huang, X.; Sun, G. Performance Assessment of Real-Time Precise Point Positioning Using BDS PPP-B2b Service Signal. *Adv. Sp. Res.* **2021**, *68*, 3242–3254. [CrossRef]
18. Wang, L.; Li, Z.; Ge, M.; Neitzel, F.; Wang, X.; Yuan, H. Investigation of the Performance of Real-Time BDS-Only Precise Point Positioning Using the IGS Real-Time Service. *GPS Solut.* **2019**, *23*, 66. [CrossRef]
19. Zhang, L.; Yang, H.; Gao, Y.; Yao, Y.; Xu, C. Evaluation and Analysis of Real-Time Precise Orbits and Clocks Products from Different IGS Analysis Centers. *Adv. Sp. Res.* **2018**, *61*, 2942–2954. [CrossRef]
20. Gao, Z.; Zhang, H.; Ge, M.; Niu, X.; Shen, W.; Wickert, J.; Schuh, H. Tightly Coupled Integration of Ionosphere-Constrained Precise Point Positioning and Inertial Navigation Systems. *Sensors* **2015**, *15*, 5783–5802. [CrossRef]
21. Gao, Z.; Ge, M.; Shen, W.; Zhang, H.; Niu, X. Ionospheric and Receiver DCB-Constrained Multi-GNSS Single-Frequency PPP Integrated with MEMS Inertial Measurements. *J. Geod.* **2017**, *91*, 1351–1366. [CrossRef]
22. Shin, E.-H. Estimation Techniques for Low-Cost Inertial Navigation. Ph.D. Thesis, The University of Calgary, Calgary, AB, Canada, May 2005.
23. Cox, D.B.J. Integration of GPS with Inertial Navigation Systems. *Navigation* **1978**, *25*, 236–245. [CrossRef]
24. Gu, S.; Dai, C.; Fang, W.; Zheng, F.; Wang, Y.; Zhang, Q.; Lou, Y.; Niu, X. Multi-GNSS PPP/INS Tightly Coupled Integration with Atmospheric Augmentation and Its Application in Urban Vehicle Navigation. *J. Geod.* **2021**, *95*, 64. [CrossRef]
25. Le, A.Q.; Lorga, J. Combining Inertial Navigation System With GPS Precise Point Positioning: Flight Test Results. In Proceedings of the ION GNSS 2006, Fort Worth, TX, USA, 26–29 September 2006; pp. 3035–3042.
26. Martell, H. Tightly Coupled Processing of Precise Point Position (PPP) and INS Data. In Proceedings of the ION GNSS 2009, Savannah, GA, USA, 22–25 September 2009; pp. 1898–1905.
27. Du, S. Integration of Precise Point Positioning and Low Cost MEMS IMU. Master's Thesis, The University of Calgary, Calgary, AB, Canada, November 2010.
28. Abd Rabbou, M.; El-Rabbany, A. Tightly Coupled Integration of GPS Precise Point Positioning and MEMS-Based Inertial Systems. *GPS Solut.* **2015**, *19*, 601–609. [CrossRef]
29. BeiDou Navigation Satellite System Signal in Space Interface Control Document: Precise Point Positioning Service Signal PPP-B2b (Version 1.0). 2020. Available online: http://en.beidou.gov.cn/SYSTEMS/ICD/202008/P020200803538771492778.pdf (accessed on 20 October 2021).
30. Gao, Z.; Zhang, H.; Ge, M.; Niu, X.; Shen, W.; Wickert, J.; Schuh, H. Tightly Coupled Integration of Multi-GNSS PPP and MEMS Inertial Measurement Unit Data. *GPS Solut.* **2017**, *21*, 377–391. [CrossRef]

31. Sun, W.; Yang, Y. BDS PPP/INS Tight Coupling Method Based on Non-Holonomic Constraint and Zero Velocity Update. *IEEE Access* **2020**, *8*, 128866–128876. [CrossRef]
32. Gao, Z.; Ge, M.; Li, Y.; Chen, Q.; Zhang, Q.; Niu, X.; Zhang, H.; Shen, W.; Schuh, H. Odometer, Low-Cost Inertial Sensors, and Four-GNSS Data to Enhance PPP and Attitude Determination. *GPS Solut.* **2018**, *22*, 57. [CrossRef]
33. Witchayangkoon, B. Elements of GPS Precise Point Positioning. Ph.D. Thesis, Spatial Information Science and Engineering, University of Maine, Orono, ME, USA, 2000.
34. Pintor, P.; González, E.; Senado, A.; Bohlig, P.; Sperl, A.; Henkel, P.; Simón, J.; Hernández, C.; de Blas, J.; Vázquez, J. Galileo High Accuracy Service (HAS) Algorithm and Receiver Development and Testing. In Proceedings of the 35th International Technical Meeting of the Satellite Division of The Institute of Navigation (ION GNSS+ 2022), Denver, CO, USA, 19–23 September 2022; pp. 836–851.
35. Cai, C.; Gao, Y. Modeling and Assessment of Combined GPS/GLONASS Precise Point Positioning. *GPS Solut.* **2013**, *17*, 223–236. [CrossRef]
36. Chen, J.; Pei, X.; Zhang, Y.; Wu, B. GPS/GLONASS System Bias Estimation and Application in GPS/GLONASS Combined Positioning. *Lect. Notes Electr. Eng.* **2013**, *244*, 323–333.
37. Choy, S.; Zhang, S.; Lahaye, F.; Héroux, P. A Comparison between GPS-Only and Combined GPS+GLONASS Precise Point Positioning. *J. Spat. Sci.* **2013**, *58*, 169–190. [CrossRef]
38. Zhou, F.; Dong, D.; Li, P.; Li, X.; Schuh, H. Influence of Stochastic Modeling for Inter-System Biases on Multi-GNSS Undifferenced and Uncombined Precise Point Positioning. *GPS Solut.* **2019**, *23*, 59. [CrossRef]
39. Jiang, N.; Xu, Y.; Xu, T.; Xu, G.; Sun, Z.; Schuh, H. GPS/BDS Short-Term ISB Modelling and Prediction. *GPS Solut.* **2017**, *21*, 163–175. [CrossRef]
40. Niu, X.; Goodall, C.; Nassar, S.; El-Sheimy, N. An efficient methodfor evaluating the performance of MEMS IMUs. In Proceedings of the Position location and Navigation Symposium, 2006 IEEE/ION, San Diego, CA, USA, 25–27 April 2016; pp. 766–771.
41. Sukkarieh, S. Low Cost, High Integrity, Aided Inertial Navigation Systems for Autonomous Land Vehicles. Ph.D. Thesis, University of Sydney, Sydney, Australia, January 2000.
42. Mohamed, A.; Schwarz, K. Adaptive Kalman Filtering for INS/GPS. *J. Geod.* **1999**, *73*, 193–203. [CrossRef]
43. Lu, C.; Li, X.; Nilsson, T.; Ning, T.; Heinkelmann, R.; Ge, M.; Glaser, S.; Schuh, H. Real-Time Retrieval of Precipitable Water Vapor from GPS and BeiDou Observations. *J. Geod.* **2015**, *89*, 843–856. [CrossRef]
44. An, X.; Meng, X.; Jiang, W. Multi-Constellation GNSS Precise Point Positioning with Multi-Frequency Raw Observations and Dual-Frequency Observations of Ionospheric-Free Linear Combination. *Satell. Navig.* **2020**, *1*, 7. [CrossRef]
45. Li, X.; Ge, M.; Dai, X.; Ren, X.; Fritsche, M.; Wickert, J.; Schuh, H. Accuracy and Reliability of Multi-GNSS Real-Time Precise Positioning: GPS, GLONASS, BeiDou, and Galileo. *J. Geod.* **2015**, *89*, 607–635. [CrossRef]
46. Huang, L.; Lu, Z.; Li, B.; Xin, G.; An, W.; Lv, H.; Wang, N.; Zhou, X. The Performance Analysis of Multi-System Integrated Precise Point Positioning (PPP). *Lect. Notes Electr. Eng.* **2016**, *390*, 317–326.

remote sensing

MDPI

Article

Some Key Issues on Pseudorange-Based Point Positioning with GPS, BDS-3, and Galileo Observations

Feng Zhou * and Xiaoyang Wang

College of Geodesy and Geomatics, Shandong University of Science and Technology, Qingdao 266590, China
* Correspondence: zhouforme@sdust.edu.cn

Abstract: Nowadays, BDS-3 and Galileo are still developing and have global service capabilities. This study aims to provide a comprehensive analysis of pseudorange-based/single point positioning (SPP) among GPS, BDS-3, and Galileo on a global scale. First, the positioning accuracy distribution of adding IGSO and GEO to the MEO of BDS-3 is analyzed. The results show that after adding IGSO and GEO, the accuracy of 3D in the Asia-Pacific region is significantly improved. Then, the positioning accuracy of the single-system and single-frequency SPP was validated and compared. The experimental results showed that the median RMS values for the GPS, Galileo, and BDS-3 are 1.10/1.10/1.30 m and 2.57/2.69/2.71 m in the horizontal and vertical components, respectively. For the horizontal component, the GPS and Galileo had better positioning accuracy in the middle- and high-latitude regions, while BDS-3 had better positioning accuracy in the Asia-Pacific region. For the vertical component, poorer positioning accuracy could be seen near the North Pole and the equator for all three systems. Meanwhile, in comparison with the single-system and single-frequency SPP, the contribution of adding pseudorange observations from the other satellite system and frequency band was analyzed fully. Overall, the positioning accuracy can be improved to varying degrees. Due to the observation of noise amplification, the positioning errors derived from dual-frequency SPP were much noisier than those from single-frequency SPP. Moreover, the positioning performance of single-frequency SPP with the ionosphere delay corrected with CODE final (COD), rapid (COR), 1-day predicted (C1P), and 2-day predicted (C2P) global ionospheric map (GIM) products was investigated. The results showed that SPP with COD had the best positioning accuracy, SPP with COR ranked second, while C1P and C2P were comparable and slightly worse than SPP with COR. SPP with GIM products demonstrated a better positioning accuracy than that of the single- and dual-frequency SPP. The stability and variability of the inter-system biases (ISBs) derived from the single-frequency and dual-frequency SPP were compared and analyzed, demonstrating that they were stable in a short time. The differences in ISBs among different receivers with single-frequency SPP are smaller than that of dual-frequency SPP.

Keywords: GNSS; GPS; BDS-3; Galileo; single point positioning (SPP); positioning performance

Citation: Zhou, F.; Wang, X. Some Key Issues on Pseudorange-Based Point Positioning with GPS, BDS-3, and Galileo Observations. *Remote Sens.* **2023**, *15*, 797. https://doi.org/10.3390/rs15030797

Academic Editor: Giuseppe Casula

Received: 28 November 2022
Revised: 17 January 2023
Accepted: 21 January 2023
Published: 31 January 2023

1. Introduction

With the fast development of the global navigation satellite system (GNSS), GNSS has been widely used in the military, disaster prevention and mitigation, engineering construction, transportation, and urban management, and can provide positioning, navigation, and timing services with different levels of accuracy [1–6]. Pseudorange-based/single point positioning (SPP), as one of the most basic modes of GNSS, is used by users in various fields for its computational power, low cost, and fast positioning. The first one that can be used for SPP is the GPS. First of all, only the GPS is capable of SPP, and due to the influence of multiple error sources, the positioning accuracy of GPS SPP has an error of tens of meters [7]. Therefore, some researchers worked on improving the C/A code measurement accuracy and the signal-in-space ranging error (SISRE) of the broadcast ephemeris and improved the accuracy of both of them to 0.3 m and 0.7 m, respectively [8].

At present, the positioning accuracy of GPS SPP in horizontal and vertical components is limited to 1–2 m and 2–3 m, respectively [9]. Galileo had four available satellites in orbit in 2012 and was able to solve SPP independently. In recent years, Galileo satellites have been increasing, and there are now 30 satellites in orbit, with a positioning accuracy of 2–3 m for Galileo SPP in the horizontal component and different frequencies in the vertical component (2–3 m for E1 and 5–7 m for E5a, E5b, and E5 (ab)) [10]. Currently, BDS-3 has 30 satellites providing navigation, positioning, and timing services, including 24 medium earth orbit (MEO), 3 inclined geosynchronous satellite orbit (IGSO), and 3 geosynchronous earth orbit (GEO) satellites [11,12]. BDS SPP, using only the B1 frequency, can provide a positioning accuracy within 10 meters [13]. The positioning accuracy of single-frequency BDS SPP is better than 2 m and 5 m in horizontal and vertical components, respectively, if group delay and code deviation correction are incorporated into the error correction [14]. In addition, the estimated TEC noise proxies (for elevations higher than 60 deg.)—a 100 s root mean square (RMS) of TEC—were: 0.05 TECU for the Galileo E5 AltBOC, 0.09 TECU for the GPS L5, 0.1 TECU for the Galileo E5a/E5b BPSK, and 0.85 TECU for the Galileo E1 CBOC. Dual-frequency combinations provide RMS values of 0.03 TECU for the Galileo E1E5 and 0.03/0.07 TECU for the GPS L1L2/L1L5. At low elevations, E5 AltBOC provides at least twice as little single-frequency TEC noise as compared with the data obtained from E5a or E5b [15].

Since the data accuracy of broadcast ephemeris itself is poor, far below the accuracy of precision ephemeris and precision clock error products, and SPP is based on the data provided by broadcast ephemeris to calculate the satellite 3D position and satellite clock error, the positioning accuracy of SPP is limited to the level of meters [16]. To improve the positioning accuracy of SPP, some GNSS researchers have contributed to the research on improving mathematical models, error model correction, and accuracy improvement for GNSS broadcast ephemeris products. Firstly, regarding the mathematical model, a weighted least squares method based on least squares, with the satellite altitude cut-off angle as the weight, has remarkably improved the positioning accuracy. In addition, iterative least squares with the user-equivalent range error (UERE) as the weight, enhances the positioning accuracy of single-frequency SPP of four single systems, GPS, GLONASS, BDS, and Galileo, to 1/3/1.5/1.5 m, 1.5/3/1.5/1.5 m, and 2.5/6/ 2.5/3 m in E, N, and U components [17,18]. In terms of the error correction model, the effect of ionospheric delay is an essential error source for single-frequency SPP. Meanwhile, single-frequency receivers can only weaken the impact of ionospheric delay by various models. The commonly used ionospheric correction model is Klobuchar, which is widely used because of its simple algorithm [19]. Some researchers have improved the basic Klobuchar model and have developed the GPS Klobuchar model (GPSK8) and the BDS Klobuchar model (BDSK8), which are able to eliminate 64.8% and 65.4% of the ionospheric delay, respectively [20]. Except for the K8 model, some scholars have studied the BDS Global Broadcast Ionospheric Delay Correction Model (BDGIM), under which the global 3D positioning accuracy of BDS SPP can reach 3.5 m, which is better than the BDSK8 model in the mid-latitude region of the Northern hemisphere [21,22]. In addition, the accuracy of the predicted GIMs generated by three different IAACs (CODE, UPC, and ESA) has been evaluated. It was found that the accuracy of the predicted GIMs from CODE was better than that of the predicted GIMs from UPC and ESA, providing a reference for single-frequency GNSS [23].

Although the positioning accuracy of SPP has improved in the case of various mathematical models and error model corrections, the available satellites for a single system have been drastically reduced in areas with complex environments and severe occlusions, where the positioning accuracy is poor or even impossible. In order to obtain the positioning performance of SPP under different environments, the positioning performance of single GPS SPP under 10°, 20°, 30° and 40° satellite cut-off angles is solved by simulating different environments with different satellite cut-off angles. The results show that when the satellite altitude angle reaches more than 30 degrees, the number of GPS satellites decreases from 7 to 4, the PDOP value increases from about 3 to more than 6, and the 3D accuracy decreases from about 2 m to more than 4 m [24]. Therefore, some scholars have conducted multi-system combination SPP. The first multi-system combination was the GPS and GLONASS combination, which increased the number of available satellites and provided more stable positioning results in complex areas [25]. With the development of Galileo and BDS, the study of the combination of GPS and Galileo or BDS has also received attention [26]. When the GPS and four Galileo In-Orbit Validation (IOV) satellites are used for simultaneous positioning, the results are improved by 2% and 10% in the horizontal and vertical components, respectively [27]. The combination of GPS and BDS was first studied in the Australian region, where the combination of BDS and GPS can produce better positioning accuracy and position stability in areas with few available GPS satellites [28]. In the Asia-Pacific region, the combination of GPS and BDS-2 has dramatically improved the positioning accuracy by about 10% better than the case of GPS-only. The combination of multiple systems SPP not only improves the accuracy of navigation and positioning but also improves the integrity of positioning [29]. However, since different navigation and positioning systems have different time and coordinate bases, and the hardware delays between systems are receiver-dependent, all these factors may lead to systematic bias when making observations. Therefore, the time, coordinate base, and receiver hardware delay must be considered when performing multi-system fusion positioning [30]. This is the inter-system bias (ISB), and this error must be dealt with when performing multi-system SPP [31]. Calculating the ISB of the GPS and Galileo is easier because the GPS and Galileo have the same frequency (e.g., L1/L5 for the GPS and E1/E5a for Galileo). The ISB of the GPS and Galileo is discussed, and it is found that the value of ISB depends on the receiver type and can be up to hundreds of seconds, according to the data provided by The Cooperative Network for GIOVE Observation (CONGO) [32]. The effect of ISB must also be considered when performing the combined GPS and BDS SPP. The intra-day stability of the ISB of the GPS and BDS was evaluated using a hypothesis testing method. The result proves that ISB is stable in a short time and can be used as an a priori value [33].

In the last two decades, numerous scholars have studied the status accuracy of single-system SPP, but there are fewer studies on the positioning excellence of different systems within different regions. Meanwhile, some scholars have studied multi-system SPP, but there are fewer studies on the combination of BDS-3 and Galileo, while there are fewer studies on the addition of other systems to the current system in different regions. Based on the above, this paper first analyzes the status of three satellite systems, GPS, BDS-3, and Galileo, regarding the number of available satellites and PDOP. The following three single systems, GPS, BDS-3, and Galileo, are solved for single-frequency SPP at an elevation cut-off of 7°, respectively, to analyze the global distribution of the positioning accuracy in horizontal and vertical components and to compare the accuracy of the three systems. Additionally, we analyzed the dual-frequency single-system SPP. The gain of the combination of GPS with BDS-3 or Galileo, BDS-3 with GPS or Galileo, and Galileo with GPS or BDS-3 in relation to the single system was then examined. Meanwhile, using several ionospheric products, we evaluated the SPP's accuracy. Finally, the stability and variability of ISB are explored.

2. Methodology of SPP

2.1. Pseudorange Observation Equations

For a specific satellite-receiver link, the general equations of pseudorange observations P can be described as:

$$P_{r,j}^{s} = \rho_r^s + c\left[\left(\delta t_r + d_{r,j}\right) - \left(\delta t^s + d_j^s\right)\right] + T_r^s + I_{r,j}^s + \varepsilon_{r,j}^s \tag{1}$$

where s, r, and j denote the satellite, receiver, and frequency band; ρ_r^s represents the geometric distance; δt_r and δt^s are the receiver and satellite clock offsets; c is the speed of light in vacuum; T_r^s is the slant tropospheric delay along the direction from satellite s to receiver r; $I_{r,j}^s$ is the slant ionospheric delay; $d_{r,j}$ and d_j^s denote the receiver- and satellite-specific pseudorange hardware delays/biases; $\varepsilon^s_{r,j}$ represents the unmodelled errors, such as the pseudorange observation noise and multipath effect, etc. It is noteworthy that δt_r and $d_{r,j}$ are high-correlated, and so are δt^s and d^s_j.

In SPP processing, the satellite position and clock offsets are calculated with broadcast ephemeris in advance. The tropospheric delays are corrected with empirical models, i.e., Saastamoinen. For single-frequency SPP, the ionospheric delays are also modeled with empirical models, such as the Klobuchar model for GPS, the BDGIM model for BDS-3, and the Nequick-G model for Galileo. For dual-frequency SPP, the ionospheric delays are usually eliminated by ionospheric-free (IF) combined observables. The corrections of the satellite-specific pseudorange hardware delays for single- and dual-frequency SPP can refer to the Interface Control Document of the specific satellite system.

2.2. *Single-Frequency SPP*

By taking the pseudorange observation on the first frequency as an example, the linearization of Equation (1) can be written as:

$$p^s_{r,1} = g^s_r \cdot x + c\delta\bar{t}_r + \varepsilon^s_{r,1} \tag{2}$$

where $p^s_{r,1}$ is observed, minus the computed value of the pseudorange observation; g^s_r is the unit vector of the component from the receiver to the satellite; x is the vector of the receiver position increments relative to the a priori position; $\delta\bar{t}_r$ is the reparameterized receiver clock offset that absorbs the receiver-specific pseudorange hardware delay, and $\delta\bar{t}_r = \delta t_r + d_{r,1}$.

2.3. *Dual-Frequency SPP*

For dual-frequency SPP, the linearized IF combined pseudorange observables are actually formed as:

$$p^s_{r,\mathrm{IF}} = g^s_r \cdot x + c\delta\tilde{t}_r + \varepsilon^s_{r,\mathrm{IF}} \tag{3}$$

with

$$\begin{cases} p^s_{r,\mathrm{IF}} = \alpha \cdot p^s_{r,1} + \beta \cdot p^s_{r,2} \\ \delta\tilde{t}_r = \delta t_r + (\alpha \cdot d_{r,1} + \beta \cdot d_{r,2}) \\ \varepsilon^s_{r,\mathrm{IF}} = \alpha \cdot \varepsilon^s_{r,1} + \beta \cdot \varepsilon^s_{r,2} \end{cases} \tag{4}$$

2.4. *The Handling of Receiver Clock Offsets in Multi-GNSS SPP*

There are two schemes for the handling of receiver clock offsets in multi-GNSS SPP. The first method is that an independent receiver clock offset per GNSS is introduced. The second one only estimates the GPS-specific receiver clock offset, while for the other satellite system, the inter-system bias (ISB) is employed. In fact, the receiver clock offset for the other system is the sum of the GPS-specific receiver clock offset and ISB. Both methods are equivalent. In this study, the first method is adopted.

3. Data Sets and Processing Strategies

3.1. *Data*

In this paper, the observation files of 145 MGEX tracking stations with a sampling interval of 30 s for a total of 28 days from 1 February 2022 to 28 February 2022 (See: ftps://gdc.cddis.eosdis.nasa.gov/pub/gnss/data/daily accessed on 27 November 2022), and the broadcast ephemeris files provided by IGS (See: ftps://gdc.cddis.eosdis.nasa.gov/pub/gnss/data/daily/2022/brdc accessed on 2 June 2022) were selected for the solution. Figure 1 displays the geographical distribution of 145 tracking stations equipped with geodetic-type receivers, all of which can receive dual-frequency observations from three navigation satellites, namely GPS (L1/L2), BDS-3 (B1/B3), and Galileo (E1/E5a). The SPP was solved by using the open-source GAMP software, and the IGS SINEX weekly solution was used as the reference coordinates [34]. The specific processing strategies are provided in Table 1.

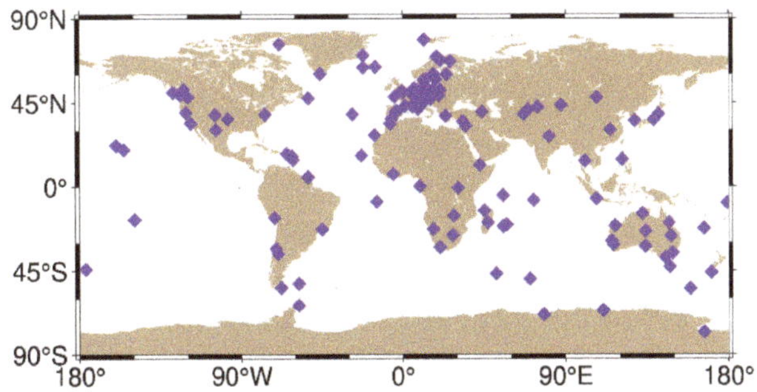

Figure 1. Geographical distribution of 145 tracking stations.

Table 1. The strategy of SPP processing.

Items	Strategies
Number of tracking stations	145
Number of satellites	GPS(32), BDS-3(30), Galileo(30)
Signal selection	GPS(L1, L2), BDS-3(B1, B3), Galileo(E1, E5a)
Sampling rate	30 s
Satellite elevation cut-off	7°
Weight of observation value	Prior standard deviation of measurement error
Tropospheric delay	Saastamoinen delay model
Ionospheric delay	Single frequency and IF combination

3.2. *Availability Analysis of GNSS Constellations*

This subsection mainly analyzes the number of visible satellites and position dilution of precision (PDOP) values for the GPS, Galileo, BDS-3-MEO, BDS-3-MEO+IGSO, and BDS-3-MEO+IGSO+GEO in any global location. Through a hypothetical simulation, we divide the world into 72×90 grids with 5° longitude and 2° latitude and assume that there is a receiver that can receive data from the GPS, BDS-3, and Galileo in the center of each grid [35]. The satellite cut-off elevation angle of the receiver is set to 7°, and the geodetic height is set to 100 m. In addition, whether a satellite can be used for positioning depends on the geometric relationship between the satellite and the receiver. This is because the satellite position accuracy calculated by the broadcast ephemeris and precision ephemeris differs by meter level, and the precision ephemeris provided by GFZ includes not only all available GPS and Galileo satellite data but also data from three types of BDS-3 satellites (MEO, IGSO, and GEO). Therefore, this paper calculates the satellite position using the precision ephemeris, with a sampling interval of 5 minutes, provided by GFZ.

Figure 2 displays the average number of visible satellites at different locations worldwide for the GPS, Galileo, BDS-3-MEO, BDS-3-MEO+IGSO, and BDS-3-MEO+IGSO+GEO on DOY 32 of 2022. The figure instructions consider that the GPS and Galileo have more visible satellites near the equator and at the poles, while fewer satellites are available in the mid-latitudes. In the case of GPS-only, there are 10–11 visible satellites at 20°–50°S and 20°–50°N and 12–13 visible satellites at other latitudes. The number of visible satellites for Galileo reached 10 in the low- and high-latitude regions. However, it is reduced to 8–9 at mid-latitudes. Since BDS-3 has three different orbits of satellites, the number of available satellites in the global region varies not only in latitude but also in the Eastern and Western hemispheres. For MEO alone, the number of visible satellites increases after an initial decrease from the low latitudes to the high latitudes, similar to the GPS and Galileo systems, with visible satellites at about eight at 20°–50°S and 20°–50°N, and about nine at other latitude regions. The visible satellites of BDS-3 will be strongly enhanced by the addition of GEO and IGSO. When IGSO is added, the number of visible satellites increases to 10–11 at high latitudes and 12 near the equator in the Eastern hemisphere, while when GEO is added, the number of visible satellites in the range of 40–180 E will increase dramatically. The number will reach 14 in the lower latitudes of the Eastern hemisphere. Thus, BDS-3 can better serve the Asia-Pacific region.

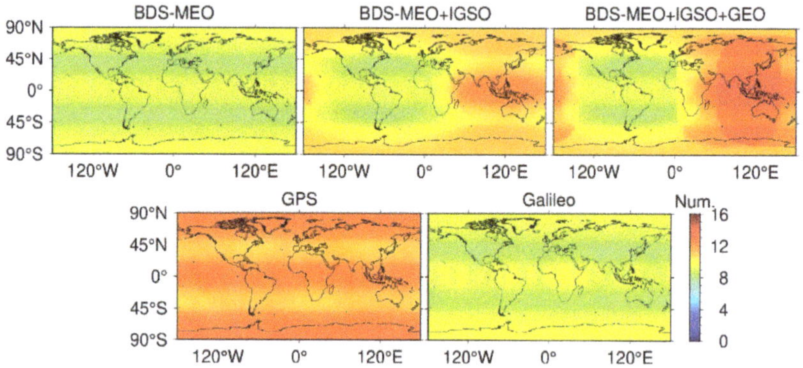

Figure 2. Number of visible satellites for GPS, Galileo, BDS-3-MEO, BDS-3-MEO+IGSO, and BDS-3-MEO+IGSO+GEO.

Figure 3 presents the global average PDOP values for the single-system situation for one day on 1 February 2022. The GPS has an average PDOP of 1.5–1.7 at the latitudes of 10°N–10°S, 40°–60°N, and 40°–60°S, and 1.7–1.9 at other latitudes. For Galileo, the average PDOP is 1.9–2.0 at 15°N–15°S, 50°–70°N, and 50°–70°S, and 2.0–2.3 at other latitudes. In the BDS-3-MEO case, similar to the GPS and Galileo, the PDOP increases from 1.9 to 2.1 and then decreases to 1.9 between 70°S and 70°N and is 2.0–2.2 at high latitudes. When the IGSO and GEO satellites are inserted, the DPOP in the Eastern hemisphere decreases dramatically between the longitudes 60° and 160°E. The PDOP ranged from 1.4 to 1.8 and even dropped to 1.3 near the equator.

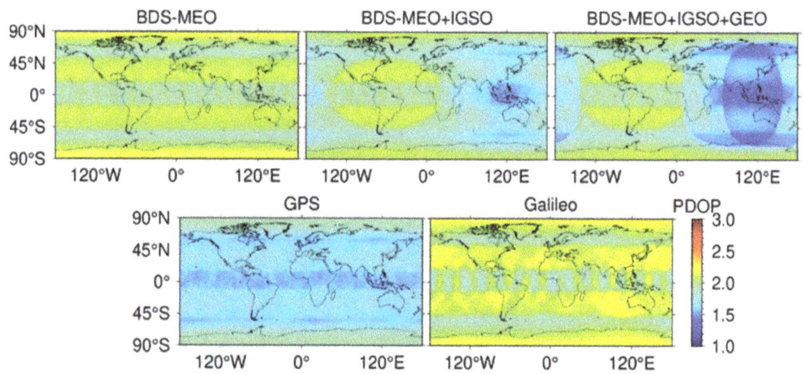

Figure 3. The PDOP for GPS, Galileo, BDS-3-MEO, BDS-3-MEO+IGSO, and BDS-3-MEO+IGSO+GEO.

4. Results Validation and Discussion

4.1. The Impact of Adding MEO, IGSO, and GEO Satellites for BDS-3 SPP

The orbital trajectories of the three satellite types of BDS-3 result in different positioning accuracies at different locations. Figure 4 shows that after adding IGSO to MEO, the positioning accuracy of the stations distributed in the Asia-Pacific region is significantly improved, and the RMS of 3D is increased from 3–4 m to 2.5–3 m. Adding GEO to MEO and IGSO, the accuracy improvement is not obvious. Table 2 lists the accuracy gains and decreases of 145 stations with IGSO in MEO and GEO in MEO and IGSO. Adding IGSO to MEO increases the accuracy by 4.7%, and adding GEO to MEO and IGSO increases the accuracy by 3.6%.

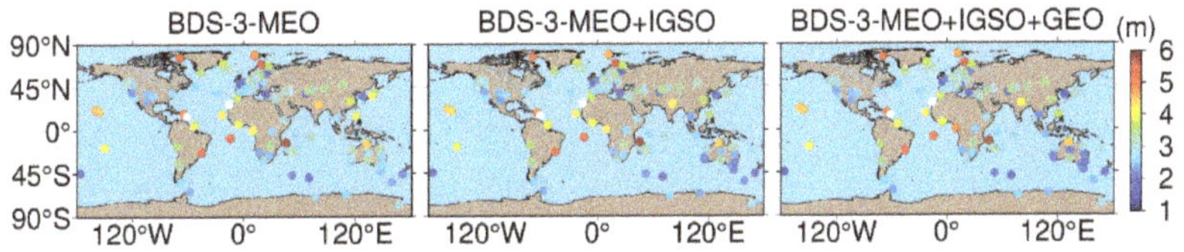

Figure 4. Geographical distribution of positioning accuracy of 145 tracking stations in 3D with IGSO and GEO satellites added to MEO.

Table 2. The number of accuracy improvements of adding IGSO in MEO and GEO in MEO and IGSO.

Satellite Type Added	Number Of Stations with Gain	Number of Stations with Reduced
MEO+IGSO	50	53
MEO+IGSO+GEO	40	30

4.2. Single-System and Single-Frequency SPP (G, C, E)

This section primarily examines the positioning performance of single-frequency single systems (G, C, and E). We first selected the BRMG in the Southern hemisphere and the ABPO tracking station in the Northern hemisphere for SPP. Figure 5 shows the single-frequency positioning errors for 2880 epochs for each station. The error varies with roughly the same trend within 1 day for all three systems, and they all have varying levels of systematic error. In addition, the error values of the GPS and Galileo are relatively close, while the error of BDS-3 fluctuates more.

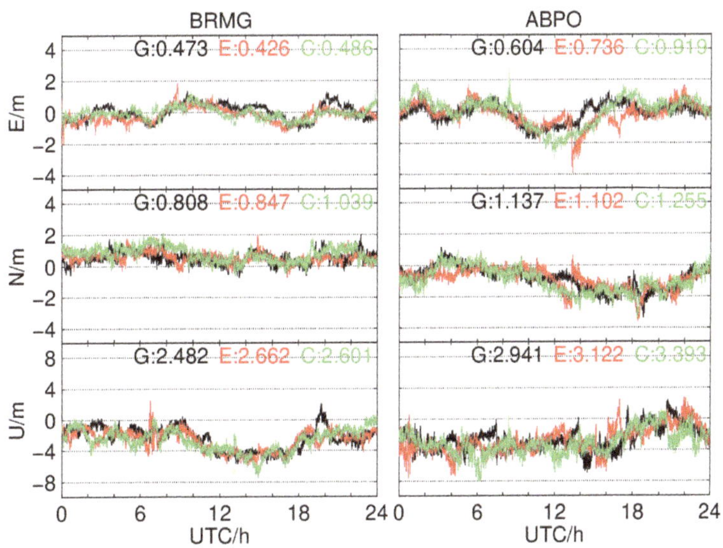

Figure 5. Single-frequency positioning errors for the E, N, and U components of the two tracking stations of BRMG and ABPO.

Figure 6 shows the number of visible satellites and PDOP values for the three systems during one day for the BRMG and ABPO tracking stations. Both stations are able to observe 7–11 GPS satellites and 5–9 Galileo satellites. BRMG is able to observe 7–19 BDS-3 satellites, while ABPO is able to observe 8–12 satellites. The PDOP values are between 2 and 3.

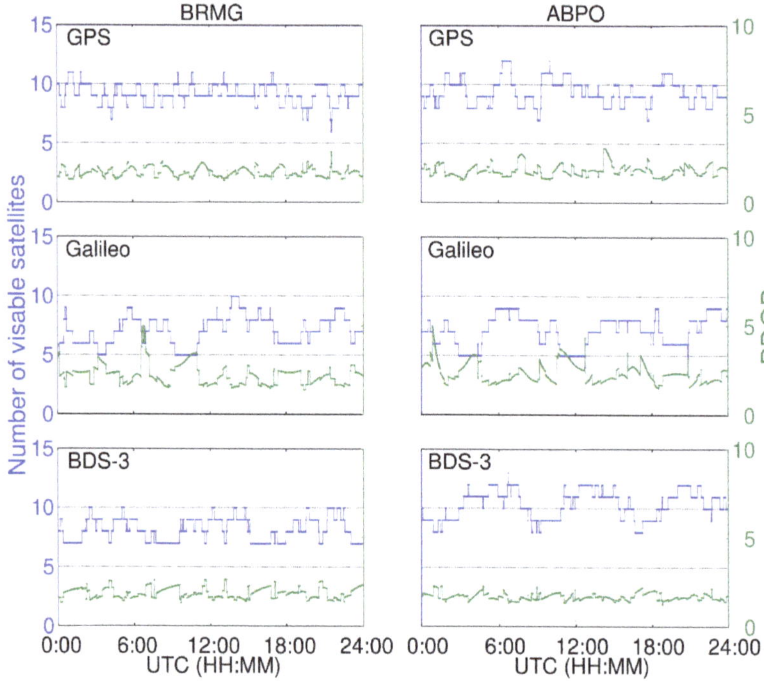

Figure 6. Number of visible satellites and PDOP values for the two tracking stations, BRMG and ABPO.

We then analyzed the overall positioning accuracy of the 145 tracking stations. Figure 7 presents the quartiles of the positioning errors (5%, 25%, 50%, 75%, and 95%) for the three single systems with an elevation cut-off of 7°. In the horizontal component, the median RMS of GPS-only and Galileo-only are the same, both at 1.10 m. The median RMS of BDS-3-only is larger than that of GPS-only and Galileo-only, at 1.30 m. In the vertical component, the GPS-only solution is the best, Galileo-only is the second best, and BDS-3-only is the worst. The median RMS of all three of them is 2.57/2.69/2.71 m. In both components, the 95% quantile RMS of the single system is less than 3.5 and 5 m.

Figure 8 exhibits the geographical distribution of the positioning accuracy for the 145 tracking stations of GPS, Galileo, and BDS-3, where white means exceeding the highest value of the ribbon, and black means less than the lowest value of the ribbon [36]. As shown in the figure, the geographic distribution of positioning accuracy is approximately the same for GPS-only and Galileo-only due to the similar distribution of the number of visible satellites around the world for these two systems. In the horizontal component, there is a significant difference in latitude between the GPS and Galileo tracking station accuracy distributions. The stations located at the middle and high latitudes in North America and Europe have a root mean square (RMS) value of about 1–1.2 m, but the RMS value increases to about 2 m for the tracking stations in the areas between 30° and 30°S and even reaches more than 3 m at low latitudes in South America and Africa. The geographical distribution of the positioning accuracy of BDS-3 varies not only in latitude but also in the Eastern and Western hemispheres. The RMS is 1–1.4 m in the middle and high latitudes of the

Western hemisphere and reaches 3–4 m in the low latitudes; however, in the low latitudes of the Eastern hemisphere, especially in the Asia-Pacific region, the RMS remains around 1.5–2 m. In the vertical component, all three systems are less accurate at low latitudes and in the Arctic, where the RMS reaches 4–6 m, and in other regions, the RMS is 1.8–3 m. The accuracy of the Arctic region is lower than that of the Antarctic region, which may be due to systematic errors or to the influence of the ionosphere, which will be further analyzed by adding dual-frequency observations to eliminate the ionospheric errors below.

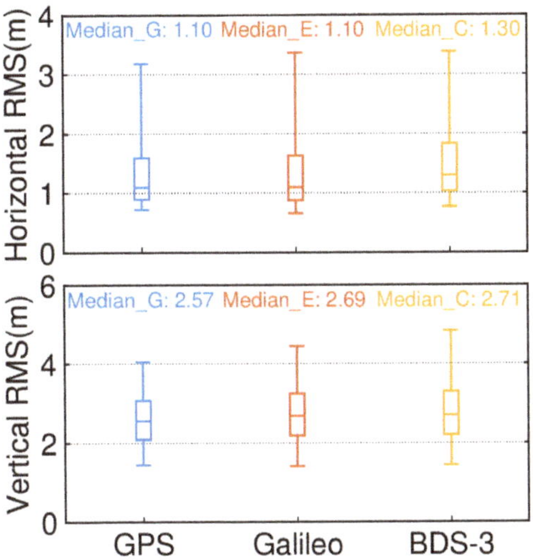

Figure 7. Boxplot of position accuracy of single-frequency SPP with an elevation cut-off of 7° for GPS, Galileo, and BDS-3.

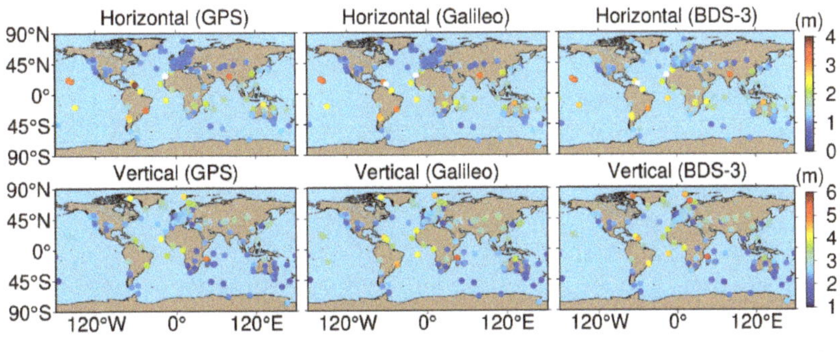

Figure 8. Geographical distribution of positioning accuracy of 145 tracking stations in horizontal and vertical components with an elevation cut-off of 7° for GPS, Galileo, and BDS-3.

The former describes the geographic distribution of the positioning accuracy for a single system. This subsection provides a more visual comparison of the geographic distribution of the positioning accuracy between the single systems (including Galileo versus GPS, BDS-3 versus GPS, and BDS-3 versus Galileo). As shown in Figure 9, the difference between the overall positioning accuracy of Galileo and the GPS is relatively small in the horizontal component, with a difference of −10% to 10%. The positioning accuracy of Galileo is better than the GPS in South America and Europe, while the GPS is slightly better than Galileo in Africa and Australia. In the vertical component, Galileo's positioning accuracy is stronger than GPS-only in Europe and low-latitude South America and slightly worse than the GPS in other regions. The geographical distribution of positioning accuracy of BDS-3 versus GPS and BDS-3 versus Galileo is similar. In the horizontal component, the positioning accuracy of BDS-3 in the Eastern hemisphere is 10–20% higher than that of the GPS or Galileo, while in the Asia–Pacific region, it is about 25% higher. In the vertical component, BDS-3 outperforms both by 10–20% in the Asia-Pacific region, and the positioning accuracy is inferior to the GPS and Galileo in other regions. In summary, the positioning accuracy of the GPS and Galileo is comparable, and BDS-3 outperforms the GPS and Galileo in the Asia-Pacific region.

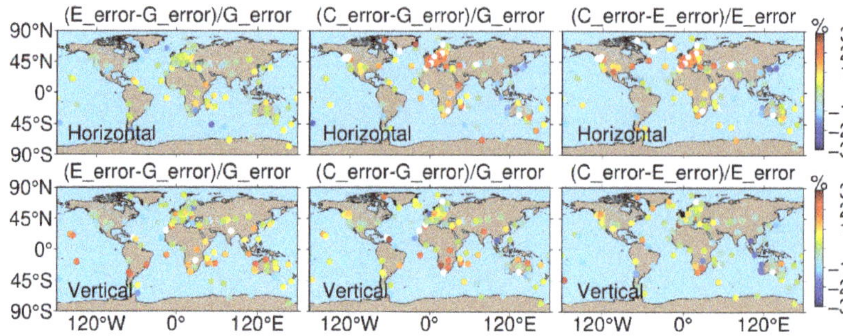

Figure 9. Geographical distribution of the single-frequency positioning accuracy of 145 tracking stations in the horizontal and vertical sections of GPS, Galileo and BDS-3.

4.3. Single-System and Dual-Frequency SPP (G, C, E)

Figure 10 shows the dual-frequency positioning errors of each station. With the addition of the dual-frequency observations, the effect of the ionosphere is eliminated, the system error is removed, and the positioning accuracy of both the N and U components is improved. The single-frequency systematic error in the E component is small, and the addition of the dual-frequency observations increases the observation noise and thus reduces the accuracy.

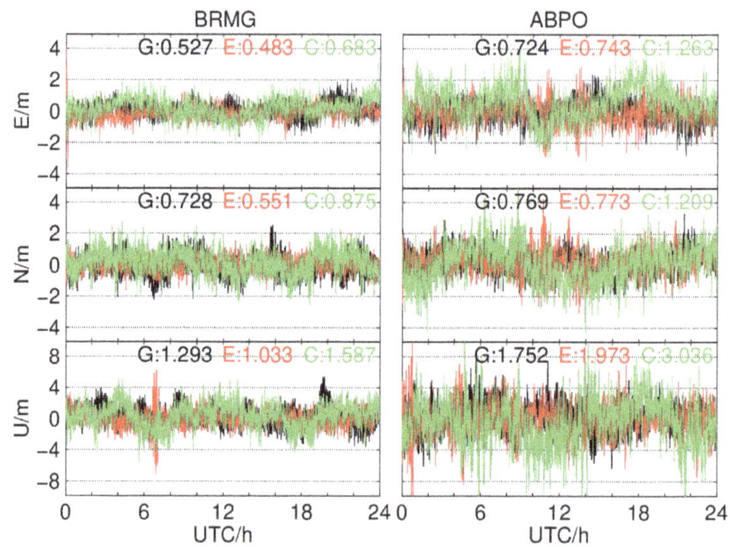

Figure 10. Dual-frequency positioning errors for the E, N, and U components of the two tracking stations of BRMG and ABPO.

The difference between the dual-frequency ionosphere-free combined SPP and single-frequency SPP is that single-frequency SPP uses an a priori model to weaken the ionospheric effects, and the commonly used a priori model is the Klobuchar model, while dual-frequency SPP uses ionosphere-free combined to eliminate the ionospheric errors. Figure 11 shows the boxplot of the positioning accuracy of the dual-frequency, ionosphere-free combined SPP. In the horizontal component, the median RMS of 145 tracking stations of GPS and Galileo dual-frequency ionosphere-free combination SPP are 0.95 m and 0.84 m, respectively, which is a smaller improvement compared to single-frequency SPP; however, the 95th percentile of the RMS is reduced from over 3 m to less than 2 m, which reduces the dispersion value and keeps the accuracy within 2 m. The positioning accuracy of the BDS-3 dual-frequency ionosphere-free SPP is not significantly improved compared to the single-frequency SPP. In the vertical component, the median and 95th percentiles of RMS for both the GPS and Galileo dual-frequency SPP were greatly decreased, from 2.57 and 2.69 m to 1.61 and 1.46 m for RMS, respectively, and from more than 4 m to about 3 m for both 95th percentiles. The median RMS of the dual-frequency ionosphere-free combination of BDS-3 is reduced from 2.71 m to 2.34 m. The addition of dual-frequency observation can effectively reduce the dispersion value and improve positioning accuracy.

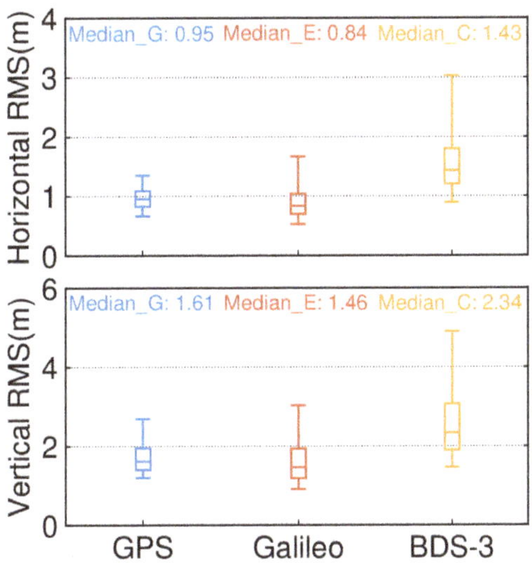

Figure 11. Boxplot of position accuracy of dual-frequency ionosphere-free combined SPP with an elevation cut-off of 7° for GPS, Galileo, and BDS-3.

Figure 12 provides the geographical distribution of the accuracy of the dual frequency. In the horizontal component, the RMS of the 145 tracking stations is between 0.8 and 1.4 m for GPS-only, and Galileo-only has a larger RMS span than the GPS, between 0.6 and 1.8 m. BDS-3-only has a larger RMS, between 1 and 3 m. In the vertical component, the RMS values of GPS-only and Galileo-only are evenly distributed with no system error, and the positioning accuracy of BDS-3-only is better in the Asia-Pacific region.

4.4. Dual-System and Single-Frequency SPP (GE, GC, CE)

Due to the few available satellites in a single system, the positioning accuracy of a single-system SPP sometimes cannot match the needs of the users, especially in areas with complex environments. Therefore, to improve the accuracy of the positioning, we added the other systems to the current system separately, using a dual system for SPP. First, we analyzed the positioning errors of the BRMG and ABPO tracking stations GE, GC, and CE for one day. As shown in Figure 13, the error trends of the single and dual systems are similar within one day, and both have system errors. The system error in the E component is not obvious regarding both stations, while the system error in the N component is about 1 m, and the system error in the U component is 2–3 m. In addition, although the trends are similar, the dual system has fewer error fluctuations and fewer coarse differences in one day compared to the single system.

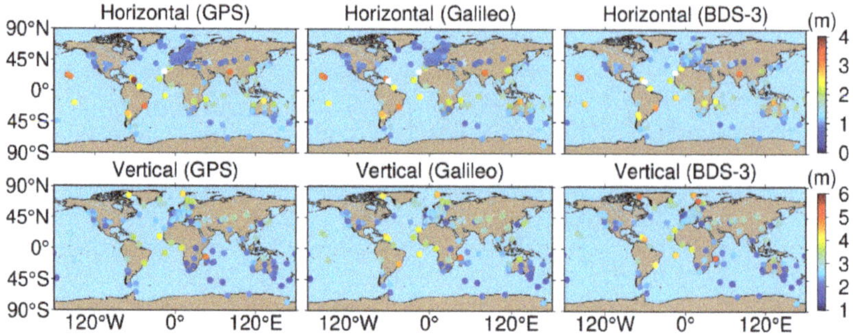

Figure 12. Geographical distribution of the dual-frequency positioning accuracy of 145 tracking stations in the horizontal and vertical sections of GPS, Galileo and BDS-3.

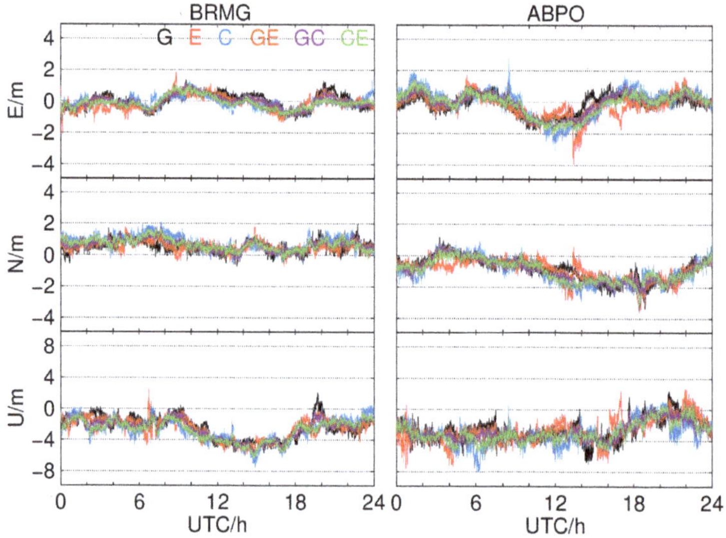

Figure 13. Dual system positioning errors for the E, N, and U components of the two tracking stations of BRMG and ABPO.

Then, we discussed the overall RMS of the dual system for the 145 tracking stations and the geographic distribution of the gain cases. The boxplot of the positioning accuracy for the dual system combination is displayed in Figure 14. Table 3 provides the number of tracking stations with the positive and negative gains from other systems for the current system, as well as the median value of the specific gain percentages for the 145 tracking stations. In the horizontal component, the addition of the GPS and Galileo to BDS-3 resulted in accuracy gains for 131 and 128 tracking stations, respectively, with median percentage gains of 14.2% and 12.2%, respectively. The stations with the largest improvements are mainly in Europe and South America. On the contrary, with the addition of BDS-3 to GPS and Galileo, the stations with greater accuracy improvements are mainly in the Asia-Pacific region, and with the addition of Galileo to GPS, only two stations have reduced accuracy, with a percentage improvement of 7.2%. In the vertical component, the GPS provides the largest contribution to BDS-3 with 125 improvements, and the median improvement is 8.3%. The improvements of BDS-3 to GPS and Galileo are mainly concentrated in the Asia-Pacific region and near the equator. In summary, GE has the best positioning accuracy, GC is the second best, and CE is the worst; the GPS has the largest gain on BDS-3, and the number of tracking stations whose accuracy has been improved by adding Galileo to GPS is the largest.

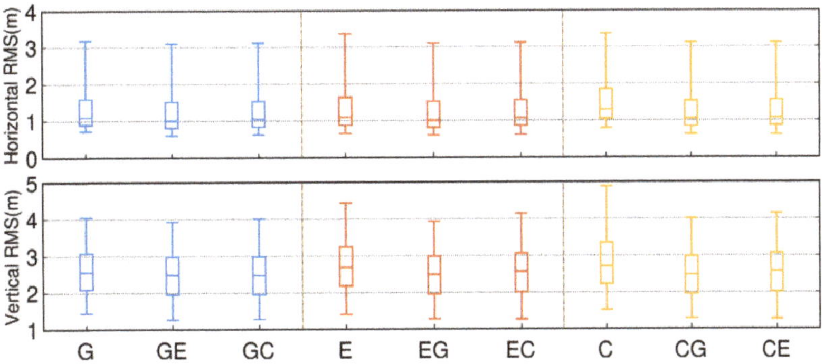

Figure 14. Boxplot of position accuracy of dual system with an elevation cut-off of 7°.

Table 3. Gain of positioning accuracy of other systems to the current system.

Component	Sys	Negative Gain	Median	Positive Gain	Median
Horizontal	GE	13	−3.2%	132	7.4%
	GC	14	−3.6%	131	14.2%
	EG	2	−2.0%	143	7.2%
	EC	17	−1.2%	128	12.2%
	CG	20	−1.6%	125	5.5%
	CE	51	−3.0%	94	5.4%
Vertical	GE	24	−2.2%	121	6.9%
	GC	20	−2.9%	125	8.4%
	EG	28	−1.0%	117	5.2%
	EC	24	−4.2%	121	7.3%
	CG	28	−1.7%	117	4.6%
	CE	30	−1.9%	115	5.2%

4.5. SPP with GIM Products

This subsection analyzes the SPP positioning accuracy using different ionospheric products. We used a total of four ionospheric products, COD is the CODE (Center for Orbit Determination in Europe, University of Berne, Switzerland) final global ionospheric map (GIM), COR is the CODE rapid GIM, and C1P and C2P are 1- and 2-day vertical TEC (VTEC) maps produced by CODE (European Centre for Orbitometry, University of Bern, Switzerland), respectively. Figure 15 shows a one-day positioning error sequence for the BRMG and ABPO tracking stations, with large error fluctuations and systematic errors for the single frequency. The error variation trends of COD, COR, C1P, and C2P are similar and have high accuracy.

Figure 16 presents the boxplot of the positioning accuracy for 145 stations using different ionospheric products. In the horizontal component, the median RMSs of SF/DF/COD/COR/C1P/C2P are 1.10/0.95/0.72/0.75/0.86/0.86 m, respectively, and several products have different degrees of median RMS improvements compared with single-frequency and dual-frequency. Additionally, the 95% quantile of C1P and C2P has decreased from more than 3 m to within 2.2 m for a single frequency. COD and COR are even reduced to within

2 m. In the vertical component, the median RMSs of SF/DF/COD/COR/C1P/C2P are 2.57/1.61/1.27/1.33/1.52/1.50 m, respectively. Furthermore, for a single frequency, the 95% quantile of SF, DF, COD, C1P, and C2P is reduced from more than 4 m to less than 3 m. In summary, the highest accuracy is achieved using COD, followed by COR, and C1P and C2 are comparable but slightly worse than COR.

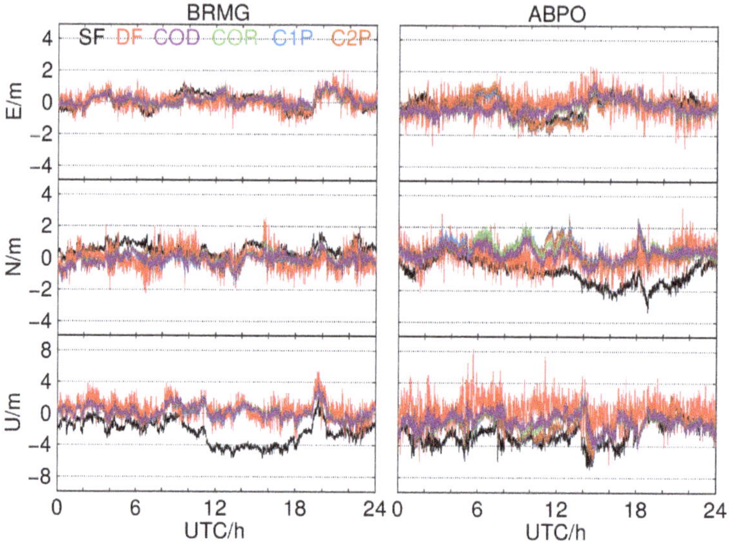

Figure 15. SPP positioning accuracy of different ionospheric products for BRMG and ABPO tracking stations.

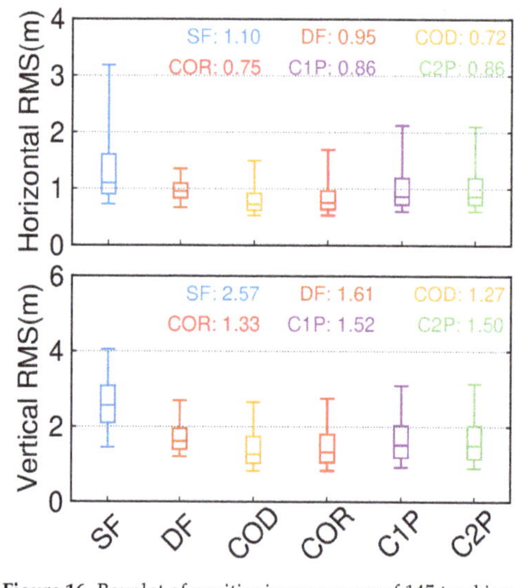

Figure 16. Boxplot of positioning accuracy of 145 tracking stations for different ionospheric products.

4.6. ISB

The value of ISB can be as high as several thousand nanoseconds, which has a powerful influence on the positioning accuracy of SPP. Therefore, the influence of ISB must be considered in multi-system fusion localization, and this subsection analyzes the stability and variability of ISB. We processed data for a total of 28 days in February 2022 to estimate the ISBs for the GPS with the Galileo combination (GE), the GPS with the BDS-3 combination (GC), and the BDS-3 with the Galileo combination (CE). A total of five receiver types—JAVAD, LEIKA, SEPT, TRIMBLE ALLOY, and TRIMBLE NETR9—were selected for the experimental analysis, and the specific receiver types and numbers are shown in Table 4.

Table 4. Type and number of receivers.

Receiver	Model	Number
JAVAD	TRE_3 TRE_3 DELTA	7 17
LEICA	LEICA GR50	8
SEPT	ASTERX4 POLARX5 POLARX5TR	4 61 19
TRIMBLE	ALLOY	23
TRIMBLE	NETR9	4

First, to analyze the stability of single-frequency and dual-frequency ISBs over a short period, five tracking stations equipped with different receiver types were selected for this paper, and the data selected were for one day on 1 February 2022. As shown in Figures 17 and 18, the single-frequency and dual-frequency ISBs of the five stations vary very little within a day, fluctuating within a certain range, and there is variability in the ISBs of different receivers. To further analyze the variability of ISB, two stations from each of the five receivers were selected to calculate the mean value as well as the standard deviation of the ISB for 28 days in February 2022, and the specific values are displayed in Tables 5 and 6.

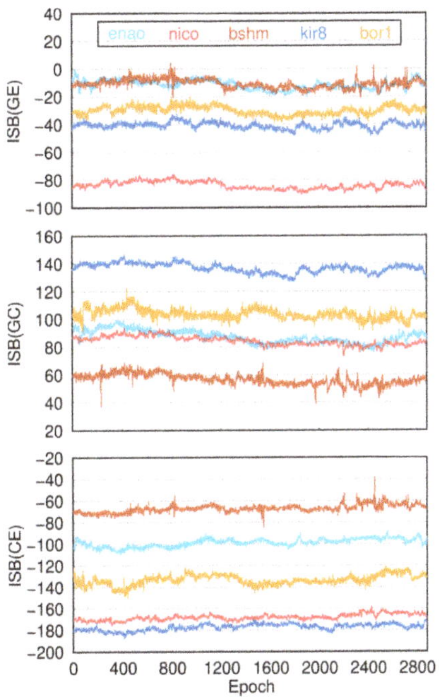

Figure 17. One-day single-frequency ISB of 5 tracking stations equipped with different receivers (1 February 2022).

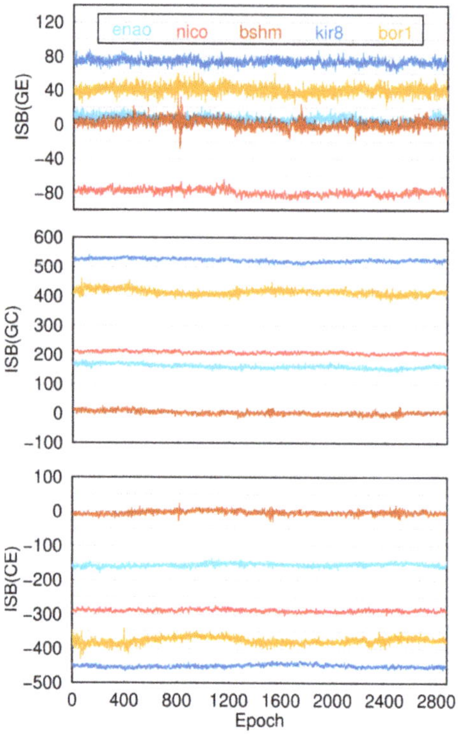

Figure 18. One-day dual-frequency ISB of 5 tracking stations equipped with different receivers (1 February 2022).

Table 5. Median and STD of the 28-day single-frequency ISB for the 10 tracking stations (1 February 2022–28 February 2022).

Station	Receiver	G-E		G-C		C-E	
		Mean	STD	Mean	STD	Mean	STD
enao	JAVAD TRE_3	−20.51	±4.90	80.24	±6.83	−100.68	±4.68
sgpo	JAVAD TRE_3	−20.49	±4.81	81.43	±6.90	−102.14	±4.51
nico	LEICA GR50	−93.23	±4.89	75.64	±6.66	−169.13	±4.61
yebe	LEICA GR50	−93.44	±4.87	73.9	±6.85	−167.84	±4.76
bshm	SEPT POLARX5	−20.16	±4.89	47.53	±6.69	−67.61	±4.44
chpi	SEPT POLARX5	−20.77	±4.64	41.93	±6.42	−62.12	±4.53
kir8	TRIMBLE ALLOY	−50.15	±4.91	129.33	±6.66	−179.35	±4.62
mayg	TRIMBLE ALLOY	−47.44	±4.41	131.27	±6.46	−178.75	±4.14
bor1	TRIMBLE NETR9	−39.5	±4.96	95.19	±6.75	−134.78	±4.76
nabg	TRIMBLE NETR9	−40.69	±5.02	88.29	±6.39	−128.8	±4.28

Figures 19 and 20 show the average ISB values of the single frequency and dual frequency for 10 tracking stations equipped with five receiver types every day of the month. The single-frequency and dual-frequency ISB values of the same receiver type are very close, which may mean that the ISB value is closely related to the hardware delay. The single-frequency ISBs of two tracking stations of the same receiver of GE, GC, and CE are less than 3 ns, 7 ns, and 6 ns, respectively. The dual-frequency ISBs of two tracking stations of the same receiver of GE, GC, and CE are less than 10 ns, 23 ns, and 11 ns, respectively. The ISB value of the dual-frequency SPP is larger, and the ISB of the KIR8 tracking station

GC reaches 517.59 ns. Additionally, the ISB difference in stations with the same receiver is also larger than that of the single-frequency ISB, and the ISB difference between the BOR1 and NABG tracking station GC reaches 23 ns; thus, we recommend using single-frequency receivers. In parallel, we used GIM products to estimate the ISB. The difference between the estimated result and the ISB value estimated by a single frequency is very small and can be controlled within 10 ns. In order to verify the stability of the ISBs in the long term, the difference in ISBs for every two adjacent days in a 28-day period is calculated for all tracking stations. The boxplot demonstrating the difference in the ISBs of GE, GC, and CE for two adjacent days is shown in Figure 21, and the difference in the ISBs of GE, GC, and CE for two adjacent days are within 4 ns, 7 ns, and 5 ns (95th percentile), respectively.

Table 6. Median and STD of the 28-day dual-frequency ISB for the 10 tracking stations (1 February 2022–28 February 2022).

Station	Receiver	G-E		G-C		C-E	
		Mean	STD	Mean	STD	Mean	STD
enao	JAVAD TRE_3	−2.58	±4.13	155.14	±5.78	−157.67	±5.76
sgpo	JAVAD TRE_3	−3.61	±3.95	157.55	±5.50	−161.17	±5.34
nico	LEICA GR50	−88.49	±3.49	201.12	±4.03	−289.62	±4.32
yebe	LEICA GR50	−98.42	±4.03	195.31	±5.09	−293.75	±5.10
bshm	SEPT POLARX5	−8.41	±5.01	−4.26	±6.60	−4.16	±6.64
chpi	SEPT POLARX5	−6.61	±4.02	−22.36	±5.50	−15.86	±5.67
kir8	TRIMBLE ALLOY	−64.94	±4.11	517.59	±5.10	−452.53	±5.01
mayg	TRIMBLE ALLOY	−69.06	±5.16	514.02	±6.33	−444.76	±6.49
bor1	TRIMBLE NETR9	−30.79	±5.57	408.88	±9.74	−377.68	±12.74
nabg	TRIMBLE NETR9	−22.25	±3.82	385.06	±8.44	−386.85	±8.21

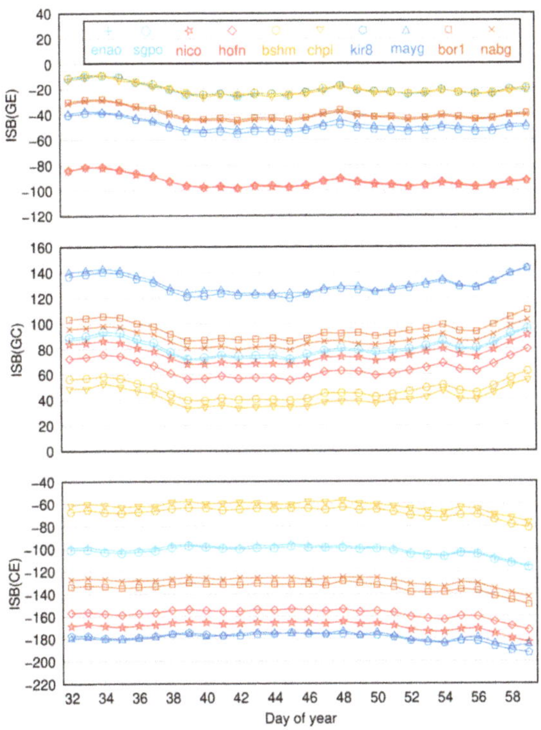

Figure 19. Average value of daily single-frequency ISB for 10 tracking stations of 5 receivers.

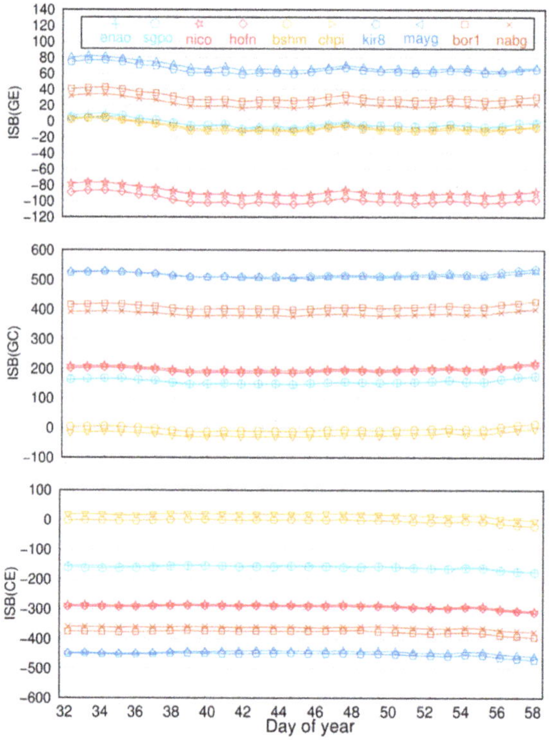

Figure 20. Average value of daily dual-frequency ISB for 10 tracking stations of 5 receivers.

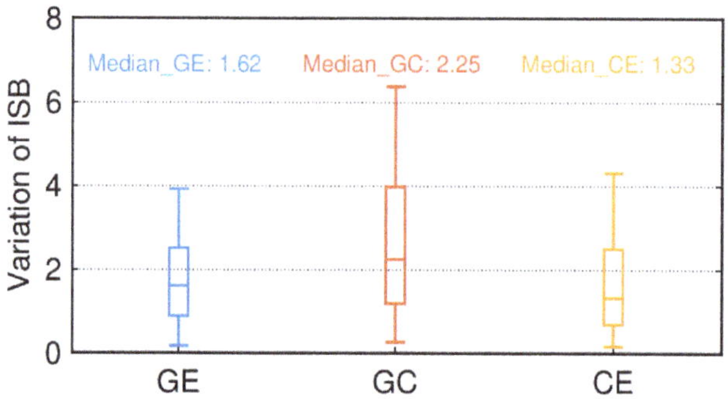

Figure 21. Boxplot of variation of ISB for two days adjacent to all tracking stations (1 February 2022 to 28 February 2022).

5. Conclusions

GNSS was an emerging field in the 20th century, among which SPP has the features of low cost, a fast solution speed, and small hardware size, which can be installed on small devices, such as cell phones and receivers, and has been widely used in car navigation, cell phone terminal positioning, earthquake mitigation and detection, agriculture and forestry, and other fields. In this paper, 145 tracking stations were selected to analyze the positioning theory of SPP and its positioning performance, and the following conclusions were obtained:

(1) First, we analyzed the satellite availability and PDOP values of the GPS, Galileo, BDS-3-MEO, BDS-3-MEO+IGSO, and BDS-3-MEO+IGSO+GEO. The results indicate that the global satellite availabilities of the GPS, Galileo, and BDS-3-MEO have similar global distributions, all of which first decrease and then increase from the equator to the poles. In the mid and low latitudes, the number of available satellites is 11/8/9, respectively. In other regions, the number of available satellites increases to 13, 10, and 10, respectively. However, when the IGSO and GEO satellites are added, the number of available satellites increases to 12 for BDS-3 and reaches 13 near the equator in the Eastern hemisphere. The number of available satellites in the Eastern hemisphere is strongly enhanced.

(2) The accuracy of the global distribution of the GPS, BDS-3, and Galileo single-frequency SPP is analyzed, while the accuracy distribution characteristics of the three systems are compared globally. The results show that in the horizontal component, the GPS and Galileo have the best positioning accuracy of 1–1.2 m in Europe and North America and 2–3 m at low latitudes, while BDS-3 has the highest positioning accuracy of about 1.2 m in the Asia-Pacific region, and the overall Eastern hemisphere is better than the Western hemisphere. In the vertical component, the accuracy of all three systems gradually deteriorates from south to north, probably due to system bias and perhaps due to the influence of the ionosphere. After the comparison of the three systems, the positioning accuracy of the GPS and Galileo is comparable, and BDS-3 outperforms the GPS and Galileo by 30% and 20% in the Asia-Pacific region on the horizontal and vertical components, respectively.

(3) We combined BDS-3 or Galileo with GPS, GPS or Galileo with BDS-3, and GPS or BDS-3 with Galileo and analyzed the percentage gain of the combined SPP with respect to the single system. The conclusions drawn are as follows: For the combination of the GPS and Galileo with BDS-3, the accuracy of the Asia-Pacific region in the horizontal and vertical components will be improved by 30% and 20%, respectively. When the GPS and Galileo are combined with BDS-3, the accuracy in North America and Europe will improve by 30–40% in the horizontal component, and the global accuracy will improve by 10–20% in the vertical component. The combination of Galileo and the GPS or BDS-3 can effectively improve the accuracy of the vertical component in the middle and low latitudes by about 15–20%.

(4) We analyzed the positioning accuracy using different ionospheric products, and the results showed that the median RMS of COD, COR, C1P, and C2P were 0.72/0.75/0.86/0.86 m and 1.27/1.33/1.52/1.50 m, respectively, with the highest accuracy using COD, followed by COR, and C1P and C2P being comparable and slightly lower than COR, thus providing a reference to facilitate the selection of products for single-frequency GNSS users.

(5) We analyzed the positioning accuracy using different ionospheric products, and the results showed that the single-frequency positioning accuracy was the worst, with the highest accuracy using COD, followed by COR, and C1P and C2P being equivalent and slightly lower than COR.

(6) Finally, we analyzed the single-frequency and dual-frequency ISBs of five receiver types, and the results showed that the ISBs are stable in a short time. At the same time, the ISB is related to the receiver type, and the difference between the single-frequency ISB of the same receiver type is smaller than that of the dual-frequency ISB; thus, it is recommended to use the single-frequency receiver.

The results of this paper clearly illustrate the positioning accuracy of three single-system SPP solutions—GPS, BDS-3, and Galileo—in different regions. As well as being able to analyze the positioning superiority of different system combinations in different regions, this provides a reference for choosing the optimal combination for different regions, while the ISB analysis provides more possibilities for the multi-system combination SPP.

Author Contributions: Conceptualization, X.W.; methodology, X.W.; software, F.Z. and X.W.; validation, F.Z.; formal analysis, X.W.; investigation, F.Z. and X.W.; writing—original draft preparation, F.Z. and X.W.; writing—review and editing, F.Z. and X.W.; visualization, X.W.; supervision, F.Z. All authors have read and agreed to the published version of the manuscript.

Funding: This research was funded by Laoshan Laboratory (LSKJ202205104, LSKJ202205104_01) and the China Postdoctoral Science Foundation (2020M673669).

Data Availability Statement: The GNSS raw observation data is available at ftps://gdc.cddis.eosdis.nasa.gov/pub/gnss/data/daily/ (accessed on 6 May 2022). The broadcast ephemeris data is available at ftps://gdc.cddis.eosdis.nasa.gov/pub/gnss/data/daily/2022/brdc/ (accessed on 6 May 2022).

Acknowledgments: The authors would like to thank the IGS Multi-GNSS Experiment (MGEX) for providing the data.

Conflicts of Interest: The authors declare no conflict of interest.

References

1. Zheng, F.; Gu, S.; Gong, X.; Lou, Y.; Fan, L.; Shi, C. Real-time single-frequency pseudorange positioning in China based on regional satellite clock and ionospheric models. *GPS Solut.* **2020**, *24*, 6. [CrossRef]
2. Hu, J.; Li, P.; Zhang, X.; Bisnath, S.; Pan, L. Precise point positioning with BDS-2 and BDS-3 constellations: Ambiguity resolution and positioning comparison. *Adv. Space Res.* **2022**, *70*, 1830–1846. [CrossRef]
3. Zhou, F.; Li, X.; Li, W.; Chen, W.; Dong, D.; Wickert, J.; Schuh, H. The Impact of Estimating High-Resolution Tropospheric Gradients on Multi-GNSS Precise Positioning. *Sensors* **2017**, *17*, 756. [CrossRef]
4. Ma, H.; Psychas, D.; Xing, X.; Zhao, Q.; Verhagen, S.; Liu, X. Influence of the inhomogeneous troposphere on GNSS positioning and integer ambiguity resolution. *Adv. Space Res.* **2021**, *67*, 1914–1928. [CrossRef]
5. Paziewski, J.; Wielgosz, P. Assessment of GPS + Galileo and multi-frequency Galileo single-epoch precise positioning with network corrections. *GPS Solut.* **2014**, *18*, 571–579. [CrossRef]
6. Liu, J.; Cao, C. Development status and trend of global navigation satellite system. *J. Navig. Position.* **2020**, *8*, 1–8. [CrossRef]
7. Yang, Z.; Xue, B. The developed procedures and developing trends of the American navigation satellites. *J. Navig. Position.* **2021**, *9*, 1–12. [CrossRef]
8. Montenbruck, O.; Steigenberger, P.; Hauschild, A. Broadcast versus precise ephemerides: A multi-GNSS perspective. *GPS Solut.* **2015**, *19*, 321–333. [CrossRef]
9. Xu, Y.; Ji, S. Data quality assessment and the positioning performance analysis of BeiDou in Hong Kong. *Surv. Rev.* **2015**, *47*, 446–457. [CrossRef]
10. Robustelli, U.; Pugliano, G. Galileo Single Point Positioning Assessment Including FOC Satellites in Eccentric Orbits. *Remote Sens.* **2019**, *11*, 1555. [CrossRef]
11. CSNO. The Current Constellation of the BDS System. China Satellite Navigation Office. 2021. Available online: http://www.csno-tarc.cn/system/constellation (accessed on 2 June 2022).
12. Yang, Z.; Xue, B. The developed procedures and developing trends of BeiDou satellite navigation system. *J. Navig. Position.* **2022**, *10*, 1–14. [CrossRef]
13. Yang, Y.; Li, J.; Wang, A.; Xu, J.; He, H.; Guo, H.; Shen, J.; Dai, X. Preliminary assessment of the navigation and positioning performance of BeiDou regional navigation satellite system. *Sci. China Earth Sci.* **2014**, *57*, 144–152. [CrossRef]
14. Guo, F.; Zhang, X.; Wang, J. Timing group delay and differential code bias corrections for BeiDou positioning. *J. Geod.* **2015**, *89*, 427–445. [CrossRef]
15. Padokhin, A.M.; Mylnikova, A.A.; Yasyukevich, Y.V.; Morozov, Y.V.; Kurbatov, G.A.; Vesnin, A.M. Galileo E5 AltBOC Signals: Application for Single-Frequency Total Electron Content Estimations. *Remote. Sens.* **2021**, *13*, 3973. [CrossRef]
16. Luo, X.; Lou, Y.; Gong, X.; Gu, S.; Chen, B. Benefit of Sparse Reference Network in BDS Single Point Positioning with Single-Frequency Measurements. *J. Navig.* **2018**, *71*, 403–418. [CrossRef]
17. Zhang, Z.; Pan, L. Current performance of open position service with almost fully deployed multi-GNSS constellations: GPS, GLONASS, Galileo, BDS-2, and BDS-3. *Adv. Space Res.* **2022**, *69*, 1994–2019. [CrossRef]
18. Blanco-Delgado, N.; Nunes, F.D. Satellite selection based on WDOP concept and convex geometry. In Proceedings of the 2010 5th ESA Workshop on Satellite Navigation Technologies and European Workshop on GNSS Signals and Signal Processing (NAVITEC), Noordwijk, The Netherlands, 8–10 December 2010; pp. 1–8. [CrossRef]
19. Klobuchar, J.A. Ionospheric Time-Delay Algorithm for Single-Frequency GPS Users. *IEEE Trans. Aerosp. Electron. Syst.* **1987**, *3*, 325–331. [CrossRef]
20. Wu, X.; Hu, X.; Wang, G.; Zhong, H.; Tang, C. Evaluation of COMPASS ionospheric model in GNSS positioning. *Adv. Space Res.* **2013**, *51*, 959–968. [CrossRef]
21. Yang, C.; Guo, J.; Geng, T.; Zhao, Q.; Jiang, K.; Xie, X.; Lv, Y. Assessment and Comparison of Broadcast Ionospheric Models: NTCM-BC, BDGIM, and Klobuchar. *Remote Sens.* **2020**, *12*, 1215. [CrossRef]
22. Li, Q.; Su, X.; Xu, Y.; Ma, H.; Liu, Z.; Cui, J.; Geng, T. Performance Analysis of GPS/BDS Broadcast Ionospheric Models in Standard Point Positioning during 2021 Strong Geomagnetic Storms. *Remote Sens.* **2022**, *14*, 4424. [CrossRef]
23. Li, M.; Yuan, Y.; Wang, N.; Li, Z.; Huo, X. Performance of various predicted GNSS global ionospheric maps relative to GPS and JASON TEC data. *GPS Solut.* **2018**, *22*, 55. [CrossRef]

24. Pan, L.; Cai, C.; Santerre, R.; Zhang, X. Performance evaluation of single-frequency point positioning with GPS, GLONASS, BeiDou and Galileo. *Surv. Rev.* **2017**, *49*, 197–205. [CrossRef]
25. Ryan, S.; Petovello, M.; Lachapelle, G. Augmentation of GPS for Ship Navigation in Constricted Water Ways. In Proceedings of the 1998 National Technical Meeting of The Institute of Navigation, Long Beach, CA, USA, 21–23 January 1998; pp. 459–467.
26. Cai, C.; Gao, Y.; Pan, L.; Dai, W. An analysis on combined GPS/COMPASS data quality and its effect on single point positioning accuracy under different observing conditions. *Adv. Space Res.* **2014**, *54*, 818–829. [CrossRef]
27. Cai, C.; Luo, X.; Liu, Z.; Xiao, Q. Galileo Signal and Positioning Performance Analysis Based on Four IOV Satellites. *J. Navig.* **2014**, *67*, 810–824. [CrossRef]
28. Odolinski, R.; Teunissen, P.J.G.; Odijk, D. First combined COMPASS/BeiDou-2 and GPS positioning results in Australia. Part I: Single-receiver and relative code-only positioning. *J. Spat. Sci.* **2014**, *59*, 3–24. [CrossRef]
29. Yang, Y.; Xu, J. GNSS receiver autonomous integrity monitoring (RAIM) algorithm based on robust estimation. *Geod. Geodyn.* **2016**, *7*, 117–123. [CrossRef]
30. Wang, J.; Knight, N.L.; Lu, X. Impact of the GNSS Time Offsets on Positioning Reliability. *J. Glob. Position. Syst.* **2011**, *10*, 165–172. [CrossRef]
31. Teng, Y.; Wang, J. New Characteristics of Geometric Dilution of Precision (GDOP) for Multi-GNSS Constellations. *J. Navig.* **2014**, *67*, 1018–1028. [CrossRef]
32. Montenbruck, O.; Hauschild, A.; Hessels, U. Characterization of GPS/GIOVE sensor stations in the CONGO network. *GPS Solut.* **2011**, *15*, 193–205. [CrossRef]
33. Zeng, A.; Yang, Y.; Ming, F.; Jing, Y. BDS–GPS inter-system bias of code observation and its preliminary analysis. *GPS Solut.* **2017**, *21*, 1573–1581. [CrossRef]
34. Zhou, F.; Dong, D.; Li, W.; Jiang, X.; Wickert, J.; Schuh, H. GAMP: An open-source software of multi-GNSS precise point positioning using undifferenced and uncombined observations. *GPS Solut.* **2018**, *22*, 33. [CrossRef]
35. Yang, Y.; Li, J.; Xu, J.; Tang, J.; Guo, H.; He, H. Contribution of the Compass satellite navigation system to global PNT users. *Chin. Sci. Bull.* **2011**, *56*, 2813–2819. [CrossRef]
36. Geng, J.; Meng, X.; Dodson, A.H.; Teferle, F.N. Integer ambiguity resolution in precise point positioning: Method comparison. *J. Geod.* **2010**, *84*, 569–581. [CrossRef]

remote sensing MDPI

Article

The Efficiency of Geodetic and Low-Cost GNSS Devices in Urban Kinematic Terrestrial Positioning in Terms of the Trajectory Generated by MMS

Filip Viler [1], Raffaela Cefalo [2], Tatiana Sluga [2], Paolo Snider [2] and Polona Pavlovčič-Prešeren [1,*]

1 Faculty of Civil and Geodetic Engineering, University of Ljubljana, Jamova cesta 2, SI-1000 Ljubljana, Slovenia
2 GeoSNav Laboratory, Department of Engineering and Architecture, University of Trieste, Via Valerio 6/2, 34127 Trieste, Italy
* Correspondence: polona.pavlovcic-preseren@fgg.uni-lj.si; Tel.: +386-1-4768-631

Abstract: The quality of geospatial data collection depends, among other things, on the reliability and efficiency of the GNSS receivers or even better integrated GNSS/INS systems used for positioning. High-precision positioning is currently not only the domain of professional receivers but can also be achieved by using simple devices, including smartphones. This research focused on the quality of 2D and 3D kinematic positioning of different geodetic and low-cost GNSS devices, using the professional mobile mapping system (MMS) as a reference. Kinematic positioning was performed simultaneously with a geodetic Septentrio AsteRx-U receiver, two u-blox receivers—ZED-F9P and ZED-F9R—and a Xiaomi Mi 8 smartphone and then compared with an Applanix Corporation GPS/INS MMS reference trajectory. The field tests were conducted in urban and non-urban environments with and without obstacles, on road sections with large manoeuvres and curves, and under overpasses and tunnels. Some general conclusions can be drawn from the analysis of the different scenarios. As expected, some results in GNSS positioning are subject to position losses, large outliers and multipath effects; however, after removing them, they are quite promising, even for the Xiaomi Mi8 smartphone. From the comparison of the GPS and GNSS solutions, as expected, GNSS processing achieved many more solutions for position determination and allowed a relevant higher number of fixed ambiguities, even if this was not true in general for the Septentrio AsteRx-U, in particular in a surveyed non-urban area with curves and serpentines characterised by a reduced signal acquisition. In GNSS mode, the Xiaomi Mi8 smartphone performed well in situations with a threshold of less than 1 m, with the percentages varying from 50% for the urban areas to 80% for the non-urban areas, which offers potential in view of future improvements for applications in terrestrial navigation.

Keywords: GNSS/INS sensors; Smartphone Xiaomi Mi8; GNSS/INS sensors; kinematic-trajectory acquisition; u-blox ZED-F9P and ZED-F9R; mobile mapping system (MMS)

Citation: Viler, F.; Cefalo, R.; Sluga, T.; Snider, P.; Pavlovčič-Prešeren, P. The Efficiency of Geodetic and Low-Cost GNSS Devices in Urban Kinematic Terrestrial Positioning in Terms of the Trajectory Generated by MMS. *Remote Sens.* **2023**, *15*, 957. https://doi.org/10.3390/rs15040957

Academic Editors: Giuseppe Casula, Zhetao Zhang and Wenkun Yu

Received: 20 January 2023
Revised: 6 February 2023
Accepted: 7 February 2023
Published: 9 February 2023

1. Introduction

The task of mapping the world and determining positions while moving has always been a challenging task in topographic surveying. With the advent of the first mobile mapping systems (MMS) in the late 1980s, where moving platforms were equipped with precise GPS (Global Positioning System) navigation and digital imaging, and later with other sensors, the systems allowed the world to be surveyed with high accuracy and productivity. As a result, several sectors have recognised the importance of high-quality georeferenced geospatial data, such as land use planning and construction, insurance and emergency services. Today, the requirements in this field have increased, especially with the development of autonomous navigation in transportation, which requires fast availability, reliability and high integrity of real-time positioning. However, the problem of GNSS navigation in urban areas remains a grey area in which positioning accuracy can be frustrating and confusing [1–3]. To this end, the performance of devices of varying quality must be tested against high-precision sensors.

Not everyone has access to highly professional MMS sensors, but there is another way to solve the problem. With the advent of low-cost sensors, it is possible to produce mobile mapping sensors at an affordable price. Today's trend in geomatics is towards mass sensing with low-cost devices that could be useful in daily life for real-time mapping of reality or its subsequent reconstruction. Dual-frequency smartphones capable of acquiring carrier-phase measurements associated with other sensors, including INS (Inertial Navigation System), present another challenge in mobile mapping today. For example, such sensors could be used in the rapid acquisition of road conditions and damage, mapping of the condition of roadside facilities and the reconstruction of car accidents. However, in order for this to be of use, the quality of their robustness and performance should be investigated and estimated.

Several authors [4–11] have already noted that modern, low-cost electronic devices can achieve high accuracy in determining kinematic GNSS trajectories. Such sensors are used, for example, in drones for the direct georeferencing of trajectories [12–17]. However, the georeferencing of trajectories for drones differs from ground-based positioning in terms of obstacles and the open sky, whereby rapid changes in the signal reception environment and measurement conditions pose a major challenge. The problem of kinematic positioning quality is also relevant to the challenges of autonomous vehicle navigation, where the requirements for integrity, reliability and quality are quite high and, in some cases, currently impossible to meet. As for smartphones, some authors [18–20] have reported that some characteristics of GNSS signals currently acquired by smartphones still have a negative impact on GNSS position accuracy. This was one of the main issues of the analyses in this study.

To meet this challenge, the reference trajectory was derived from a professional MMS equipped with a POS/LV (Position and Orientation System for Land Vehicles) from Applanix Corporation (Applanix, Richmond Hill, ON, Canada). A detailed explanation of the system from the year 2000 can be found, for example, in [21,22]. Among other studies, Applanix measurements have already been used to evaluate the extraction of road geometry parameters [23], for studies comparing the kinematic performance of different satellite systems (GPS, GLONASS, Galileo) in terrestrial measurements using single frequencies [24] and multi-frequencies [25] at the University of Trieste. However, this study was extended to the observation of multi-constellation systems and the use of a Xiaomi Mi8 Android smartphone, two low-cost u-blox receivers—ZED-F9P and ZED-F9R—and a Septentrio AsteRx-U geodetic GNSS receiver in terms of the reference trajectory of the Applanix Corporation GPS/INS MMS system.

Thus far, this article has shown the importance of positioning quality in geospatial data collection, using low-cost sensors from u-blox or smartphones that are affordable for all. The main contribution of this article is the comparison of trajectories acquired by using different GNSS devices and a comparison with the reference trajectory. Although the authors are aware that both smartphones and u-blox devices can receive both GNSS and INS measurements, the focus of this study was on the performance of GNSS receivers in urban and semi-urban environments. The goal was to gain information on the quality of GNSS positioning and to compare the performance of GNSS-based geodetic receivers and the much cheaper u-blox and dual-frequency smartphones operating in kinematic mode. All the findings obtained in this study should be extended and improved in the future to also provide deeper insights into GNSS/INS observations and post-processing quality.

1.1. Low-Cost Devices in High-Precision Kinematic Terrestrial Positioning

The group of low-cost devices that currently provide high-precision positioning based on carrier-phase GNSS or GNSS/INS measurements includes devices from u-blox (Thalwil, Switzerland), Septentrio (Leuven, Belgium), VectorNav (Dallas, TX, USA), ComNav (Surveying Hub B.V., Hilversum, The Netherlands), and others. These devices are already installed in various road vehicles, drones, boats and satellites, and they have dramatically improved processing chains in real-world data collection. The group of low-cost devices also includes some smartphones, tablets and wristwatches with a production date of 2018 and beyond.

As for testing low-cost GNSS devices in kinematic terrestrial surveys, many studies have already been conducted since the introduction of multi-constellation and dual-frequency smartphones in 2018. Broekman and Gräbe showed that a low-cost real-time mobile geolocation service provides centimetre accuracy up to 15 km from the base station [13]. The problem was further investigated by Janos and Kuros [14], who used ZED-F9P for stop-and-go topographic real-time kinematic (RTK) surveys under difficult measurement conditions. Sana et al. (2022) [26] followed up on previous studies by testing the kinematic performance of RTK and post-processing of two vehicles in two different environments: once in a non-urban area with a long baseline of about 30 km with varying visibility and screened locations and once in an urban area within a radius of about 10 km with some buildings and open areas. Their experiences, especially the conclusions that the real-time kinematic solutions matched well with the post-processed solutions, served as the basis for this research. In addition, the authors' previous research relating to the interference of ZED-F9P receivers should be noted when several ZED-F9P receivers were tested under vertical and horizontal L1/E1 chirp jamming [27,28], while the study by Dimc et al. [29] was extended to the L1/E1 interference of the kinematic vehicle with several geodetic receivers and two u-bloxes on board, namely ZED-F9P and ZED-F9R.

The year 2016 represented the first milestone in the use of smartphones in high-precision GNSS positioning. At that time, the extraction of raw observation data for post-processing (i.e., pseudorange, carrier phase, Doppler shift, and carrier-to-noise ratio (C/N0)) was introduced under the Android Nougat operating system [30]. Until then, the outputs of GNSS chipsets in smartphones were only PVT (position, velocity and time). The very next year, the first two smartphones—the Samsung S8 and the Huawei P10—were launched that could perform single-frequency carrier phase measurements. However, the most important milestone was the launch of the Xiaomi Mi8 in May 2018 with the new Broadcom BCM47755 GNSS chipset [31], which enabled the acquisition of dual-frequency carrier phase data on L1/L5 and E1/E5a for GPS and Galileo, respectively.

These advances led to several studies being carried out on raw GNSS measurements on smartphones, aiming at a comprehensive analysis of the observations of the positioning quality and performance of various dual-frequency smartphones, starting with the Xiaomi Mi8 [4,5,7,9,18,30,32–37]. Zeng et al. [38] reported that the carrier-to-noise density ratio (C/N0) of a smartphone is about 10 dB-Hz lower than that of a geodetic receiver, but the multipath is much higher. They reported that frequent cycle-slips and loss of lock already limit the accuracy of high-precision positioning in static mode but even more so in kinematic mode. In addition, the authors proposed that the duty cycles should be turned off, which resulted in better carrier phase measurement accuracy and much higher positioning accuracy.

Liu et al. [39] followed up this study in 2019 and investigated the quality of raw GNSS observations from smartphones in terms of C/N0, noise, carrier-phase tracking and velocity estimation. The authors suggested that an altitude-based weighting algorithm is not suitable for low-cost receivers, including smartphones, while a C/N0 weighting would be a better choice for these devices. Later, Banville et al. [40], Paziewski et al. [41], and Robustelli et al. [9] confirmed the suitability of this processing approach. The fact that this study is based on u-blox instruments operating in the L1/E1 and L2 frequency bands, smartphones operating in the L1/E1 and L5/E5a frequency bands, as well as a geodetic instrument receiving observations at all frequencies, was a good starting point for further analysis. A valuable study on real-time kinematic positioning with smartphones was recently published by Li et al. [42]. The study used two Huawei Mate30 and two Huawei P40 smartphones with two installation modes: vehicle roof mode, in which the smartphones were mounted on the roof outside the vehicle, and dashboard mode, in which the smartphones were stabilised inside the vehicle. The authors confirmed that the installation on the vehicle roof provided better results, especially in terms of RMS (Root Mean Square) for the vertical component. Therefore, in the current experiment, the authors used the aforementioned method, i.e., the smartphone was mounted on the roof of the vehicle.

As part of this brief overview of the use of low-cost devices and smartphones in kinematic terrestrial surveying, it is also worth mentioning the 2021 Google Smartphone Decimeter Challenge experiment that inspired Everett et al. [43] to optimise the use of RTKLIB, specifically for measurements with Android smartphones, which highlighted the changes needed given the low quality of measurements on these platforms.

1.2. Paper Focus and Outline

In the context of the state-of-the-art overview, the major motivation for the research described in this paper was to conduct field experiments and analyse the performance of GNSS devices of different quality operating simultaneously. The goal was to gain knowledge to characterise the behaviour of each device during kinematic positioning and to gain information to improve trajectory determination by GNSS and INS in post-processing. The main objective of the research was to compare the positions in terms of the reference positions based on double difference processing and to determine the positioning quality, especially under difficult signal reception conditions. The focus of the research, meanwhile, was to identify and investigate anomalies in smartphone Xiaomi Mi8 and u-blox receivers.

The main research question of this study was: 'Is there a significant difference in performance between low-cost, i.e., u-blox receivers and smartphones, and professional GNSS receivers in kinematic vehicle positioning?' More specifically, this study focuses on three different questions related to the following:

- Determining the quality of georeferencing in purely kinematic GNSS mode;
- Distinguishing between quality of observations and kinematic results from several GNSS devices under consideration; and
- Evaluating the performance of GNSS devices compared to the mobile mapping system (MMS).

The innovation of this study lies in the reference trajectory being generated using a professional MMS. As far as the authors are aware, there are no examples of kinematic comparisons of smartphone and u-blox positioning performance against positioning systems using the acquisition of an accurate MMS as a benchmark. The remainder of this document is organised as follows: Section 2 describes the experimental setup in the study area with the locations where the measurements were performed, including a description of the hardware used. The processing aspects that were followed in performing this analysis are described in Section 3. The experimental analyses are described in Section 4 and the final conclusions are drawn in Section 5.

2. Materials and Methods

2.1. Survey Experimental Design

For this study, a vehicle from the GeoSNav laboratory of the University of Trieste was used, which was already equipped with a POS/LV (Position and Orientation System/Land Vehicles) Mobile Mapping System from Applanix Corporation, namely Applanix POS LV-420, Version 4 (Applanix, Richmond Hill, ON, Canada). For the purposes of this experiment, the vehicle was additionally equipped with two u-blox devices, namely ZED-F9P and ZED-F9R, an Xiaomi Mi8 smartphone and a Septentrio AsteRx-U geodetic receiver (Septentrio, Leuven, Belgium) with a PolaNt-x MF antenna (Figure 1).

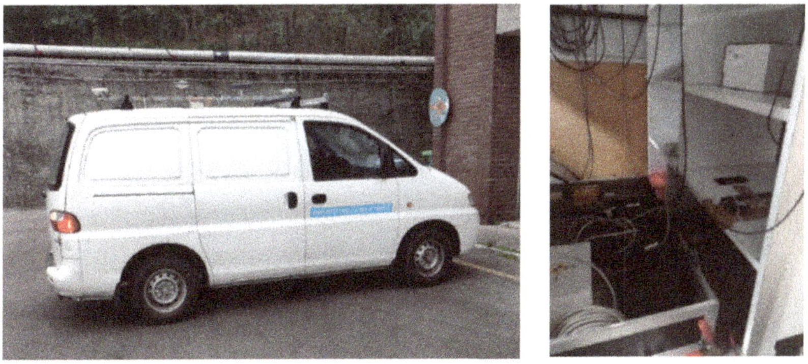

(**a**) (**b**)

Figure 1. (**a**) The MMS of the GeoSNav Lab vehicle, University of Trieste, with the GNSS antennas mounted on the roof; (**b**) the setup of the instruments inside the vehicle.

2.2. *MMS POS/LV and GNSS Receivers Used in the Study*

The antennas of the receivers were located at various positions on the iron shaft mounted on the roof and axial to the vehicle, thus making it easy to measure the horizontal distances between them. They served as a reference for the evaluation positioning results (Figures 2 and 3).

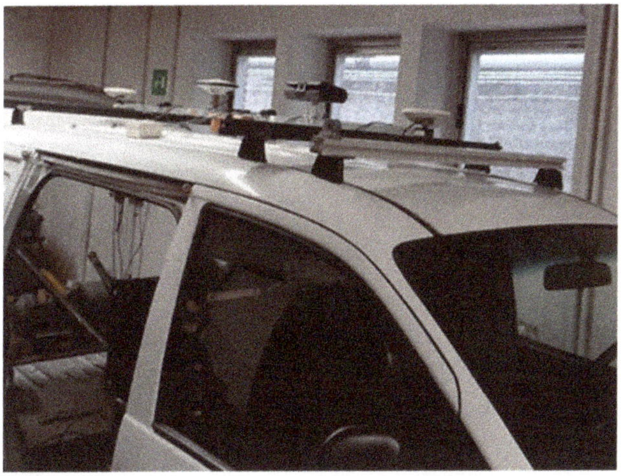

Figure 2. The setup of GNSS antennas, u-bloxes and smartphone on the roof of the vehicle.

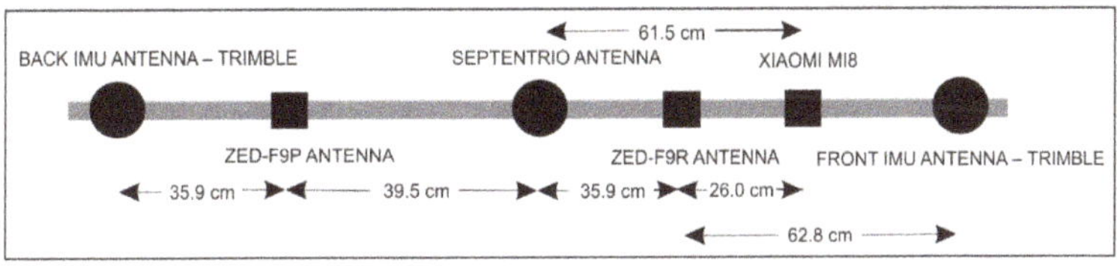

Figure 3. The setup of the Trimble antennas of the MMS (at the front and rear of the vehicle), Septentrio PolaNt-x MF multi-frequency GNSS antenna, Xiaomi Mi8 smartphone and low-cost antennas ANN-MB-00 from ZED-F9P and ZED-F9R with the horizontal distances.

As shown in Figure 2, the central element of the Applanix system, the PCS (POS Computer System), which can process data from various sensors for real-time positioning and stores the data for post-processing analyses (as in the case in the current study), is in the vehicle. The integrated inertial system is a Litton LN-200 Inertial Measurements Unit (IMU) with three accelerometers and three solid-state fibre-optic gyros (Northrop Grumman, Falls Church, VA, USA). Two Trimble Zephyr GPS antennas are mounted on the car roof and connected to two different BD950 GPS cards in the PCS of the Applanix system. The rear antenna is GPS L1/L2 and is used for positioning; the front is GPS L1 and is used for attitude determination. Trimble's geodetic receivers send positioning and heading data to the PCS, the latter using GAMS (GPS Azimuth Measurement Subsystem). By using a carrier-phase double-difference algorithm to measure the relative position vector between the two antennas, GAMS calibrates IMU and ensures that the azimuth does not drift (when GPS/GNSS coverage exists). During the survey, data from the GPS receivers had an acquisition rate of 1 Hz, while the inertial system sent data to the CPU system at a rate of 200 Hz. The integrated measurements from each sensor were processed using the Kalman filter.

In addition to the Applanix professional system, two SimpleRTK2B boards (Ardusimple, Lleida, Spain) were used for this research, one with a ZED-F9P receiver and the other with a ZED-F9R. The boards were connected to a laptop to provide power to both panels and to store the kinematic observations. The reason for storing the observations directly on the computer with the installed open-source programme from U-Center [44] rather than on the SD card was that based on previous experiments, the memory unit was often overloaded when performing measurements in multi-constellations, thus resulting in gaps in the measurements. The two boards and the laptop were inside the vehicle next to the Applanix system components. Two low-cost antennas ANN-MB-00 were mounted on the roof of the vehicle on a special iron structure already used for GPS Trimble antennas and the Septentrio PolaNt-x antenna (Figure 3).

Another device, the Xiaomi Mi8 smartphone, was mounted parallel to the roof of the vehicle (Figure 1) in the direction of travel. The smart device was in a specially made plastic box so that it was safe, protected and fixed during the measurements. The duty cycle mode was turned off in the smartphone settings because it determines the time the GNSS chip is actively tracking signals and can affect the quality of the GNSS data. This is because the longer the duty cycle, the better the quality of the GNSS observation.

2.3. MMS POS/LV and GNSS Receivers Used in the Study

The kinematic tests were conducted on 11 February 2022 in the city centre of Trieste, Italy, and on the outskirts of the nearby region. The measurements in the city were performed only a few metres above sea level, while for the non-urban area, the measurements were taken on a Karst plateau at an altitude of about 350 metres (Figure 4). To further differentiate the measurement conditions and the behaviour of the different receivers under these conditions, the overall trajectory was divided into different cases: tunnel (Case A), city (Case B) with tall buildings representing a typical urban canyon area, a serpentine road (Case C), and a GNSS vegetation-barrier area in the Karst plateau above Trieste (Case D). The results were compared separately for each of these four areas. To obtain a better idea of the measurement conditions, some characteristic points from the trajectory of the vehicle are shown in Figure 5.

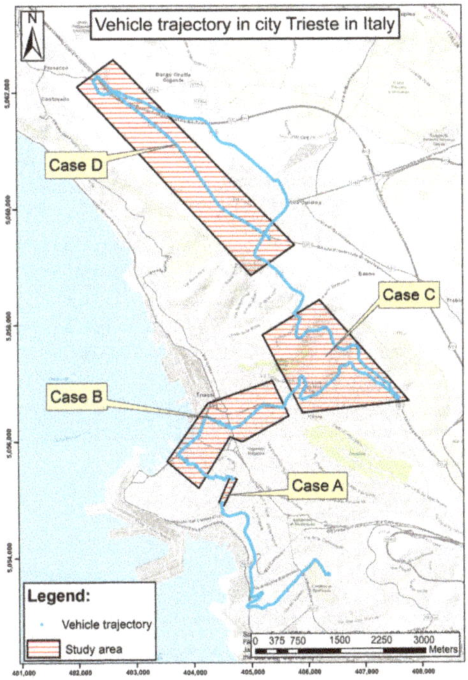

Figure 4. The surveyed area, the city of Trieste and nearby region, Italy.

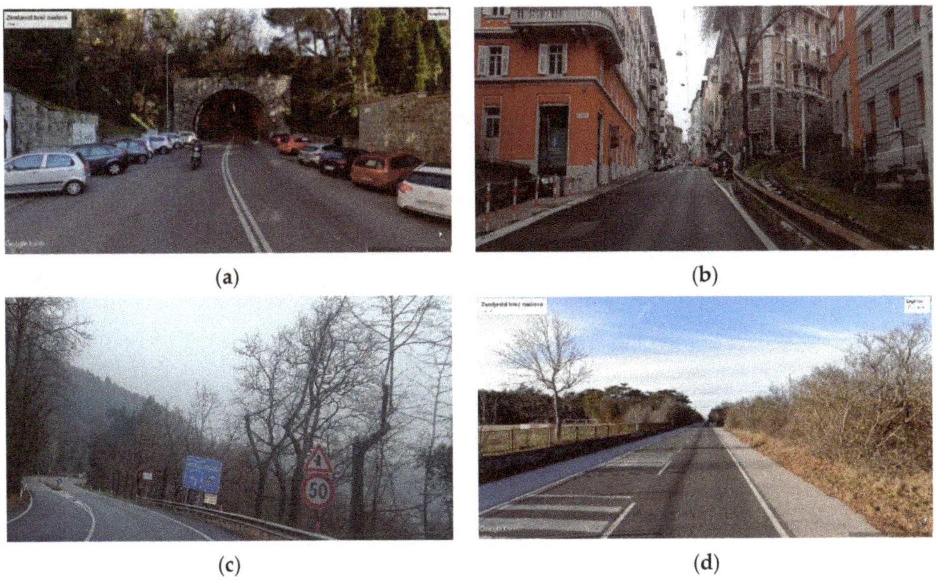

Figure 5. (**a**) Case A: the entrance into the tunnel (45.642059°N, 13.7744820°E); (**b**) Case B: urban canyons in the city centre (45.655748°N, 13.771247°E); (**c**) Case C: serpentines (45.657905°N, 13.813880°E); (**d**) Case D: straight road with vegetation as only barrier (45.690686°N, 13.769906°E).

3. Data Processing

The data processing chain implemented in this paper can be described as follows:

- Analysis of data quality from different receivers operating in GNSS mode;
- Reference trajectory computation from MMS measurements;
- Trajectory computation from low-cost devices, namely u-blox ZED-F9P, ZED-F9R, and Xiaomi Mi8, and Septentrio AsteRx-U professional geodetic receiver; and
- Evaluation of the performances of low-cost GNSS/INS devices compared to the professional MMS system and the Septentrio AsteRx-U geodetic receiver.

The trajectories from all the receivers, including MMS, were computed by the post-processing relative double-difference carrier-phase technique. For the MMS, the Position and Orientation System Post-Processing Package (POSPac™) Mobile Mapping Suite (MMS™) was used [45], which provides a deep level of sensor integration and error modelling and processing modes. For relative double difference positioning, the nearby Trieste reference station (the distance from the study area was around 11 km) belonging to the Antonio Marussi Friuli Venezia Giulia Region GNSS network was used [46]. Observations for geodetic and low-cost receivers, namely u-bloxes and Xiaomi Mi8, were processed using RTKLIB software, demo 5, version b34d [47,48].

RTKLIB is an open-source program package for standard and precise positioning and consists of a portable program library and several APs (application programs). It supports: (1) standard and precise positioning algorithms with GPS, GLONASS, Galileo, QZSS, BeiDou, and SBAS; (2) single, DGPS/DGNSS, kinematic, static, PPP-kinematic and PPP-static positioning modes with GNSS for both real-time and post-processing. A number of features may contribute to its superior performance compared to the software, namely: (a) it is free, while commercial software comes with a cost that can be significant for some users; (b) flexibility in adjusting the processing parameters so that better results can be achieved; (c) possibilities to improve the current version of RTKLIB also by changing some algorithms.

Since the MMS relay on GPS-only constellations, the initial processing strategy for all the used receivers was based only on GPS; successively, a multi-constellation solution using GPS, GLONASS, and Galileo was applied. For static mode ambiguity resolution, the "fix-and-hold" mode was used, which allows a GNSS receiver to maintain positioning even if the number of visible satellites decreases or the measurement quality degrades. In contrast, the "continuous" ambiguity mode was used for kinematic positioning because in this mode, ambiguities are resolved in real time and solutions are continuously updated as new observations become available. Since there were many obstacles along the roads, the lower obstacles were bypassed at the cut-off angle at 15°.

Observations were processed using broadcast ephemerides, iono-free mode, and Saastamoinen's tropospheric delay model. Unfortunately, there are no available antenna calibrations for the patch antennas of the u-blox receivers ZED-F9P and ZED-F9R and for the Xiaomi smartphones, as there were for the Septentrio AsteRx-U. However, in this research, an approximated antenna phase centre offset was calculated based on the procedure proposed by Netthonglang et al. (2019) [49], which is based on averaging post-processed coordinates in northing and easting (reference was added). This study was performed prior to this particular experiment in the static mode. All additional processing settings used are listed in Table 1.

To investigate the benefits of multi-constellation, observations from the Septentrio AsteRx-U, the ZED-F9P and ZED-F9R u-bloxes and the Xiaomi Mi8 were also processed for GPS, GLONASS and Galileo, but without BeiDou, since the Marussi Network reference station, configured to collect observations from only three constellations, was used for relative positioning. Finally, the provided geographic coordinates were projected into the reference system UTM ETRS.

Table 1. Processing parameters used in RTKLIB.

Parameters	RTKLIB
Constellations	GPS/GPS + GLONASS + Galileo
Observations	code and carrier phase (L1/E1 + L2/E5b + L5/E5a)
Ambiguity	fix-and-hold/continuous
Ephemeris	broadcast
Elevation angle	15°
Ionospheric delay	iono-free (LC)
Tropospheric delay	Saastamoinen

3.1. *Quality of GNSS Observations*

To compare the performance of the Xiaomi Mi8, the ZED-F9P u-blox, the ZED-F9R u-blox and the Septentrio AsteRx-U receiver, several pre-analyses were performed, namely:

- Satellite visibility during measurements;
- Multipath; and
- Cycle slips for the received signals.

Figure 6 shows the available number of GNSS satellites and the metric quantification of the multipath effect on the L1 frequency of the total trajectory for the Septentrio AsteRx-U geodetic receiver. The figures illustrate the areas with worse measurement conditions, which refer to urban street canyons and areas with lush roadside vegetation, as shown in Figure 6.

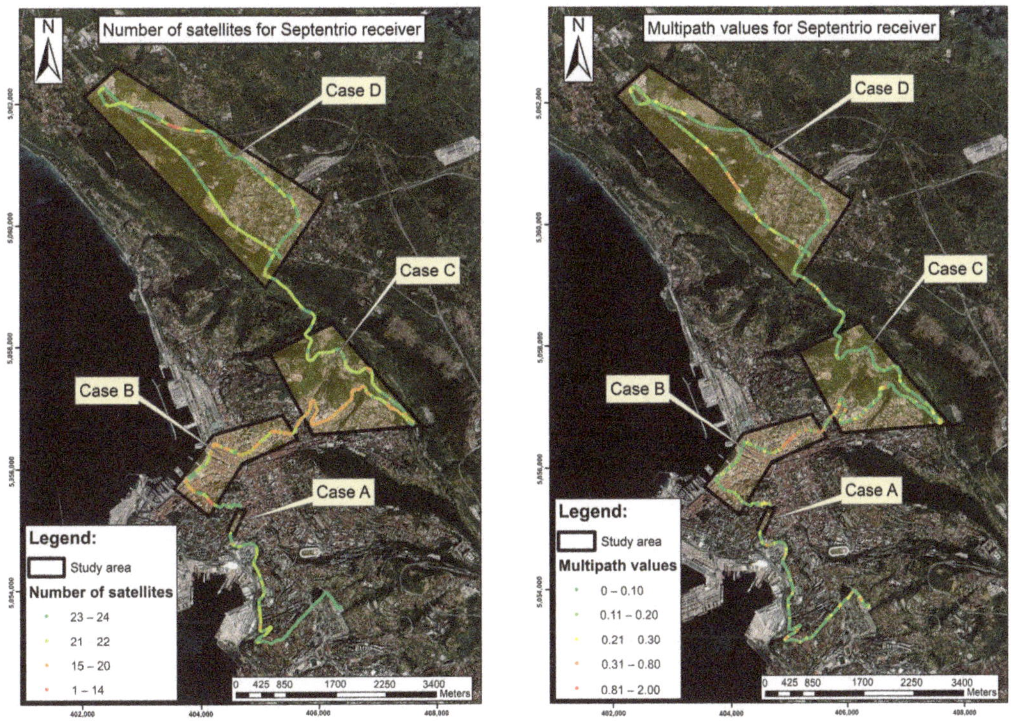

(**a**) (**b**)

Figure 6. (**a**) Number of available satellites along the route, and (**b**) multipath in metres for the Sep tentrio AsteRx-U receiver.

Initially, the multipath was estimated; however, furthermore in RTKLIB, a module was used to resolve LLI by using several algorithms, including integer least squares and LAMBDA (Least-Squares Ambiguity Decorrelation). For static sessions, the "fix-and-hold-algorithm was used", while for kinematic, the "continuous mode" was used (Table 1).

As shown in Table 2, a smartphone was capable of acquiring observations only at the L1 and L5 frequencies, while the u-bloxes were capable only at the L1 and L2 frequencies. From the results, it can be seen that the Septentrio AsteRx-U geodetic receiver, which is equipped with an active geodetic GNSS antenna with built-in low-noise amplifiers that amplify the signals before further processing, and which has a higher gain, performed much better under difficult conditions. The root mean square error (RMSE) for the multipath along the entire track is 0.3554 m, 0.4320 m, and 0.3313 for frequencies L1, L2, and L5, respectively. The smartphone's multipath estimate is much higher on the L1 frequency band with respect to the Septentrio and two of the u-bloxes, but it reaches the u-bloxes' results from L1 and L2 on the L5 frequency band. It is also evident that the L5's Septentrio observations were much less affected by the multipath. The results confirm the fact that L5 signals are less susceptible to multipath because they rely on a higher frequency and wider bandwidth compared to other GNSS signals.

Table 2. Statistics for GNSS multipath on different frequencies for the receivers for the whole track.

Parameters	Type of GNSS Device			
	Septentrio AsteRx-U	**u-blox ZED-F9P**	**u-blox ZED-F9R**	**Smartphone Xiaomi Mi8**
RMSE for multipath L1 (m)	0.3554	0.5129	0.4848	0.7394
RMSE for multipath L2 (m)	0.4320	0.5292	0.5581	-
RMSE for multipath L5 (m)	0.3313	-	-	0.5515

The advantages of the Septentrio AsteRx-U receiver's active antenna are also evident in Figure 7, which shows the carrier-to-noise ratio (C/N0) in colours and the loss-of-lock indicator (LLI) phenomenon in red lines. LLI is an indicator that shows when the receiver has lost the link or signal to a satellite, resulting in a cycle slip event (Figure 7). This information is particularly useful for GNSS post-processing to improve position accuracy and repeatability. As can be seen in Figure 7, LLIs occur for all receivers during kinematic positioning (from 13:55 to 14:45), but they are less frequent for the Septentrio AsteRx-U antenna. LLIs are also common and quite frequent with the u-bloxes and the Xiaomi Mi8 smartphone in static mode. Overall, the measurements with the Septentrio AsteRx-U were the least affected by LLIs even in kinematic measurements; however, interestingly, the problem was greatest with the ZED-F9R u-blox. Moreover, it can be confirmed that the signal quality of the overall measurements with the Xiaomi Mi8 receiver is significantly worse compared to the u-bloxes, as previously demonstrated by other authors [50], albeit with different smartphones.

LLI indicators show a temporary loss of signal coherence, which could be caused by interference, non-line-of sight conditions, or multipath, and they often produce cycle slips in the receiver. Cycle slips have a significant negative impact on GNSS positioning, but fortunately, they can be recovered during post-processing. The method used by RTKLIB compares the time series of carrier phase measurements from a GNSS receiver with the predicted values calculated from code and carrier phase measurements. The occurrence of a cycle slip is evident by a sudden change in the carrier phase observation, which does not match the predicted value. The software uses the carrier phase residuals or an advanced Kalman filter—the method used in our case—to estimate the carrier phase bias. When the problem is detected, the software adjusts the carrier phase prediction to account for the cycle deviation, thus improving the GNSS position accuracy.

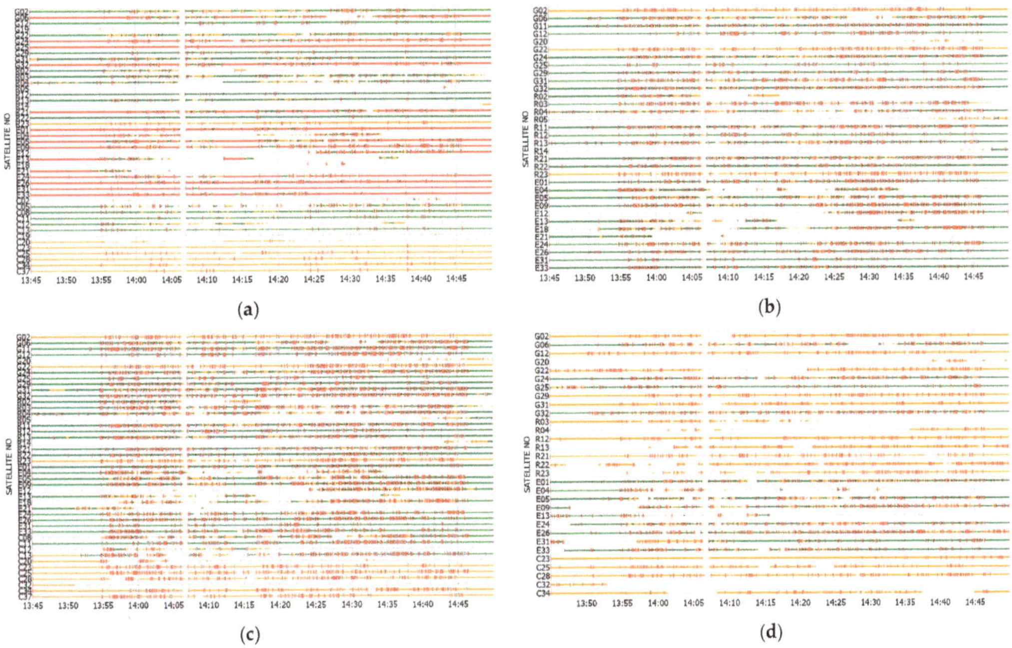

Figure 7. Satellite visibility, cycle slips (vertical red lines), and carrier-to-noise ratio (C/N0) for (**a**) Septentrio AsteRx-U receiver, (**b**) ZED-F9P u-blox receiver, (**c**) ZED-F9R u-blox receiver, and (**d**) Xiaomi Mi8. C/N0 values are shown in a colour scale from green (45 dB-Hz), orange, purple, blue, red (25 dB-Hz) to grey (<25 dB-Hz).

3.2. Quality of GNSS Positioning

The goal of processing the observations was to test the quality of the positions over time, especially in challenging conditions for GNSS. In the following analyses, all the comparisons were made under the hypothesis that the MMS solution is more accurate compared to the other receivers. Since the Applanix system only allows the computation of trajectories in the GPS constellation, all the comparisons were initially performed only in the context of the trajectories generated by GPS; then, the multi-constellation results for Septentrio, u-bloxes, and the smartphone were computed. It could be said that the reference trajectory from Applanix was at an advantage due to operating in GPS/INS mode, while the trajectories from the other receivers were at an advantage due to operating only in GNSS mode. The GPS and GNSS solutions were analysed for the trajectory under consideration, with the percentage determined by the number of total solutions obtained from the MMS. Similar comparisons were then made for the individual cases, i.e., cases A–D, as shown in Figure 4.

The reference coordinates (e_0, n_0, h_0) at each epoch t were obtained from the post-processing results of the MMS and computed for the position of the Septentrio AsteRx-U, ZED-F9P, ZED-F9R, and Xiaomi Mi8 smartphone. Since in GNSS positioning, the vertical accuracy can be significantly lower than the horizontal accuracy, especially under difficult conditions such as urban canyons or dense vegetation, the spatial distances d_{3D} were estimated separately in further analyses. In each epoch t, they were computed between the considered receivers $(e_{rec}, n_{rec}, h_{rec})$ and the reference receiver (e_0, n_0, h_0), namely:

$$\begin{aligned} d_{2D}(t) &= \sqrt{(e_0(t) - e_{rec}(t))^2 + (n_0(t) - n_{rec}(t))^2} \\ d_{3D}(t) &= \sqrt{(d_{2D}(t))^2 + (h_0(t) - h_{rec}(t))^2} \end{aligned} \tag{1}$$

In the quality analysis, the first criterion for the quality of the positioning was the ability to resolve the carrier phase ambiguities, while the second criterion was based on the comparison of the calculated distance ($d_{2D,computed}(t)$ and $d_{3D,computed}(t)$) between the Applanix and the specific receiver at the specific epoch (Equation (1)) with the measured horizontal distance between the Applanix and each of the receiver d_0 (see Figure 3). The measured height difference should be zero, since all instruments were placed on the same levelled car mount. The calculation of the deviation of the distances (horizontal and spatial) for each specific epoch t was made separately based on the following equation:

$$dev\_d(t) = d_{computed}(t) - d_0 \tag{2}$$

where d stands for either d_{2D} or d_{3D}. In addition, since many positions turned out to be outliers, the deviations were analysed for the given threshold. Deviations greater than 1 m, 30 cm, and 10 cm were identified as outliers and removed from the analysis.

3.3. Statistical Testing

To test the equality of variances between two populations, namely GPS and GNSS solutions for each of the receivers, the Fisher–Snedecor test (test F) was used. The null and alternative hypotheses were as follows:

$$H_0 : \sigma^2_{GPS} = \sigma^2_{GNSS}, H_1 : \sigma^2_{GPS} \neq \sigma^2_{GNSS} \tag{3}$$

based on the assumption that for each test, there were two independent random samples of the deviations in horizontal distances (the same procedure was used for the spatial distances) and the reference ones from Equation (2), namely $dev\_d_{GPS_i}$, and $i = 1, \ldots, n$, and $dev\_d_{GNSS_i}$, $dev\_d_{GNSS_i}$, and $i = 1, \ldots, m$, drawn from two normal distributions, $N(\mu_{GPS}, \sigma_{GPS})$ and $N(\mu_{GNSS}, \sigma_{GNSS})$. To perform the test, a random sample of n should be obtained from one population and a sample of m should be obtained from the second population. The means and variances of the sample followed the equation for GPS or GNSS:

$$\overline{dev\_d} = \frac{1}{n}\sum_{i=1}^{n} dev\_d_i,\ s^2_{dev\_d} = \frac{\sum_{i=1}^{n}\left(dev\_d_i - \overline{dev\_d}\right)^2}{n-1}, \tag{4}$$

while the statistics F for different receivers were based on the equation:

$$F = \frac{s^2_{dev\_d_{GPS}}}{s^2_{dev\_d_{GNSS}}} \tag{5}$$

followed F-distribution with $n-1$ and $m-1$ degrees of freedom. If the calculated F-score is close to one, the null hypothesis that the samples come from the same populations could be accepted; the more the ratio from Equation (5) deviates from 1, the stronger the evidence for unequal population variances exists. If the F-score is greater than the defined $F_{\alpha/2,n-1,m-1}$ or less than $F_{1-\alpha/2,n-1,m-1}$ for the significance level α, that leads to the conclusion that the null hypothesis (Equation (3)) could be rejected, and it could be stated that the variances from the two different solutions were not the same in both groups.

4. Results and Discussion

4.1. Entire Trajectory: Static and Kinematic Sessions

This scenario consisted of a static session (10 min) at the beginning of the survey, a kinematic session that lasted approximately 50 min, and a final 10 min static session after calibration of the Applanix system. The comparison of the solutions of GPS and GNSS (Tables 3 and 4, Figure 8) shows that in the case of GNSS, many more positions could be obtained on the whole trajectory when the observations were processed in the multi-constellation mode. For the Xiaomi Mi8 phone, almost all ambiguities were determined as floats, although the range of deviations in distances was in the same range as for the u-bloxes, albeit it is still slightly worse.

Table 3. GPS solutions, solutions with fixed ambiguities, and deviation of distances below the given threshold. Situation for the entire trajectory: total solutions 5148).

Parameters	Septentrio AsteRx-U	u-blox ZED-F9P	u-blox ZED-F9R	Smartphone Xiaomi Mi8
	Type of GNSS Device			
GPS solutions	3628 (70.5%)	3174 (61.7%)	3125 (60.1%)	3286 (63.8%)
Fixed ambiguities	2320 (64.0%)	644 (20.3%)	1204 (38.5%)	33 (1.0%)
dev_d_{2D} < 1 m	3168 (61.5%)	2869 (55.7%)	2832 (55.0%)	2287 (44.4%)
dev_d_{2D} < 30 cm	2907 (56.5%)	1344 (26.1%)	1537 (29.9%)	912 (17.7%)
dev_d_{2D} < 10 cm	2443 (47.5%)	906 (17.6%)	1146 (22.3%)	31 (0.6%)
dev_d_{3D} < 1 m	2889 (56.1%)	2611 (50.7%)	2428 (47.2%)	1651 (32.1%)
dev_d_{3D} < 30 cm	2712 (53.7%)	1294 (25.1%)	1085 (21.1%)	473 (9.2%)
dev_d_{3D} < 10 cm	2300 (44.7%)	130 (2.5%)	849 (16.5%)	0 (0%)

Table 4. GNSS solutions, solutions with fixed ambiguities, and deviation of distances below the given threshold. Situation: entire trajectory (static and kinematic: total solutions 5148).

Parameters	Septentrio AsteRx-U	u-blox ZED-F9P	u-blox ZED-F9R	Smartphone Xiaomi Mi8
	Type of GNSS Device			
GNSS solutions	4171 (81.0%)	3831 (74.4%)	3764 (73.1%)	3951 (76.8%)
Fixed ambiguities	2176 (52.2%)	1995 (52.1%)	2095 (55.7%)	30 (0.8%)
dev_d_{2D} < 1 m	3369 (65.4%)	3279 (63.7%)	3264 (63.4%)	2177 (42.3%)
dev_d_{2D} < 30 cm	2940 (57.1%)	1798 (34.9%)	1963 (38.1%)	719 (14.0%)
dev_d_{2D} < 10 cm	2420 (47.0%)	1032 (20.0%)	1370 (26.6%)	243 (4.7%)
dev_d_{3D} < 1 m	3123 (60.7%)	2215 (60.7%)	2814 (54.7%)	1326 (25.8%)
dev_d_{3D} < 30 cm	2832 (55.0%)	1702 (33.1%)	1318 (25.6%)	261 (5.1%)
dev_d_{3D} < 10 cm	2233 (43.4%)	335 (6.5%)	943 (18.3%)	0 (0.0%)

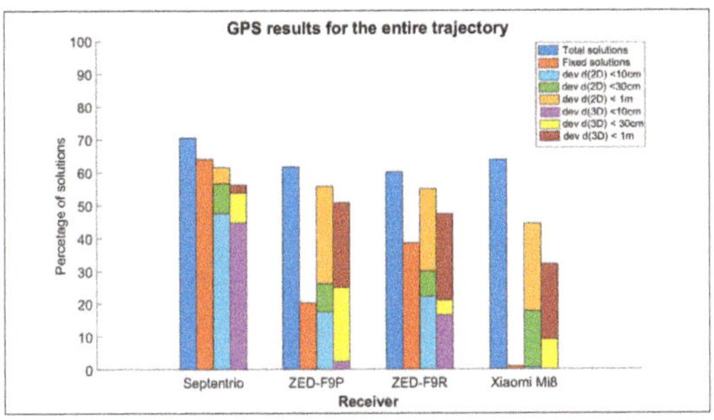

(a)

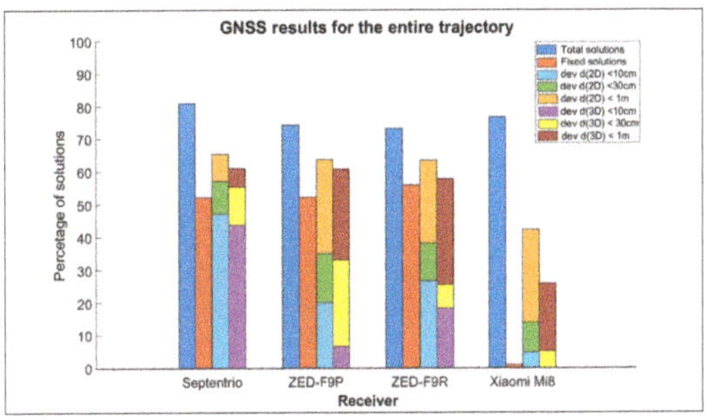

(b)

Figure 8. Entire trajectory (static and kinematic): (**a**) GPS solutions; (**b**) GNSS solutions. Each bar in a group shows: available solutions (1st bar), percentage of fixed solutions (2nd bar), deviation at horizontal distances (3rd bar, where light blue is for thresholds less than 10 cm, green is for thresholds less than 30 cm, and brown is for thresholds less than 1 m), and 4th bar shows deviation at spatial distances (yellow, green, and dark red for 10 cm, 30 cm, and 1 m thresholds, respectively).

From Figure 8, it can be seen that the Septentrio AsteRx-U receiver outperformed all the other receivers both in terms of the number of total solutions and the number of fixed solutions. From GPS, 62% of all the available solutions were those where the deviations of the calculated distances from the (measured) reference distances were less than 1 m, and they were 57% and 48% where the deviations were less than 30 cm and 10 cm, respectively. The results are slightly better for the GNSS solutions.

From the overall trajectory statistics, it can be seen that the Xiaomi Mi8 allowed more usable solutions than the u-blox receivers; however, in 99% of cases, the ambiguities were detected as floats. For the phone, 44% of the possible GPS solutions were those where the deviation of the distances from the reference positions was less than 1 m, while 18% and 1% of the total positions on the trajectory were those where the deviations were less than 30 cm and 10 cm, respectively. For the GNSS solutions, only 14% and 5% of the GNSS solutions were those where the deviations of the distances were less than 30 cm and 10 cm, respectively. The results from the u-bloxes are interesting, wherein ZED-F9R seemed to perform better than ZED-F9P both in the number of fixed solutions and in the quality of positioning.

Results with Deviation in Distances and Heights below 30 cm

The statistics for the filtered deviation of distances (Tables 5 and 6) that were less than 30 cm show that the Septentrio AsteRx-U receiver performed better than others in the number of available solutions (47.5% for the GPS and 57.1% for the GNSS), with deviations in distances calculated with a mean of −5.8 cm ± 6.5 cm for the GPS solutions (Table 5) and −6.0 cm ± 8.4 cm for the GNSS solutions (Table 6). One of the possible reasons for the larger standard deviations in the GNSS solutions could be the fact that Applanix worked in GPS/INS mode; therefore, the results from the GPS are more appropriate for comparisons. Nevertheless, it is noticeable that despite the percentage of GNSS being higher (81.0%) in terms of the GPS-only solutions (70.5%), the number of solutions calculated with fixed ambiguities is correspondingly lower (2176 compared to 2320), which is probably due to the possible insufficient performance of one of the systems used in the multi-constellation solution.

Table 5. Statistics (mean, standard deviation, and RMSE) for GPS solutions for filtered values of deviation in distances that were below 30 cm. Situation: entire trajectory (static and kinematic: 5148 total epochs).

Parameters	Type of GNSS Device			
	Septentrio AsteRx-U	**u-blox ZED-F9P**	**u-blox ZED-F9R**	**Smartphone Xiaomi Mi8**
GPS solutions	2443 (47.5%)	1344 (26.1%)	1537 (29.9%)	912 (17.8%)
$\overline{dev\_d_{2D}}$ (m)	−0.058	−0.095	0.026	0.207
s (m)	0.065	0.097	0.101	0.061
RMSE (m)	0.087	0.135	0.105	0.215
$\overline{dev\_d_{3D}}$ (m)	−0.016	−0.165	0.058	0.251
s (m)	0.087	0.100	0.082	0.027
RMSE (m)	0.089	0.193	0.100	0.252

It is also even clearer from the graphical representation of the statistics for the deviation of distances of less than 30 cm for all receivers that the Septentrio AsteRx-U receiver was much better at determining the correct position. For the Xiaomi Mi8, the median of the receiver deviates the most compared to the others. The overall scatter is smallest for the Septentrio and largest for the ZED-F9R; the width between each quartile and the scatter is much smaller for the Septentrio than for the other receivers. The situation is the same for both GPS (Table 5, Figure 9a) and GNSS processing (Table 6, Figure 9b). The interquartile range between the GPS and the GNSS for the ZED-F9R and the Xiaomi Mi8 changes drastically. Since there is no overlap between the boxes in any of the Figure 9 plots, it can be concluded that there is a clear difference between the data sets from different receivers. An *F*-test was used to detect differences in the variances for the GPS and GNSS solutions. The results are shown in Table 7.

Table 6. Statistics (mean, standard deviation, and RMSE) for GNSS solutions for filtered values of deviation in distances that were below 30cm. Situation: entire trajectory (static and kinematic: 5148 total epochs).

Parameters	Type of GNSS Device			
	Septentrio AsteRx-U	**u-blox ZED-F9P**	**u-blox ZED-F9R**	**Smartphone Xiaomi Mi8**
GNSS solutions	2940 (57.1%)	1798 (34.9%)	1963 (38.1%)	719 (14.0%)
$\overline{dev\_d_{2D}}$ (m)	−0.060	−0.100	0.047	0.159
s (m)	0.084	0.095	0.101	0.181
RMSE (m)	0.100	0.138	0.111	0.204
$\overline{dev\_d_{3D}}$ (m)	−0.025	−0.112	0.066	0.229
s (m)	0.098	0.140	0.083	0.034
RMSE (m)	0.101	0.179	0.106	0.231

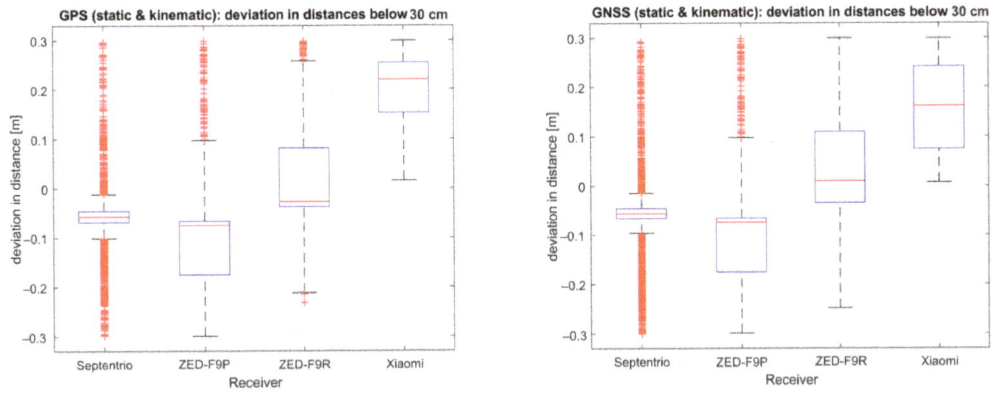

Figure 9. Boxplot for deviation of horizontal distances below 30 cm: (**a**) GPS solution; (**b**) GNSS solution.

Table 7. *F*-test ($\alpha = 5\%$) for deviations in distances of less than 30 cm; comparison between GPS and GNSS solutions for each of the receivers.

Parameters	Type of GNSS Device			
	Septentrio AsteRx-U	**u-blox ZED-F9P**	**u-blox ZED-F9R**	**Smartphone Xiaomi Mi8**
F-test (2D, 3D)	H_0 cannot be rejected	H_0 cannot be rejected	H_0 cannot be rejected	H_0 cannot be rejected

For the comparison of the variances between the two groups for each of the receivers separately, i.e., the first group represents the GPS solutions and the second represents the GNSS solutions, in both cases, the null hypothesis H_0 that means and variances are equal cannot be rejected. However, the number of possible solutions from the GPS processing in terms of the GNSS is much lower (see Tables 5 and 6).

4.2. *Case A—Tunnelling*

For the topographic survey at the tunnel exit (Figure 10), the main goal was to investigate when the receivers can recover their position after leaving the tunnel and how much they deviate from the reference position. This information can be found in Table 7.

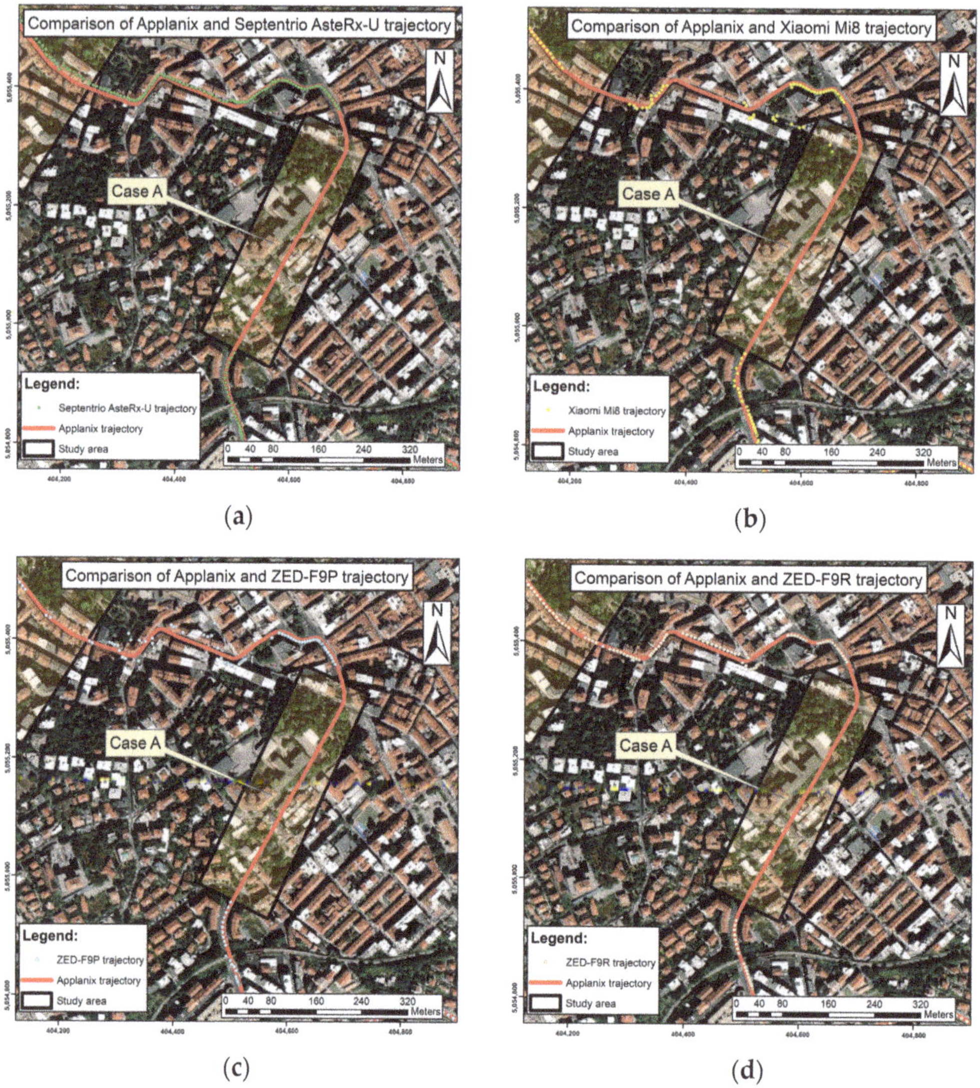

Figure 10. Case A—tunnel: performance of (**a**) Septentrio AsteRx-U, (**b**) Xiaomi Mi8, (**c**) ZED-F9P, and (**d**) ZED-F9R.

After exiting the tunnel, the Xiaomi Mi8 was the last device to recover its position (d_{along}) at about 57 m, while the other devices were able to recover their position slightly earlier—after 30 m in the case of the ZED-F9P (Table 8). The performance of the Xiaomi Mi8 and the ZED-F9P in the segment after leaving the tunnel was the worst among all the receivers tested (see Figure 10b). Interestingly, the Septentrio and ZED-F9R receivers recovered their position much better than the others, which deviated up to 2 m ($d_{cross-track}$)

from the Applanix trajectory (Figure 10). However, these are the results for this case; therefore, the findings cannot be generalised to other situations at this stage.

Table 8. Time of reacquisition of position (t), distance along track from tunnel exit (d_{along}), and deviations of first point (from tunnel) from MMS trajectory ($d_{cross-track}$): Case A—Tunnel.

Parameter	Type of GNSS Device			
	Septentrio AsteRx-U	**u-blox ZED-F9P**	**u-blox ZED-F9R**	**Smartphone Xiaomi Mi8**
t	5 s	4 s	6 s	9 s
$d_{2D,along}$	34.21 m	30.22 m	40.02 m	57.07 m
$d_{2D,cross-track}$	0.79 m	2.16 m	0.56 m	2.08 m

4.3. Case B—Urban Canyons

In urban areas, there should be a total of 700 solutions; however, due to the obstacles, far less appropriate solutions could be obtained (Tables 9 and 10, Figure 11).

Table 9. GPS solutions, solutions with fixed ambiguities, and deviation of distances below the given threshold. Situation: Case B—urban area (no. of total solutions: 700).

Parameters	Type of GNSS Device			
	Septentrio AsteRx-U	**u-blox ZED-F9P**	**u-blox ZED-F9R**	**Smartphone Xiaomi Mi8**
GPS solutions	525 (74.9%)	434 (61.9%)	460 (65.6%)	474 (67.6%)
Fixed ambiguities	253 (48.3%)	1 (0.2%)	6 (1.3%)	5 (1.1%)
dev_d_{2D} < 1 m	326 (46.6%)	292 (41.7%)	318 (45.4%)	61 (5.7%)
dev_d_{2D} < 30 cm	324 (46.3%)	157 (22.4%)	29 (4.1%)	27 (3.9%)
dev_d_{2D} < 10 cm	243 (34.7%)	2 (0.3%)	14 (2.0%)	2 (0.3%)
dev_d_{3D} < 1 m	287 (41.0%)	197 (28.1%)	315 (45.0%)	0 (0%)
dev_d_{3D} < 30 cm	260 (37.1%)	58 (8.3%)	27 (3.9%)	0 (0%)
dev_d_{3D} < 10 cm	49 (7.0%)	1 (0.1%)	13 (1.9%)	0 (0%)

Table 10. GNSS solutions, solutions with fixed ambiguities, and deviation of distances below the given threshold. Situation: Case B—urban area (no. of total solutions: 700).

Parameters	Type of GNSS Device			
	Septentrio AsteRx-U	**u-blox ZED-F9P**	**u-blox ZED-F9R**	**Smartphone Xiaomi Mi8**
GNSS solutions	700 (100%)	647 (92.3%)	615 (87.7%)	605 (86.3%)
Fixed ambiguities	223 (31.9%)	238 (38.8%)	211 (34.4%)	6 (1.0%)
dev_d_{2D} < 1 m	477 (68.1%)	342 (48.9%)	349 (49.9%)	348 (49.7%)
dev_d_{2D} < 30 cm	376 (53.7%)	319 (45.6%)	324 (46.3%)	160 (22.9%)
dev_d_{2D} < 10 cm	262 (37.4%)	60 (8.6%)	265 (37.9%)	2 (0.3%)
dev_d_{3D} < 1 m	361 (51.6%)	329 (47.0%)	331 (47.3%)	2 (0.3%)
dev_d_{3D} < 30 cm	265 (37.9%)	257 (36.7%)	41 (5.9%)	0 (0%)
dev_d_{3D} < 10 cm	48 (6.9%)	56 (8.0%)	24 (3.4%)	0 (0%)

Figure 11 shows that the Septentrio was able to determine the position for Case B (surveying in urban areas) most of the time, while the other receivers had difficulty determining the position in urban street canyons, especially the Xiaomi Mi8 phone receiver, where some of the positions deviated greatly from the reference positions. Both the u-bloxes and the Xiaomi Mi8 were only able to delineate 1.1% and 1.3% of the possible solutions with fixed ambiguities, respectively. As for the deviation of the distances, 46.3% of the

Septentrio were below 30 cm and 34.7% were below 10 cm, while the others, especially the ZED-F9R and the Xiaomi Mi8, were less successful, with only 4.1% of the positions where the deviations were below 30 cm and small percentages (0.3%, 2.0% and 0.3% for the ZED-F9P, ZED-F9R and Xiaomi Mi8, respectively) where the deviations from the reference positions were below 10 cm.

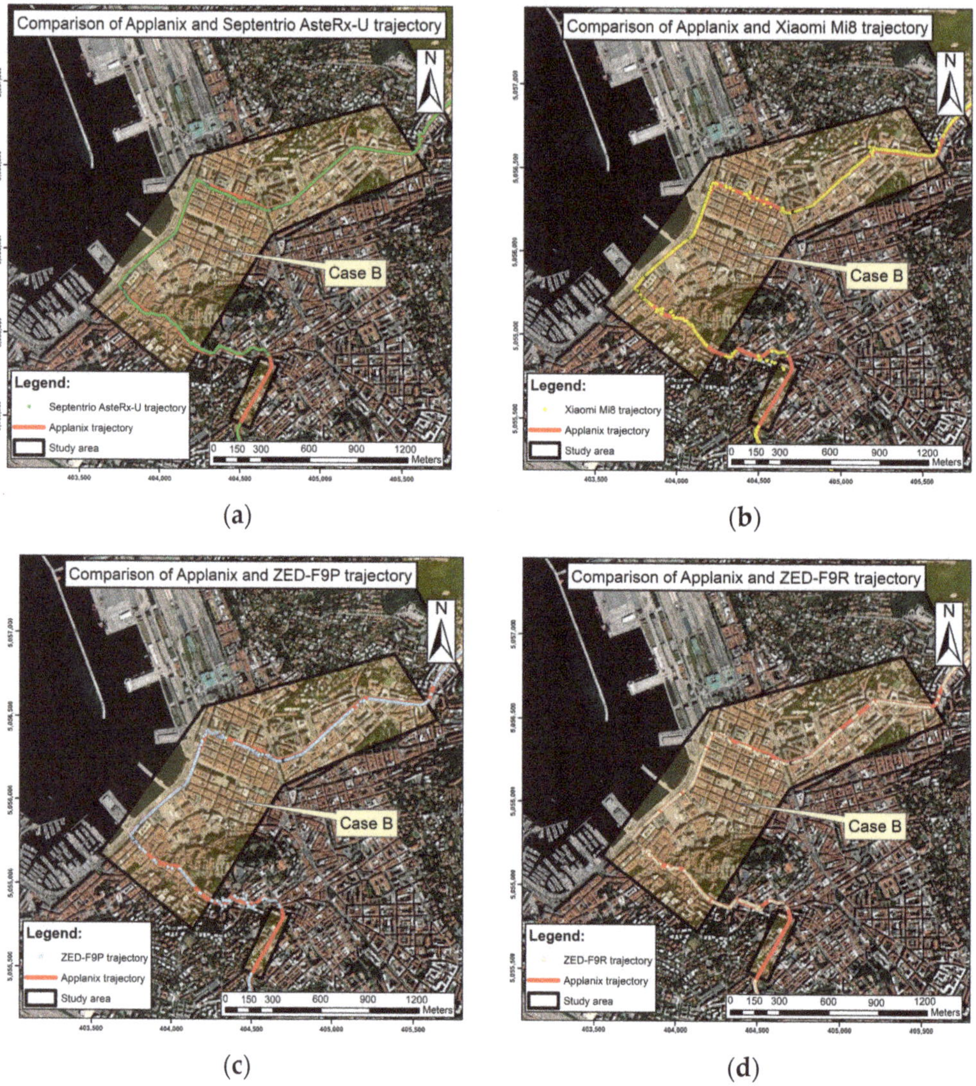

Figure 11. Case B: performance of: (**a**) Septentrio AsteRx-U, (**b**) Xiaomi Mi8, (**c**) u-blox ZED-F9P, and (**d**) u-blox ZED-F9R.

Comparing the positioning of the GPS and GNSS (Tables 9 and 10 and Figure 12), it is clear that a larger number of satellites (e.g., when using the GNSS) leads to a higher number of calculated positions compared to the GPS and often to better quality positioning, as can be seen in particular for the two u-blox receivers and the Xiaomi Mi8 smartphone. For those devices, there is also a very significant improvement in the determination of the fixed ambiguities in the case of the GNSS, while for the Septentrio AsteRx-U receiver, there is a decrease; this mainly leads to an improvement in accuracy for the deviations below 1 m but less for the other thresholds (30 cm and 10 cm). For positioning, it is also seen that much more data can be acquired in GNSS mode than in GPS mode. The same is true for positioning performance. The GNSS solutions show a significant improvement in deviations between the receivers and the Applanix of less than 1 m or 30 cm for all the low-cost receivers tested and 10 cm and an improvement in deviations below 10 cm for the two u-blox devices, especially for the ZED-F9R. From Figure 12a,b, it can be seen that for the entire trajectory the horizontal positioning was clearly better compared to the spatial one. Unfortunately, the positioning of the Xiaomi Mi8 smartphone in urban areas has not proven to be adequate, as few GNSS solutions have managed to be positioned below the threshold of qualitative accuracy; all of them refer to adequate horizontal positioning, while spatial positioning has failed in all epochs for Case B.

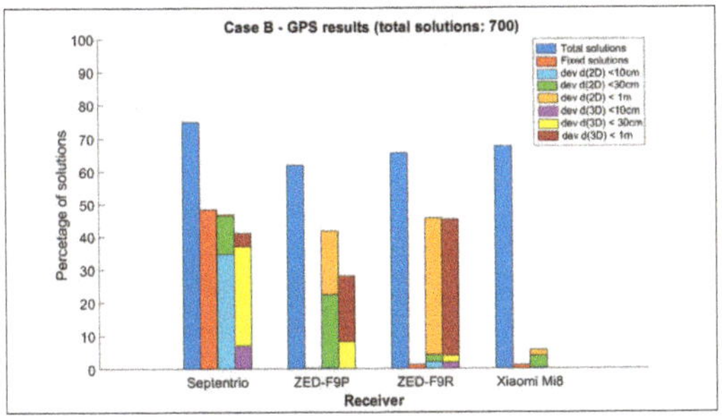

(**a**)

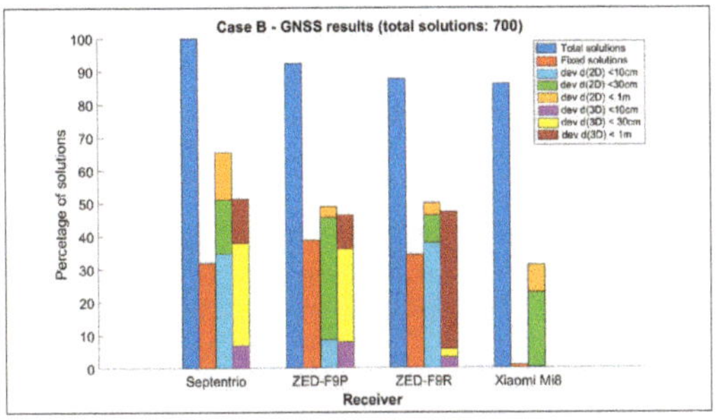

(**b**)

Figure 12. Percentage of solutions for Case B for (**a**) GPS solutions; (**b**) GNSS solutions. Each bar in a group shows: available solutions (1st bar), percentage of fixed solutions (2nd bar), deviation at horizontal distances (3rd bar, where light blue is for thresholds less than 10 cm, green is for thresholds less than 30 cm, and brown is for thresholds less than 1 m), and 4th bar shows deviation at spatial distances (yellow, green, and dark red for 10 cm, 30 cm, and 1 m thresholds, respectively).

4.4. Case C—Area with Curves and Serpentines

The example refers to a situation with fixed barriers along the road, although occasional vegetation barriers were present, but they were not vegetated due to the winter season. In this case C, the elevation of the trajectory gradually increased, resulting in a road shape with multiple curves. Figure 13 illustrates how the receivers in this area performed in GNSS positioning.

From the graphical representation (Figures 13 and 14), it is already clear that the Septentrio AsteRx-U geodetic receiver outperforms the other receivers in the number of detected positions. Interestingly, however, the Xiaomi Mi8 also detected more positions than the two u-bloxes. However, the quality of the position determination for the smartphone is significantly worse compared to the u-bloxes (Tables 11 and 12, Figure 14).

As in case B, the u-bloxes were able to determine many more fixed solutions in GNSS mode; however, as in the other cases, this was not true for the Septentrio AsteRx-U. The majority of all the available solutions were achieved with a deviation of the distances from the (measured) reference of less than 1 m for the Septentrio and 84.0%, 79.0% and 79.4% of the solutions for ZED-F9P, ZED-F9R, and Xiaomi Mi8, respectively. At the 30 cm threshold, the Septentrio was superior to the other receivers, as 80.0% of the solutions were less than 30 cm and 37.9% of the solutions were less than 10 cm. The GNSS solutions were significantly better for all the receivers with respect to the GPS.

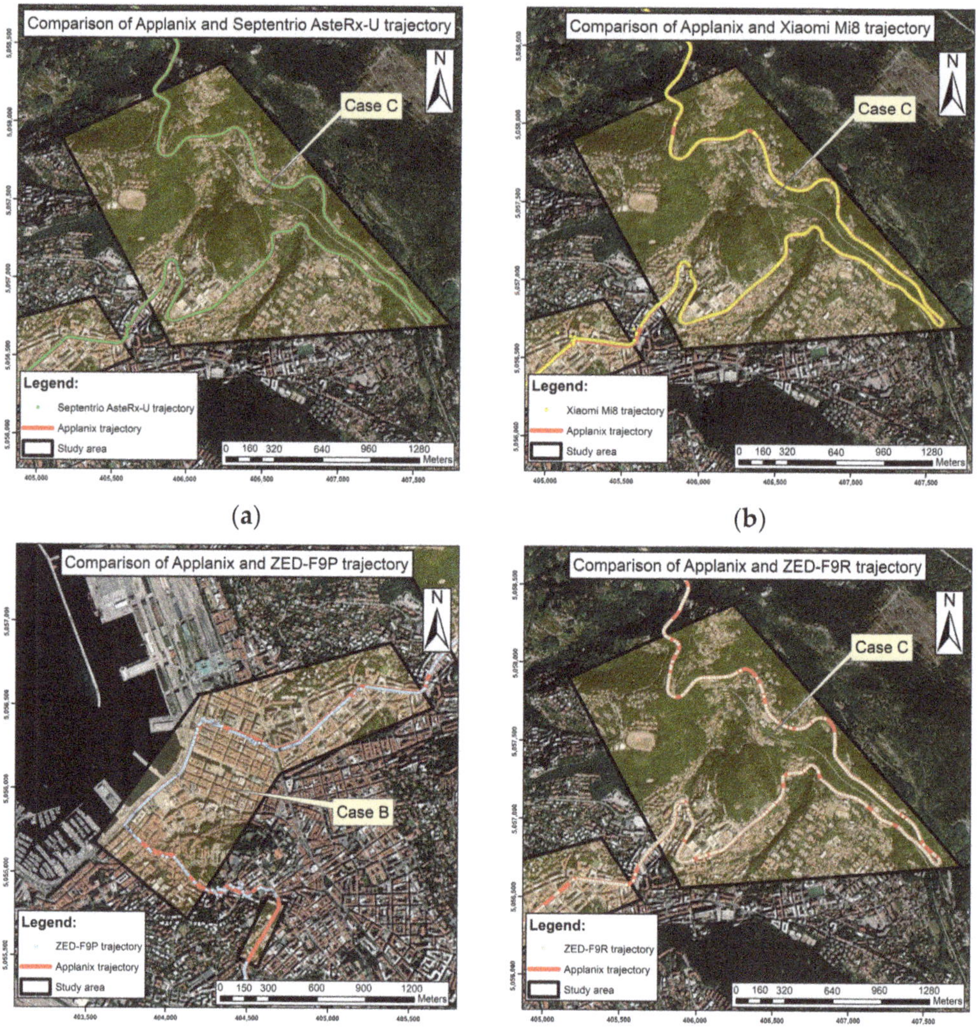

Figure 13. Case C: performance of: (**a**) Septentrio AsteRx-U, (**b**) Xiaomi Mi8, (**c**) u-blox ZED-F9P, and (**d**) u-blox ZED-F9R.

4.5. *Case D—Area with Vegetation by the Roadside*

The last case D involved a section of the trajectory that was kept flat, with some vegetation barriers near the roadway, which were less leafy due to it being winter. The measurement took place on the outskirts of the city, on the karst plateau, mostly at an altitude of about 300 m above sea level. Figure 15 shows that all the receivers except the ZED-F9R were able to determine the position in most of the section. As in the previous cases, the quality of position determination with the Xiaomi Mi8 was significantly worse than that of the other receivers (Tables 13 and 14, Figure 16).

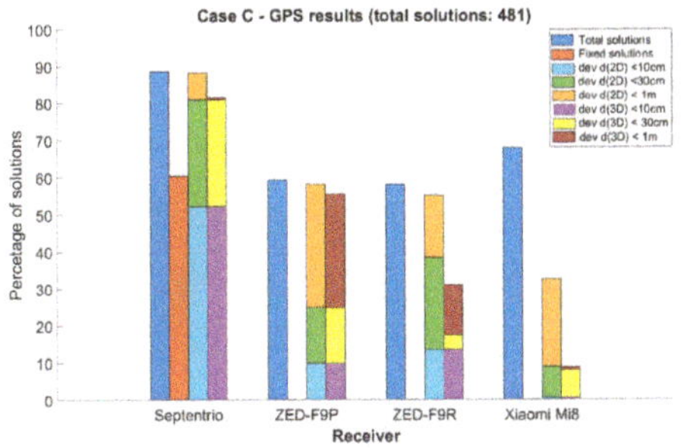

(a)

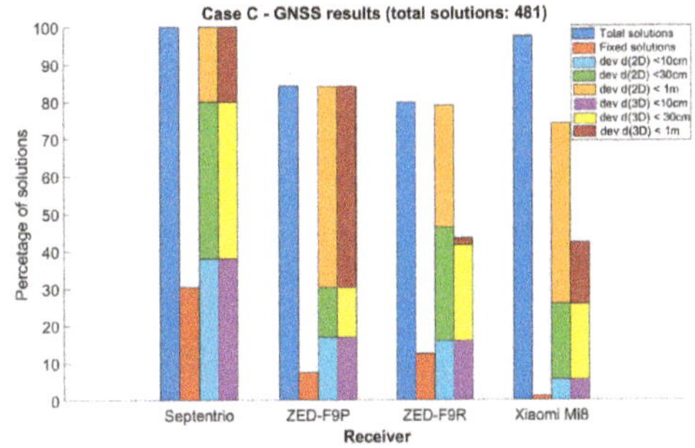

(b)

Figure 14. Percentage of solutions for Case C for (**a**) GPS solutions; (**b**) GNSS solutions. Each bar in a group shows: available solutions (1st bar), percentage of fixed solutions (2nd bar), deviation at horizontal distances (3rd bar, where light blue is for thresholds less than 10 cm, green is for thresholds less than 30 cm, and brown is for thresholds less than 1 m), and 4th bar shows deviation at spatial distances (yellow, green, and dark red for 10 cm, 30 cm, and 1 m thresholds, respectively).

Table 11. GPS solutions, solutions with fixed ambiguities, and deviation of distances below the given threshold. Situation: Case C—serpentines (no. of total solutions: 481).

	Type of GNSS Device			
Parameters	**Septentrio AsteRx-U**	**u-blox ZED-F9P**	**u-blox ZED-F9R**	**Smartphone Xiaomi Mi8**
GPS solutions	426 (88.6%)	285 (59.3%)	279 (58.0%)	326 (67.8%)
Fixed ambiguities	257 (60.3%)	0 (0%)	0 (0%)	0 (0%)
$dev\_d_{2D} < 1$ m	424 (88.2%)	278 (57.8%)	265 (55.1%)	157 (32.6%)
$dev\_d_{2D} < 30$ cm	389 (80.9%)	118 (24.5%)	184 (38.3%)	43 (8.9%)
$dev\_d_{2D} < 10$ cm	251 (52.2%)	47 (9.8%)	65 (13.5%)	2 (0.4%)
$dev\_d_{3D} < 1$ m	392 (81.5%)	265 (55.1%)	149 (31.0%)	38 (7.9%)
$dev\_d_{3D} < 30$ cm	389 (80.9%)	118 (24.5%)	84 (17.5%)	37 (7.9%)
$dev\_d_{3D} < 10$ cm	251 (52.2%)	47 (9.8%)	65 (13.5%)	2 (0.4%)

Table 12. GNSS solutions, solutions with fixed ambiguities, and deviation of distances below the given threshold. threshold Situation: Case C—serpentines (no. of total solutions: 481).

	Type of GNSS Device			
Parameters	**Septentrio AsteRx-U**	**u-blox ZED-F9P**	**u-blox ZED-F9R**	**Smartphone Xiaomi Mi8**
GNSS solutions	481 (100%)	405 (84.2%)	384 (79.8%)	469 (97.5%)
Fixed ambiguities	146 (30.3%)	30 (7.4%)	48 (12.5%)	5 (1.1%)
$dev\_d_{2D} < 1$ m	481 (100%)	404 (84.0%)	380 (79.0%)	353 (73.4%)
$dev\_d_{2D} < 30$ cm	385 (80.0%)	145 (30.2%)	222 (46.2%)	121 (25.2%)
$dev\_d_{2D} < 10$ cm	134 (37.9%)	81 (16.8%)	76 (15.8%)	27 (5.6%)
$dev\_d_{3D} < 1$ m	481 (100%)	400 (83.2%)	210 (43.7%)	199 (41.4%)
$dev\_d_{3D} < 30$ cm	385 (80.0%)	145 (30.2%)	199 (41.6%)	120 (25.2%)
$dev\_d_{3D} < 10$ cm	134 (37.9%)	80 (16.8%)	76 (15.8%)	27 (5.6%)

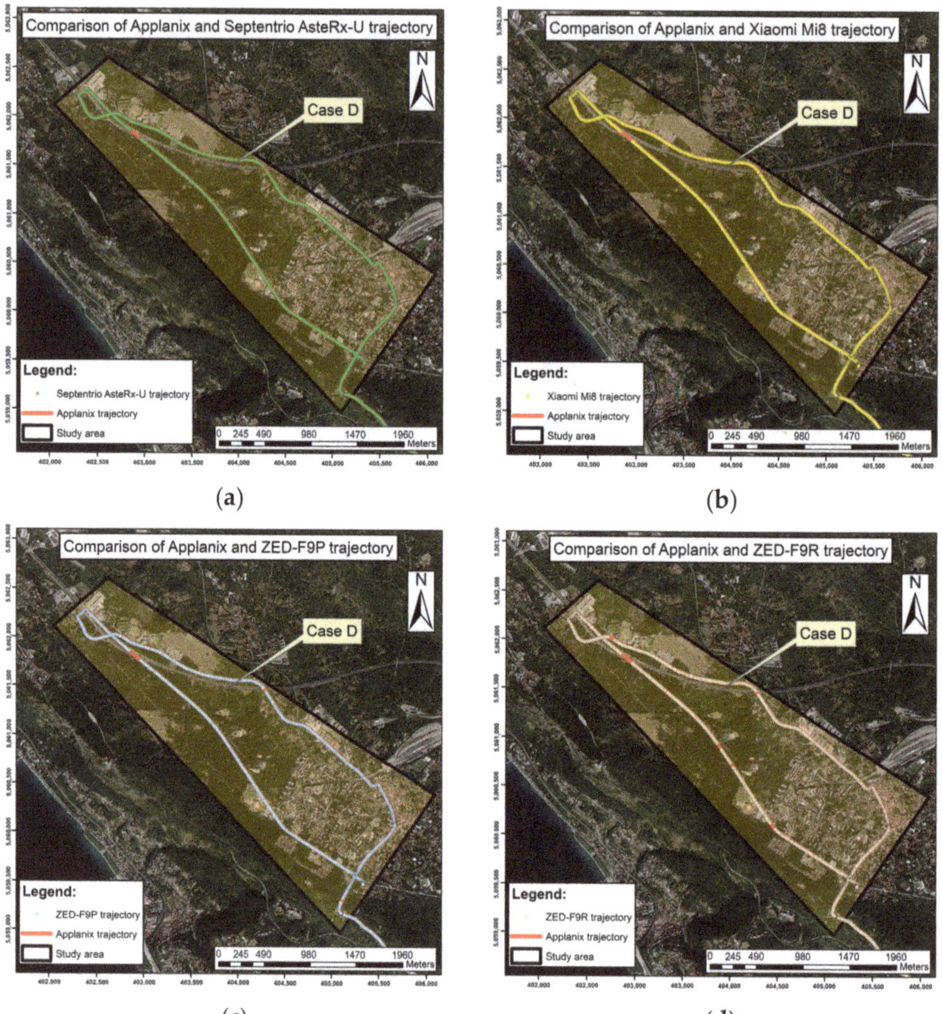

Figure 15. Case D: performance of: (**a**) Septentrio AsteRx-U, (**b**) Xiaomi Mi8, (**c**) u-blox ZED-F9P, and (**d**) u-blox ZED-F9R.

In addition, in terms of possible positioning and in situations with fixed ambiguities, the results using the GNSS are much better than with the GPS only. Again, with the Xiaomi Mi8 smartphone, it was not possible to define the ambiguities as fixed. The geodetic receiver also performed much better in this case than the other receivers in terms of the quality of the position determination. Interestingly, fixed solutions have prevailed in GSS processing with respect to GPS, especially for ZED-F9P and ZED-F9R. The deviation of the calculated horizontal distances from the reference distances was less than 10 cm in 91.9% of the total epochs. Interestingly, the Xiaomi Mi8 smartphone performed well in situations with a threshold of less than 1 m, but in all the situations, the u-bloxes were superior to the Xiaomi Mi 8. The ZED -F9P and Xiaomi Mi8 only achieved positioning quality below 10 cm in a few situations; in this case, the ZED-F9R was slightly better, but it was definitely not comparable to the highly successful Septentrio AsteRx-U.

Table 13. GPS solutions, solutions with fixed ambiguities, and deviation of distances below the given threshold. Situation Case D—open sky conditions with vegetation by the roadside (no. of total: 1001).

Parameters	Type of GNSS Device			
	Septentrio AsteRx-U	**u-blox ZED-F9P**	**u-blox ZED-F9R**	**Smartphone Xiaomi Mi8**
GPS solutions	963 (95.3%)	875 (86.5%)	860 (85.0%)	914 (90.4%)
Fixed ambiguities	669 (69.5%)	46 (5.3%)	261 (30.4%)	1 (0.1%)
$dev\_d_{2D} < 1$ m	943 (94.3%)	871 (87.1%)	849 (84.9%)	786 (78.6%)
$dev\_d_{2D} < 30$ cm	911 (91.1%)	167 (16.7%)	358 (35.8%)	176 (17.6%)
$dev\_d_{2D} < 10$ cm	793 (79.3%)	40 (4.0%)	141 (14.1%)	11 (1.1%)
$dev\_d_{3D} < 1$ m	895 (89.5%)	783 (78.3%)	788 (78.8%)	492 (49.2%)
$dev\_d_{3D} < 30$ cm	895 (89.5%)	167 (16.7%)	358 (35.8%)	176 (17.6%)
$dev\_d_{3D} < 10$ cm	793 (79.3%)	40 (4.0%)	141 (14.1%)	11 (1.1%)

Table 14. GNSS solutions, solutions with fixed ambiguities, and deviation of distances below the given threshold. Situation Case D—open sky conditions with vegetation by the roadside (no. of total: 1001).

Parameters	Type of GNSS Device			
	Septentrio AsteRx-U	**u-blox ZED-F9P**	**u-blox ZED-F9R**	**Smartphone Xiaomi Mi8**
GNSS solutions	990 (98.0%)	971 (96.1%)	946 (93.6%)	989 (97.9%)
Fixed ambiguities	711 (71.8%)	648 (66.7%)	750 (79.3%)	8 (0.8%)
$dev\_d_{2D} < 1$ m	972 (97.2%)	970 (97.0%)	945 (94.5%)	644 (64.4%)
$dev\_d_{2D} < 30$ cm	967 (96.7%)	288 (28.8%)	348 (34.8%)	141 (14.1%)
$dev\_d_{2D} < 10$ cm	919 (91.9%)	45 (4.5%)	115 (11.5%)	11 (1.1%)
$dev\_d_{3D} < 1$ m	972 (97.2%)	951 (95.1%)	945 (94.5%)	577 (57.7%)
$dev\_d_{3D} < 30$ cm	967 (96.7%)	288 (28.8%)	348 (34.8%)	141 (14.1%)
$dev\_d_{3D} < 10$ cm	919 (91.9%)	45 (4.5%)	115 (11.5%)	11 (1.1%)

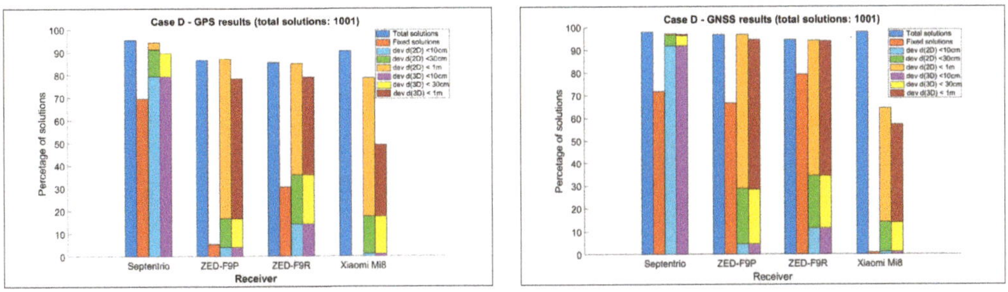

(**a**) (**b**)

Figure 16. Percentage of solutions for Case D for (**a**) GPS solutions; (**b**) GNSS solutions. Each bar in a group shows: available solutions (1st bar), percentage of fixed solutions (2nd bar), deviation at horizontal distances (3rd bar, where light blue is for thresholds less than 10 cm, green is for thresholds less than 30 cm, and brown is for thresholds less than 1 m), and 4th bar shows deviation at spatial distances (yellow, green, and dark red for 10 cm, 30 cm, and 1 m thresholds, respectively).

5. Conclusions

In this paper, the observations and position quality of a Septentrio AsteRx-U geodetic receiver, two u-blox receivers –ZED-F9P and ZED-F9R– and a Xiaomi Mi8 smartphone operating in GPS only and GNSS modes were analysed in relation to the Applanix MMS trajectory results. The study was carried out in urban and non-urban areas and in GNSS favourable measurement conditions. The results obtained in this research lead to the following conclusions, especially in terms of the low-cost used devices:

- From the comparison of the GPS only and GNSS solutions for geodetic and low-cost receivers, it can be concluded that in terms of the GPS, GNSS processing achieved many more solutions for position determination and determined ambiguities in many more cases with fixed values, even if this is not true in general for the Septentrio AsteRx-U and in particular in case C: a non-urban area with curves and serpentines characterised by a reduced signal acquisition. Comparing the means and standard deviations of the deviation of the distances from the reference, it can be stated that the variances for the GPS and GNSS solutions were comparable in this case.
- There is a significant difference in quality between kinematic positioning with geodetic and low-cost receivers. The geodetic receiver is much more stable in ambiguity solutions, especially when compared to a smartphone. The same is true for solutions where the threshold for positioning quality is set at 30 cm and even more clearly at 10 cm. The differences certainly also relate to the design of the receiving antenna. The geodetic GNSS antenna enables the elimination of some multipath signals and obviously outperforms the small antenna of a smartphone and the patch antenna of the u-bloxes. Therefore, it can be said that caution should be exercised when using low-cost receivers in terrestrial measurements in urban environments, and further studies and research are required in order to eliminate observations that are loaded with effects (these can be multipaths or interferences).
- Care should also be taken when moving out of shaded areas that make GNSS positioning impossible, especially with low-cost receivers. In the future, when processing combined GNSS and INS observations in the Kalman filter for low-cost receivers, it would be useful to include the new GNSS resolution obtained from the transient in the final solution, since the solutions—especially for the low-cost receivers—deviated significantly from the Applanix reference solutions immediately after repositioning.
- In situations with many curves and manoeuvres on the road, the low-cost receivers also performed worse compared to the geodetic receiver; in these situations, they determined significantly fewer positions than the Septentrio AsteRx-U receiver. Since this portion of the trajectory may have also been affected by the poor survey conditions due to vegetation, further research is required in order to determine how the receivers respond in curves and turns in the open sky.
- In GNSS mode, the Xiaomi Mi8 smartphone performed well in situations with a threshold of less than 1 m, with the percentages varying from 50% for the urban areas to 80% for the non-urban areas, which offers potential in view of future improvements for applications in terrestrial navigation.
- The general conclusion is that even low-cost devices, especially u-blox receivers and in particular those operating in GNSS/INS mode, are suitable for kinematic terrestrial positioning; however, their use in problematic positioning areas and in urban environments with obstacles should be treated with greater caution than is the case when using professional geodetic receivers.

Future studies will focus on the fusion of GNSS and MEMS (Micro-Electro-Mechanical System) sensors in smartphones and u-bloxes to enable more accurate and continuous positioning using the Kalman filter, especially in problematic environments such as street canyons, dense vegetation and under conditions of intentional interference, namely jamming and spoofing.

Author Contributions: Conceptualization, R.C. and P.P.-P.; methodology, R.C., P.P.-P. and F.V.; software and validation F.V. and P.P.-P.; formal analysis, P.P.-P. and F.V.; field tests: F.V., R.C. and T.S., P.S. and P.P.-P.; writing—original draft preparation, P.P.-P. and F.V.; writing—review and editing, R.C.; visualization, F.V. and P.P.-P.; project administration, R.C. and P.P.-P. All authors have read and agreed to the published version of the manuscript.

Funding: The authors acknowledge the financial support of the Slovenian Research Agency (research core funding No. P2-0227 Geoinformation Infrastructure and Sustainable Spatial Development of Slovenia).

Data Availability Statement: The data supporting the results of this study are available at https://unilj-my.sharepoint.com/:f:/g/personal/ppavlovc_fgg_uni-lj_si/EnrZ6jQl_DtIg4rVMZVNMUkBCqJdv5NC2oyrBsQpI2wiew?e=ukq78M (accessed on 7 January 2023).

Conflicts of Interest: The authors declare no conflict of interest.

References

1. Groves, P.D. Shadow matching: A new GNSS positioning technique for urban canyons. *J. Navig.* **2011**, *64*, 417–430. [CrossRef]
2. Ben-Moshe, B.; Elkin, E.; Levi, H.; Weissman, A. Improving accuracy of GNSS devices in urban canyons. In Proceedings of the 23rd Annual Canadian Conference on Computational Geometry, Toronto, ON, Canada, 10–12 August 2011.
3. Xi, C.; Cheng-Dong, X. Performance analysis of multi-constellation GNSS in urban canyons based on fuzzy comprehensive evaluation. In Proceedings of the 29th Chinese Control And Decision Conference (CCDC), Chongqing, China, 28–30 May 2017; pp. 3040–3045. [CrossRef]
4. Fortunato, M.; Ravanelli, M.; Mazzoni, A. Real-time geophysical applications with Android GNSS raw measurements. *Remote Sens.* **2019**, *11*, 2113. [CrossRef]
5. Magalhães, A.; Bastos, L.; Maia, D.; Gonçalves, J.A. Relative positioning in remote areas using a gnss dual frequency smartphone. *Sensors* **2021**, *21*, 8354. [CrossRef]
6. Angrisano, A.; Gaglione, S. Smartphone GNSS Performance in an Urban Scenario with RAIM Application. *Sensors* **2022**, *22*, 786. [CrossRef]
7. Li, Y.; Cai, C.; Xu, Z. A Combined Elevation Angle and C/N0 Weighting Method for GNSS PPP on Xiaomi MI8 Smartphones. *Sensors* **2022**, *22*, 2804. [CrossRef]
8. Paziewski, J.; Fortunato, M.; Mazzoni, A.; Odolinski, R. An analysis of multi-GNSS observations tracked by recent Android smartphones and smartphone-only relative positioning results. *Meas. J. Int. Meas. Confed.* **2021**, *175*, 109162. [CrossRef]
9. Robustelli, U.; Paziewski, J.; Pugliano, G. Observation quality assessment and performance of GNSS standalone positioning with code pseudoranges of dual-frequency android smartphones. *Sensors* **2021**, *21*, 2125. [CrossRef]
10. Bakuła, M.; Uradziński, M.; Krasuski, K. Performance of DGPS Smartphone Positioning with the Use of P(L1) vs. P(L5) Pseudorange Measurements. *Remote Sens.* **2022**, *14*, 929. [CrossRef]
11. Uradziński, M.; Bakuła, M. Assessment of static positioning accuracy using low-cost smartphone GPS devices for geodetic survey points' determination and monitoring. *Appl. Sci.* **2020**, *10*, 5308. [CrossRef]
12. Rabah, M.; Basiouny, M.; Ghanem, E.; Elhadary, A. Using RTK and VRS in direct geo-referencing of the UAV imagery. *NRIAG J. Astron. Geophys.* **2018**, *7*, 220–226. [CrossRef]
13. Broekman, A.; Gräbe, P.J. A low-cost, mobile real-time kinematic geolocation service for engineering and research applications. *HardwareX* **2021**, *10*, e00203. [CrossRef]
14. Janos, D.; Kuras, P. Evaluation of low-cost gnss receiver under demanding conditions in rtk network mode. *Sensors* **2021**, *21*, 5552. [CrossRef]
15. Van Nguyen, N.; Cho, W.; Hayashi, K. Performance evaluation of a typical low-cost multi-frequency multi-GNSS device for positioning and navigation in agriculture—Part 1: Static testing. *Smart Agric. Technol.* **2021**, *1*, 100004. [CrossRef]
16. Semler, Q.; Mangin, L.; Moussaoui, A.; Semin, E. Development of a low-cost centimetric GNSS positioning solution for android applications. *Int. Arch. Photogramm. Remote Sens. Spat. Inf. Sci. ISPRS Arch.* **2019**, *42*, 309–314. [CrossRef]
17. Wielgocka, N.; Hadas, T.; Kaczmarek, A.; Marut, G. Feasibility of using low-cost dual-frequency gnss receivers for land surveying. *Sensors* **2021**, *21*, 1956. [CrossRef]
18. Wu, Q.; Sun, M.; Zhou, C.; Zhang, P. Precise point positioning using dual-frequency GNSS observations on smartphone. *Sensors* **2019**, *19*, 2189. [CrossRef]
19. Zhu, H.; Xia, L.; Wu, D.; Xia, J.; Li, Q. Study on multi-gnss precise point positioning performance with adverse effects of satellite signals on android smartphone. *Sensors* **2020**, *20*, 6447. [CrossRef]
20. Benvenuto, L.; Cosso, T.; Delzanno, G. An Adaptive Algorithm for Multipath Mitigation in GNSS Positioning with Android Smartphones. *Sensors* **2022**, *22*, 5790. [CrossRef]

21. Lithopoulos, E. The Applanix Approach to GPS/INS Integration. In *Photogrammetric Week 1999*; Fritsch, D., Spiller, R., Eds.; Wichmann Verlag: Heidelberg, Germany, 1999; pp. 53–57.
22. Mostafa, M.; Hutton, J.O.E.; Reid, B.; Hill, R. GPS/IMU products—The Applanix approach. *Photogramm. Week* **2001**, 63–82.
23. Cefalo, R.; Grandi, G.; Roberti, R.; Sluga, T. Extraction of road geometric parameters from high resolution remote sensing images validated by GNSS/INS geodetic techniques. *Lect. Notes Comput. Sci.* **2017**, *10407*, 181–195. [CrossRef]
24. Tarantino, E.; Novelli, A.; Cefalo, R.; Sluga, T.; Tommasi, A. Single-frequency kinematic performance comparison between Galileo, GPS, and GLONASS satellite positioning systems using an MMS-generated trajectory as a reference: Preliminary results. *ISPRS Int. J. Geo-Inf.* **2018**, *7*, 122. [CrossRef]
25. Bastos, L.; Buist, P.; Cefalo, R.; Goncalves, J.A.; Ivan, A.; Magalhaes, A.; Pandele, A.; Porretta, M.; Radutu, A.; Sluga, T.; et al. Kinematic Galileo and GPS Performances in Aerial, Terrestrial, and Maritime Environments. *Remote Sens.* **2022**, *14*, 3414. [CrossRef]
26. Sanna, G.; Pisanu, T.; Garau, S. Behavior of Low-Cost Receivers in Base-Rover Configuration with Geodetic-Grade Antennas. *Sensors* **2022**, *22*, 2779. [CrossRef]
27. Pavlovčič-Prešeren, P.; Dimc, F.; Bažec, M. A comparative analysis of the response of GNSS receivers under vertical and horizontal L1/E1 chirp jamming. *Sensors* **2021**, *21*, 1446. [CrossRef]
28. Bažec, M.; Dimc, F.; Pavlovčič-Prešeren, P. Evaluating the vulnerability of several geodetic GNSS receivers under chirp signal L1/E1 Jamming. *Sensors* **2020**, *20*, 814. [CrossRef]
29. Dimc, F.; Pavlovčič Prešeren, P.; Bažec, M. Robustness against Chirp Signal Interference of On-Board Vehicle Geodetic and Low-Cost GNSS Receivers. *Sensors* **2021**, *21*, 5257. [CrossRef]
30. Robustelli, U.; Baiocchi, V.; Pugliano, G. Assessment of dual frequency GNSS observations from a Xiaomi Mi 8 android smartphone and positioning performance analysis. *Electron.* **2019**, *8*, 91. [CrossRef]
31. European Global Navigation Satellite Systems Agency (GSA). *White Paper on Using GNSS Raw Measurements on Android Devices*; EUSPA: Prague, Czech Republic, 2017; ISBN 9789292060336.
32. Gao, C.; Chen, B.; Lui, Y.; Sun, P. Real-time Precise Point Positioning with a Xiaomi MI 8 Android Smartphone. *Sensors* **2019**, *19*, 2835.
33. Fortunato, M. GMV NSL-Testing the Dual Frequency GNSS Smartphone. Available online: https://gmvnsl.com/about-nsl/nsl-blog/15-products-and-services/55-xiaomi-mi8 (accessed on 29 September 2022).
34. Guo, L.; Wang, F.; Sang, J.; Lin, X.; Gong, X.; Zhang, W. Characteristics analysis of raw multi-GNSS measurement from Xiaomi Mi 8 and positioning performance improvement with L5/E5 frequency in an urban environment. *Remote Sens.* **2020**, *12*, 744. [CrossRef]
35. Shinghal, G.; Bisnath, S. Conditioning and PPP processing of smartphone GNSS measurements in realistic environments. *Satell. Navig.* **2021**, *2*, 10. [CrossRef]
36. Wu, D.; Xiong, W.; Guo, J. Establishment and Repetition Survey of Primary GNSS Control Network of Hong Kong–Zhuhai–Macao Bridge. *J. Surv. Eng.* **2022**, *148*, 05021006. [CrossRef]
37. Elmezayen, A.; El-Rabbany, A. Precise point positioning using world's first dual-frequency GPS/galileo smartphone. *Sensors* **2019**, *19*, 2593. [CrossRef]
38. Zhang, X.; Tao, X.; Zhu, F.; Shi, X.; Wang, F. Quality assessment of GNSS observations from an Android N smartphone and positioning performance analysis using time-differenced filtering approach. *GPS Solut.* **2018**, *22*, 70. [CrossRef]
39. Liu, W.; Shi, X.; Zhu, F.; Tao, X.; Wang, F. Quality analysis of multi-GNSS raw observations and a velocity-aided positioning approach based on smartphones. *Adv. Sp. Res.* **2019**, *63*, 2358–2377. [CrossRef]
40. Banville, S.; Lachapelle, G.; Ghoddousi-Fard, R.; Gratton, P. Automated processing of low-cost GNSS receiver data. In Proceedings of the 32nd International Technical Meeting of the Satellite Division of The Institute of Navigation (ION GNSS+ 2019), Miami, FL, USA, 16–20 September 2019; pp. 3636–3652. [CrossRef]
41. Paziewski, J.; Sieradzki, R.; Baryla, R. Signal characterization and assessment of code GNSS positioning with low-power consumption smartphones. *GPS Solut.* **2019**, *23*, 1–12. [CrossRef]
42. Li, Z.; Wang, L.; Wang, N.; Li, R.; Liu, A. Real-time GNSS precise point positioning with smartphones for vehicle navigation. *Satell. Navig.* **2022**, *3*, 19. [CrossRef]
43. Everett, T.; Taylor, T.; Lee, D.K.; Akos, D.M. Optimizing the Use of RTKLIB for Smartphone-Based GNSS Measurements. *Sensors* **2022**, *22*, 3825. [CrossRef]
44. U-Center. Available online: https://www.u-blox.com/en/product/u-center (accessed on 12 December 2021).
45. Applanix Corporation. *PUBS-MAN-001768-POSPacTM MMSTM GNSS-Inertial Tools Software, Version 7.2*; Revision 12—User Guide; Applanix Corporation: Richmond Hill, ON, Canada, 2016.
46. Rete GNSS FVG—A. Marussi. Available online: https://rem.regione.fvg.it/ (accessed on 11 February 2022).
47. Takasu, T. RTKLIB: An Open source program package for RTK-GPS. Available online: http://rtklib.com/ (accessed on 4 December 2020).
48. Everett, T. RTKLIB Demo5_b34d. Available online: https://rtkexplorer.com/downloads/rtklib-code/ (accessed on 5 May 2022).

49. Netthonglang, C.; Thongtan, T.; Satirapod, C. GNSS Precise Positioning Determinations Using Smartphones. In Proceedings of the 2019 IEEE Asia Pacific Conference on Circuits and Systems (APCCAS), Bangkok, Thailand, 11–14 November 2019; pp. 401–404. [CrossRef]
50. Retscher, G.; Weigert, T. Assessment of a dual-frequency multi-GNSS smartphone for surveying applications. *Appl. Geomatics* **2022**, *14*, 765–784. [CrossRef]

remote sensing

MDPI

Article

Multi-GNSS Differential Inter-System Bias Estimation for Smartphone RTK Positioning: Feasibility Analysis and Performance

Rui Shang [1], Chengfa Gao [1,*], Lu Gan [1], Ruicheng Zhang [1], Wang Gao [2] and Xiaolin Meng [3]

1 School of Transportation, Southeast University, Nanjing 211189, China
2 School of Instrument Science and Engineering, Southeast University, Nanjing 211189, China
3 Faculty of Architecture, Civil and Transportation Engineering, Beijing University of Technology, Beijing 100124, China
* Correspondence: gaochfa@seu.edu.cn

Abstract: An inter-system model for multi-GNSSs (global navigation satellite systems) makes the interoperability among different GNSS constellations possible. In recent years, inter-system models for geodetic receivers have been extensively studied. However, the precise positioning of smartphones suffers from various problems, and the current research mostly focuses on how to achieve the GNSS ambiguity resolution. Based on the research of receiver channel-dependent bias, in this study, we will research the temporal behaviors of differential inter-system bias (DISB) and implement an inter-system model for smartphones. A representative Huawei P40 (HP40) smartphone was used in the experiments, and the results show the following: (1) For the HP40, the frequencies of Code Division Multiple Access (CDMA) systems are free of receiver channel-dependent phase bias, which provides the chances for further interoperability among these systems. However, the code observations of the HP40 are influenced by receiver channel-dependent code bias; it is therefore suggested to set a large initial standard deviation (STD) value for code observations in the positioning. (2) GPS L1/QZSS L1 and BDS-2 B1I /BDS-3 B1I are free of phase DISB, and there is obvious phase DISB between GPS L1 and Galileo E1; even so, the valuations are sufficiently stable and the STD is close to 0.005 cycles. However, the phase DISB of GPS L1/BDS B1I is unstable. (3) For kinematic positioning, when the stable phase DISB is introduced, a 3–38.9% improvement in the N/E/U directions of the positioning accuracies in the inter-system differencing is achieved compared with the intra-system differencing.

Keywords: Android; multi-GNSS; real-time kinematic (RTK) positioning; receiver channel-dependent bias; differential inter-system bias (DISB)

Citation: Shang, R.; Gao, C.; Gan, L.; Zhang, R.; Gao, W.; Meng, X. Multi-GNSS Differential Inter-System Bias Estimation for Smartphone RTK Positioning: Feasibility Analysis and Performance. *Remote Sens.* **2023**, *15*, 1476. https://doi.org/10.3390/rs15061476

Academic Editors: Giuseppe Casula, Zhetao Zhang and Wenkun Yu

Received: 3 February 2023
Revised: 4 March 2023
Accepted: 5 March 2023
Published: 7 March 2023

1. Introduction

The global navigation satellite system (GNSS) provides an important support for social development [1,2], and the popularity of smart devices has further promoted the technical and theoretical innovation of GNSS navigation applications. At present, there are more than 5 billion satellite navigation and positioning devices in the world, more than 75% of which are smartphones, and more than 50% of the applications downloaded through the app store are related to location services [3]. With the development of science and technology, the Internet of Things and artificial intelligence puts forward increasing demands for location services related to GNSS navigation and positioning [4]. In 'Google I/O of 2016', Google announced that they would open the data interface to obtain original code, phase and Doppler data in the Android Nougat system [5]. The availability of original GNSS measurements in smartphones makes it possible to achieve popular precise positioning [6–9]. Compared with geodetic GNSS receivers, there are two major issues in the relative precise positioning of a smartphone. One is the serious multipath errors caused by the omnidirectional linearly polarized antennas used in current smartphones, while the other is several anomalous errors presented in smartphone GNSS observation, referred to as the receiver channel-dependent bias, which is caused by the budget GNSS chip with technical flaws [10].

To mitigate these multipath problems for smartphones, two main approaches are involved at present [11]. One is using an external geodetic antenna with an advanced multipath rejection ability. For instance, an external antenna was used for the Xiaomi Mi 8 smartphone by Geng and Li [12], and after that, a reliable GPS ambiguity-fixed solution can be achieved. The studies of Darugna et al. [13] and Bochkati et al. [14] also obtained ambiguity-fixed solutions for GPS/Galileo by using a choke ring antenna platform. The other approach is processing the data with advanced algorithm processing. For instance, the sensors-aided multipath detection and exclusion algorithm [15], ray tracing, Vondrak filtering and support vector regression [16–18] can be applied. The study of receiver channel-dependent bias mainly includes the investigation of the characteristics and methods to avoid bias. For the investigated characteristics, Humphreys et al. [19] found that for a Samsung Galaxy S5, there is an approximate linear growth error in the phase observation. Riley et al. [20] also point out that there is a random offset in the DD phase residual of the Nexus 9. The research also points out that for the Huawei P30, receiver channel-dependent bias was found in the frequencies of GPS L5, Galileo E1/E5a and BDS B1I while not being detected in GPS L1 [21–23]. Li and Geng [24] extracted the receiver channel-dependent phase bias of the Nexus 9 through the zero-baseline experiment and found that the phase bias is different for each satellite. In addition, although the receiver channel-dependent phase bias changes randomly with the tracking state, it is stable under the smooth tracking of the phase observation. Recently, Li et al. [25] examined the characterization of the receiver channel-dependent bias for a Huawei Mate 20 with an Android 9.0.1 operating system and found that the bias not only results from low-cost GNSS chips but is also affected by the operating system. Thus, in order to achieve an integer ambiguity resolution, Paziewski et al. [23] only used phase observations without receiver channel-dependent bias for GNSS positioning. Li et al. [25] proposed an online receiver channel-dependent bias correction method to try to solve the problem.

Through the continuing effort to address these challenges, the GNSS relative positioning accuracy of smartphones has improved enormously, and some systems have even reached the centimeter level. However, the above methods all adopt the intra-system model of selecting a reference satellite in the multi-system combination positioning. In recent years, with the maturity and improvement in GNSS data processing theory, the inter-system combination has become a research hotspot in multi-GNSS positioning [26]. Compared with intra-system biases in the intra-system model, including the differential code biases (DCBs) and differential phase biases (DPBs) between different frequencies in a single GNSS constellation, the inter-system biases (ISBs) between different GNSSs should be considered in the inter-system model [27]. In addition, the following two main aspects need to be considered when building an inter-system model [28]: The first is different navigation systems. Due to the different signal channels adopted by GNSSs, the hardware delay is usually difficult to eliminate, so the influence of the differential inter-system bias (DISB) needs to be fully considered in the tight combination relative model [29,30]. The second is different signal frequencies. It is difficult to directly combine the non-overlapping frequencies to obtain the integer-estimable DD ambiguity. In addition to the phase DISB, the single-difference (SD) ambiguity after the combination of the observations should also be considered during the fusion processing of different frequency signals. In high-precision inter-system positioning for a geodesic receiver, according to the different processing methods, the research results show that some of the phase and code DISBs between are zero [31–35], and although the phase and code DISBs of some other combinations are not zero, the time-varying characteristics are stable, so they can be corrected in advance or estimated as a parameter [36–41]. In theory, introducing the DISB in the positioning model can improve the strength of the positioning model by using the time-varying stability of the DISB. The positioning experiments show that the performance of the ambiguity resolution and positioning is improved by using the inter-system model after the correct phase and code DISBs, especially in a poor observation environment such as under a high cut-off angle, and the speed of ambiguity resolution is also significantly improved.

Through the analysis of the current situation regarding the high-precision positioning of smartphones, it can be seen that with the arrival of the era of the parallel GNSS multi-system, it is imperative to realize the inter-system positioning of smartphones. However, in studies of the inter-system difference between smartphones, in addition to the DISB, the receiver channel-dependent bias should also be considered. To extend the inter-system model in smartphone GNSS positioning, in this paper, we investigate the characteristics of the receiver channel-dependent bias and the DISB for smartphones, especially for phase observations. It is noteworthy that according to the research of Li et al. [24], unlike the geodetic receiver, the inter-frequency bias (IFB) of GLONASS does not have a linear relationship with the frequency number of GLONASS satellites, and it is difficult to separate. Therefore, this paper mainly focuses on the CDMA systems. An HP40 is used as a representative in this manuscript, the phase measurements of which are not interfered with by duty-cycle issues, and an RF shielding box was employed to reduce the multipath effects. In the following, the receiver channel-dependent bias estimation method and inter-system model are formulated in Section 2. The static datasets and processing methods used in this paper, as well as the receiver channel-dependent bias and DISB characteristics analyses, are carefully addressed in Section 3. The kinematic experiments under numeric combinations of satellites are explained in Section 4 to demonstrate the improvement in the positioning performance of the inter-system differencing compared with intra-system differencing. The discussion is given in Section 5 and the research conclusions are summarized in Section 6.

2. Methodology

We focus solely on the estimation of receiver channel-dependent bias and DISB; so, in the following, the external survey-grade GNSS antenna was used as a surrogate for embedded smartphone GNSS antenna to avoid multipath influence. Furthermore, zero or short baselines are mainly adopted, so we do not consider the atmospheric delay. Based on the above settings, we first introduce the inter-system model, as well as receiver channel-dependent bias correction method. Furthermore, we then discuss the statistical hypothesis test of DISB.

2.1. *Inter-System Model for Smartphone*

Through inter-satellite single difference, some delays related to the station can be eliminated, such as the satellite clock error and the satellite hardware delays. Assuming that the antenna phase center deviation, earth rotation effect, tidal effect, relativistic effect and other errors have been corrected by the model, for a CDMA system A, the SD observation equation between receivers a and b can be obtained:

$$\begin{cases} \Delta L_{ab,j}^{A_s} = \Delta \rho_{ab,j}^{A_s} + \Delta dt_{ab,jl}^{A} + \lambda_j^A \left[\Delta N_{ab,j}^{A_s} + \Delta \phi_{ab,j}^{A_s} \right] + \Delta \varepsilon_{ab,j}^{A_s} \\ \Delta P_{ab,j}^{A_s} = \Delta \rho_{ab,j}^{A_s} + \Delta dt_{ab,jp}^{A} + \Delta \zeta_{ab,j}^{A_s} \end{cases} \tag{1}$$

where L and P are the phase and code observable in meters, respectively. The superscript $s = 1, 2, \ldots, m$ refers to the index of satellites, the subscript j represents the frequency band. The symbol ρ represents the geometric distance between the satellite and receiver. $\Delta dt_{ab,jl}^{A}$ and $\Delta dt_{ab,jp}^{A}$ are the phase and code receiver clock bias in meters, respectively. $\Delta \phi_{ab,j}^{A_s}$ is the receiver channel-dependent phase bias for the satellite in cycle. $\lambda_j^A$is the wavelength, and $\Delta N_{ab,j}^{A_s}$ is the ambiguity. $\Delta \varepsilon_{ab,j}^{A_s}$ is the phase observation noise, and $\Delta \zeta_{ab,j}^{A_s}$ is the code observation noise containing both the noise and the initial code bias. Therefore, for CDMA systems, the intra-system model is as follows:

$$\begin{cases} \Delta L_{ab,j}^{A_1} = \Delta\rho_{ab,j}^{A_1} + \Delta\overline{dt}_{ab,jl}^{A} + \Delta\varepsilon_{ab,j}^{A_1} \\ \Delta P_{ab,j}^{A_1} = \Delta\rho_{ab,j}^{A_1} + \Delta dt_{ab,jp}^{A} + \Delta\xi_{ab,j}^{A_1} \\ \Delta L_{ab,j}^{A_2} = \Delta\rho_{ab,j}^{A_2} + \Delta\overline{dt}_{ab,jl}^{A} + \lambda_j^A\left[\Delta\nabla N_{ab,j}^{A_1A_2} + \Delta\phi_{ab,j}^{A_2}\right] + \Delta\varepsilon_{ab,j}^{A_2} \\ \Delta P_{ab,j}^{A_2} = \Delta\rho_{ab,j}^{A_2} + \Delta dt_{ab,jp}^{A} + \Delta\xi_{ab,j}^{A_2} \\ \Delta L_{ab,j}^{A_3} = \Delta\rho_{ab,j}^{A_3} + \Delta\overline{dt}_{ab,jl}^{A} + \lambda_j^A\left[\Delta\nabla N_{ab,j}^{A_1A_3} + \Delta\phi_{ab,j}^{A_3}\right] + \Delta\varepsilon_{ab,j}^{A_3} \\ \Delta P_{ab,j}^{A_3} = \Delta\rho_{ab,j}^{A_3} + \Delta dt_{ab,jp}^{A} + \Delta\xi_{ab,j}^{A_3} \\ \vdots \end{cases} \tag{2}$$

where

$$\Delta\overline{dt}_{ab,jl}^{A} = \Delta dt_{ab,jl}^{A} + \lambda_j^A\left[\Delta N_{ab,j}^{A_1} + \Delta\phi_{ab,j}^{A_1}\right] \tag{3}$$

For receiver channel-dependent phase bias estimate, if we fix the positions of both receivers to the known value, only the DD phase bias $\Delta\nabla N_{ab,j}^{A_1A_s} + \Delta\phi_{ab,j}^{A_s}$ needs to be estimated in Equation (3). As demonstrated by Li and Geng [10], using the fact that the ambiguity of the continuously tracked phase observations is kept as an integer constant, after the recombination, a simplified DD receiver channel-dependent phase bias can be written as

$$\Delta\phi_{ab,j}^{A_s}(t) = \Delta\nabla N_{ab,j}^{A_1A_2}(t) - \left[\Delta\nabla N_{ab,j}^{A_1A_2}(t_0)\right] \tag{4}$$

where $[\cdot]$ is an integer rounding operation, $\Delta\nabla N_{ab,j}^{A_1A_2}(t0)$ is the estimated DD phase bias at the first epoch t_0 of this continuously tracked phase arc and t is the current epoch.

Once the receiver channel-dependent bias is addressed, when estimating the clock bias for each system, the intra-system model for multi-systems can be obtained. Furthermore, if all systems estimate a common clock bias, as shown in Equation (5), through parameter reorganization, we can obtain the phase DISB between systems, as well as the inter-system model. It is worth noting that the equation for code inter-system model is relatively simple, and the users can use phase equation as samples to write.

$$\begin{cases} \Delta L_{ab,j}^{A_1} = \Delta\rho_{ab,j}^{A_1} + \Delta\overline{dt}_{ab,jl}^{A} + \Delta\varepsilon_{ab,j}^{A_1} \\ \Delta L_{ab,j}^{A_2} = \Delta\rho_{ab,j}^{A_2} + \Delta\overline{dt}_{ab,jl}^{A} + \lambda_j^A\Delta\nabla N_{ab,j}^{A_1A_2} + \Delta\varepsilon_{ab,j}^{A_2} \\ \Delta L_{ab,j}^{A_3} = \Delta\rho_{ab,j}^{A_3} + \Delta\overline{dt}_{ab,jl}^{A} + \lambda_j^A\Delta\nabla N_{ab,j}^{A_1A_3} + \Delta\varepsilon_{ab,j}^{A_3} \\ \vdots \\ \Delta L_{ab,j}^{B_1} = \Delta\rho_{ab,j}^{B_1} + \Delta\overline{dt}_{ab,jl}^{A} + \lambda_j^A\delta_{ab,jl}^{AB} + \Delta\varepsilon_{ab,j}^{B_1} \\ \Delta L_{ab,j}^{B_2} = \Delta\rho_{ab,j}^{B_2} + \Delta\overline{dt}_{ab,jl}^{A} + \lambda_j^A\delta_{ab,jl}^{AB} + \lambda_j^B\Delta\nabla N_{ab,j}^{B_1B_2} + \Delta\varepsilon_{ab,j}^{B_2} \\ \Delta L_{ab,j}^{B_3} = \Delta\rho_{ab,j}^{B_3} + \Delta\overline{dt}_{ab,jl}^{A} + \lambda_j^A\delta_{ab,jl}^{AB} + \lambda_j^B\Delta\nabla N_{ab,j}^{B_1B_3} + \Delta\varepsilon_{ab,j}^{B_3} \\ \vdots \end{cases} \tag{5}$$

where

$$\lambda_j^A\delta_{ab,jl}^{AB} = \Delta\overline{dt}_{ab,jl}^{A} - \Delta\overline{dt}_{ab,jl}^{B} = \Delta dt_{ab,jl}^{A} - \Delta dt_{ab,jl}^{B} + \lambda_j^B\left(\Delta N_{ab,j}^{B_1} - \frac{\lambda_j^A}{\lambda_j^B}\Delta N_{ab,j}^{A_1}\right) \tag{6}$$

where B is another CDMA system, $\delta_{ab,jl}^{AB}$ donates the phase DISB in cycles. As described by Equations (5) and (6), this inter-system model can estimate the phase DISB for both overlapping and non-overlapping frequencies combinations, but for non-overlapping frequencies, the phase DISB contains the initial ambiguity of the reference satellite. Therefore, in order to ensure the stability of the phase DISB, the phase DISB should be estimated in real time.

2.2. Statistical Hypothesis Test of DISB

After obtaining the DISB estimation through Equation (5), the stability of the DISB can be evaluated using the hypothesis test method [42,43]. Assume that n DISB $\delta_{ab,jl}^{AB}(1)$, $\delta_{ab,jl}^{AB}(2), \ldots, \delta_{ab,jl}^{AB}(n)$ can be obtained in the observation period, then the weighted average value of DISB, accuracy and stability of the test quantity T can be obtained:

$$\bar{\delta}_{ab,jl}^{AB} = \frac{\sum_{t=1}^{n} \delta_{ab,jl}^{AB}(t) / \sigma_{\delta_{ab,jl}^{AB}}^{2}(t)}{\sum_{t=1}^{n} 1/\sigma_{\delta_{ab,jl}^{AB}}^{2}(t)} \tag{7}$$

$$\sigma_{\bar{\delta}_{ab,jl}^{AB}}^{2} = \frac{1}{\sum_{i=1}^{n} 1/\sigma_{\delta_{ab,jl}^{AB}}^{2}(t)} \tag{8}$$

$$T = \sum_{i=1}^{n} \frac{(\delta_{ab,jl}^{AB}(t) - \bar{\delta}_{ab,jl}^{AB})^{2}}{\sigma_{\delta_{ab,jl}^{AB}}^{2}(t)} \tag{9}$$

where $\sigma_{\delta_{ab,jl}^{AB}}^{2}$ is the posterior variance of $\delta_{ab,jl}^{AB}(t)$, under the assumption that $\delta_{ab,jl}^{AB}(t)$ is normally distributed, T is the Chi-square test with $n-1$ degrees of freedom. If $T < \chi_{\alpha}^{2}(n-1,0)$, it goes to show that the DISB valuation has not changed significantly in test time; α is the significance level, and the value of α is 5% in this paper.

3. Receiver Channel-Dependent Bias and DISB Characteristics

In this part, we investigate the characteristics of the receiver channel-dependent bias and DISB from CDMA systems for smartphone and geodetic receivers in detail to validate the feasibility of the inter-system model.

3.1. Experimental Setup and Processing Methods

A zero baseline formed by a smartphone and geodetic receiver was used to study the temporal properties of smartphone receiver channel-dependent bias and DISB. Furthermore, a 4m short-baseline CUT0-CUTB (equipped with both Trimble NetR9 receiver and TRM59800.00 antenna designed by Trimble company) from Curtin GNSS Research Centre was added to compare the receiver channel-dependent bias and DISB characteristics between the smartphone and geodetic receiver. Figure 1 shows the layout of smartphone and base stations for static experiments. As shown in the upper part of Figure 1, the datasets for the zero baseline were collected at Southeast University. A representative HP40 smartphone was used as the rover station, and the JSJN station (one of the Continuously Operating Reference Stations (CORS) of Southeast University) was set as base station. JSJN is composed of the geodetic GNSS receivers CHCNAV P5 (one of the smart GNSS geodetic reference receivers of CHC Navigation company) and a HI-TARGET AT-53501 antenna (one of the 3D choke ring antennas designed by Hi-Target company). To create the zero baseline with JSJN, a signal splitter was used, and the CHCNAV P5 and a sending antenna were connected to two output ports of the splitter. The sending antenna was fixed in a small radio frequency (RF) shielding box, and the HP40 was also placed in the RF shielding box. This experimental method can block the GNSS signals obtained by the antenna of HP40. Meanwhile, we placed the antenna of JSJN on an open roof; thus, these data were not interfered with by multipath signals. Geo++ RINEX (Receiver INdependent EXchange format) Logger (Version 2.1.6) was used to obtain the raw GNSS code and phase data from HP40.

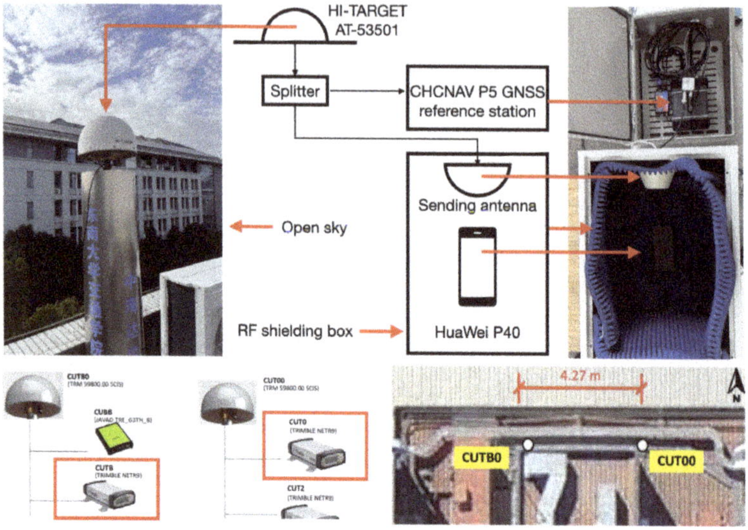

Figure 1. Layout of the smartphone and receiver for two static baselines.

Detailed information and the observation duration of two baselines are presented in Table 1. For JSNJ-HP40, 1h data were collected with sampling rate of 1 s (8:00–9:00, 18 October 2022), and for CUT0-CUTB, 24 h data were collected with sampling rate of 30 s (31 May 2020). From Table 1, we can also see that the code observation on B2a/B3I of HP40 cannot be obtained; thus, only the first frequency observations of all GNSS systems were analyzed in the following experiment. Figure 2 shows the tracked satellites of two baselines with 15° cut-off, and the experimental settings are as follows. Kalman filtering with kinematic positioning mode is used in the experiments, and the commonly used elevation angle-dependent stochastic model was employed [44–46]. For AR, the LAMBDA (Least-squares AMBiguity Decorrelation Adjustment) method was adopted to search ambiguity. Furthermore, the threshold for the ratio test was strictly set as 3.0.

Table 1. The information and details of static datasets.

Baseline	Device	Antenna	Duration (UTC Time)	Systems and Frequencies	
				L	P
JSJN-HP40	CHCNAV P5 (base)	HI-TARGET AT-53501	1 h (08:00–09:00, 18 December 2022)	G:L1/L5; E:E1/E5a; C:B1I B1C B2a; J:L1/L5	G:L1/L5; E:E1/E5a; C:B1I; J:L1/L5
	HP40 (rover)	Embedded antenna		G:L1/L5; E:E1/E5a; C:B1I B1C B2a; J:L1/L5	G:L1/L5; E:E1/E5a; C:B1I B1C B2a; J:L1/L5
CUT0-CUTB	Trimble NetR9 (base)	TRM59800.00	24 h (31 May 2020)	G:L1/L5; E:E1/E5a; C:B1I B1C B2a; J:L1/L5	G:L1/L5; E:E1/E5a; C:B1I B1C B2a; J:L1/L5
	Trimble NetR9 (rover)	TRM59800.00		G:L1/L5; E:E1/E5a; C:B1I B1C B2a; J:L1/L5	G:L1/L5; E:E1/E5a; C:B1I B1C B2a; J:L1/L5

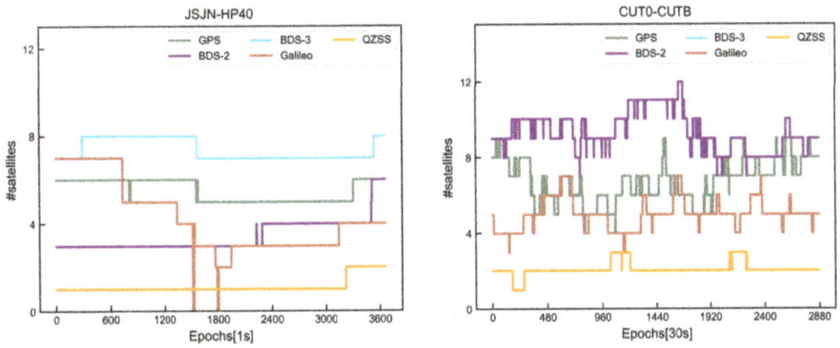

Figure 2. Total number of tracked GPS/BDS-2/BDS-3/Galileo/QZSS satellites for the two static baselines datasets.

3.2. *Temporal Properties of Receiver Channel-Dependent Bias*

The DD receiver channel-dependent phase bias estimates of the JSJN-HP40 and CUT0-CUTB are shown in Figure 3. For convenience in analyzing performance of each system, the DD phase bias of GPS/BDS-2/BDS-3/Galileo/QZSS for the JSJN-HP40 is displayed separately. According to Figure 2, the CUT0-CUTB was unable to receive the BDS-3 signals, so only the results of GPS/BDS-2/Galileo/QZSS were analyzed. Furthermore, in each panel, different colors denote different satellite DD phase bias. Based on earlier finds, the DD phase bias of each satellite signals received by geodetic receivers is identical, as shown in the right panel of Figure 3, and the results of CUT0-CUTB show this conclusion as well. In addition, according to the research of Geng et al. [12] and Li et al. [25], for some smartphones, such as XiaoMi Mi8, the receiver channel-dependent phase bias is nonzero and each satellite has a different value. In addition, for Huawei Mate 20, all satellite systems suffer from receiver channel-dependent phase bias, and the offsets differ from satellite systems and frequency bands. These smartphones cannot meet the integer ambiguity resolution (IAR) unless the DD phase bias is handled correctly. Fortunately, it can be seen from the left part of Figure 3 that for HP40, the phase bias of all visible satellites is close to zero. Furthermore, the phase bias standard deviation (STD) of all satellites is basically less than 0.002 cycles, being even smaller than that of the CUT0-CUTB (JSJN-HP40 is a zero baseline, while CUT0-CUTB is a short baseline). We thus demonstrate that the DD IAR of HP40 will not be affected by receiver channel-dependent phase bias. In conclusion, the DISB estimate and inter-system differencing for HP40 can be implemented without worrying about the DD receiver channel-dependent phase bias.

Figure 4 shows the DD code bias estimates of the two baselines; it can be seen that there is obvious systematic fluctuation for the JSJN-HP40. In addition, we carried out GPS L1/Galileo E1 DD code positioning in Figure 5. Again, here we can see the obvious systematic fluctuation in N/E/U directions for the JSJN-HP40, which means that the code observations of HP40 is perhaps strained by the receiver channel-dependent code bias. Furthermore, from Figure 4, we can also see that influenced by smartphone code noise, the STD of the DD code bias of the JSJN-HP40 is greater than CUT0-CUTB. Because the code observations make less of a contribution compared to the phase observations in the RTK positioning, we thus suggested setting a large initial STD value for code observations (such as 0.6m). The receiver channel-dependent code bias can therefore be neglected and their effects will show up in the code residuals [47]. Considering the influence of receiver channel-dependent code bias, during the next sections, we will not describe the code DISB.

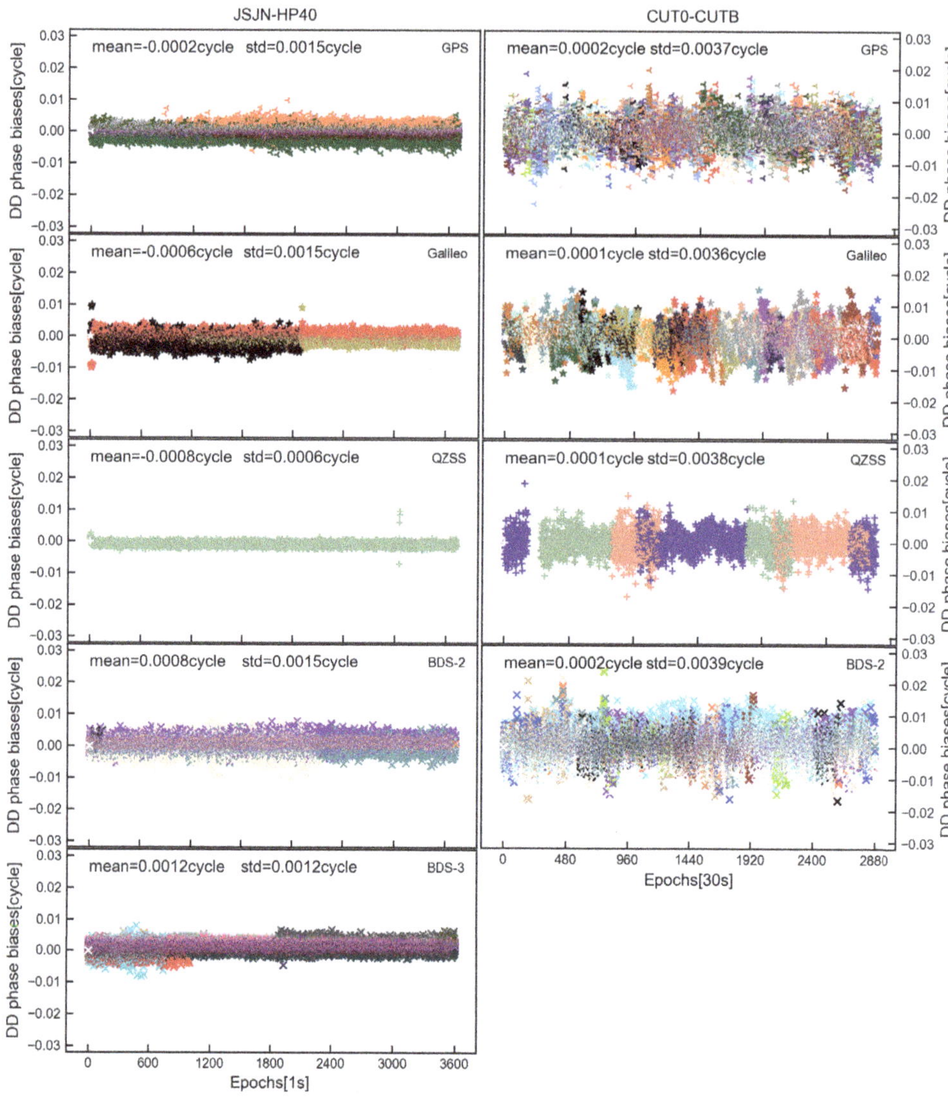

Figure 3. The DD receiver channel−dependent phase bias (cycle) of different systems and satellites for JSJN−HP40 and CUT0−CUTB.

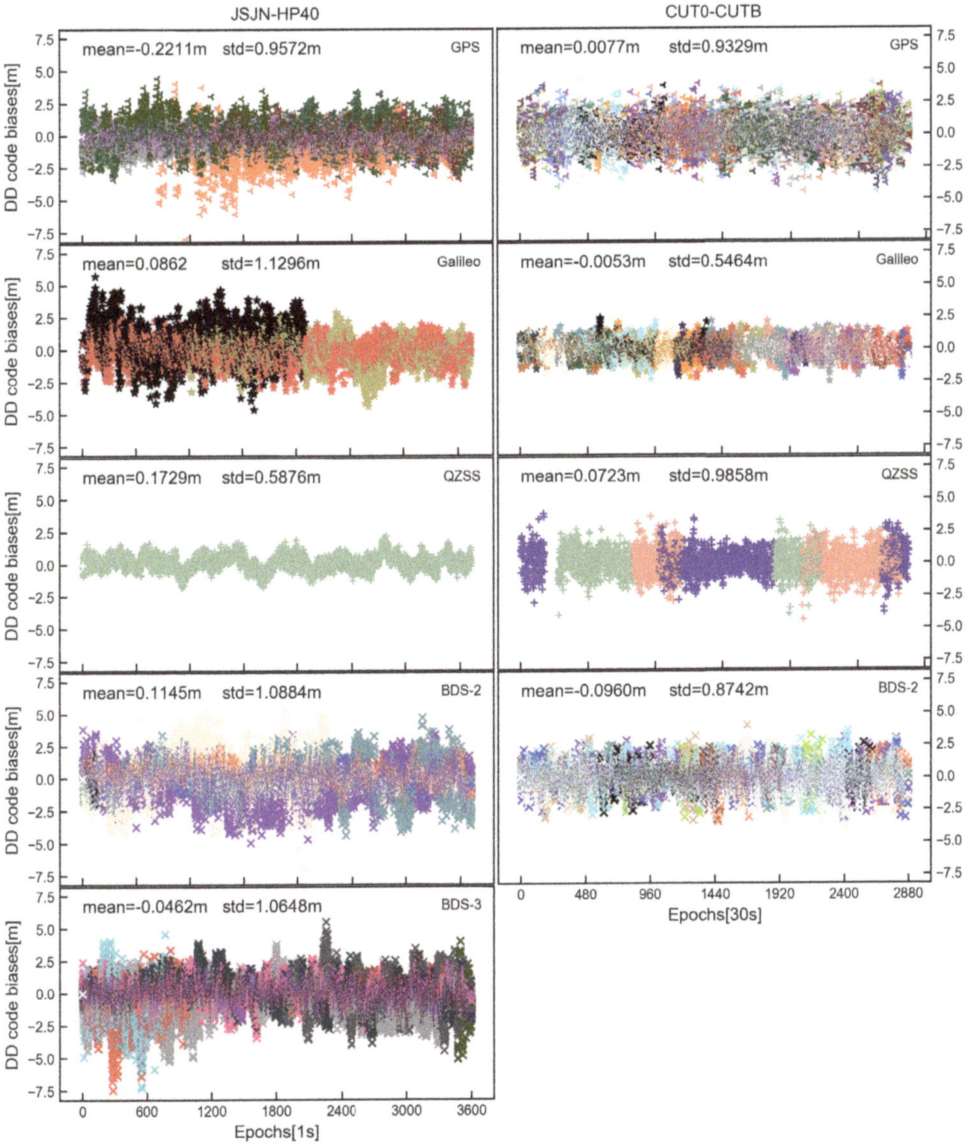

Figure 4. The DD receiver channel−dependent code bias (m) of different systems and satellites for JSJN−HP40 and CUT0−CUTB.

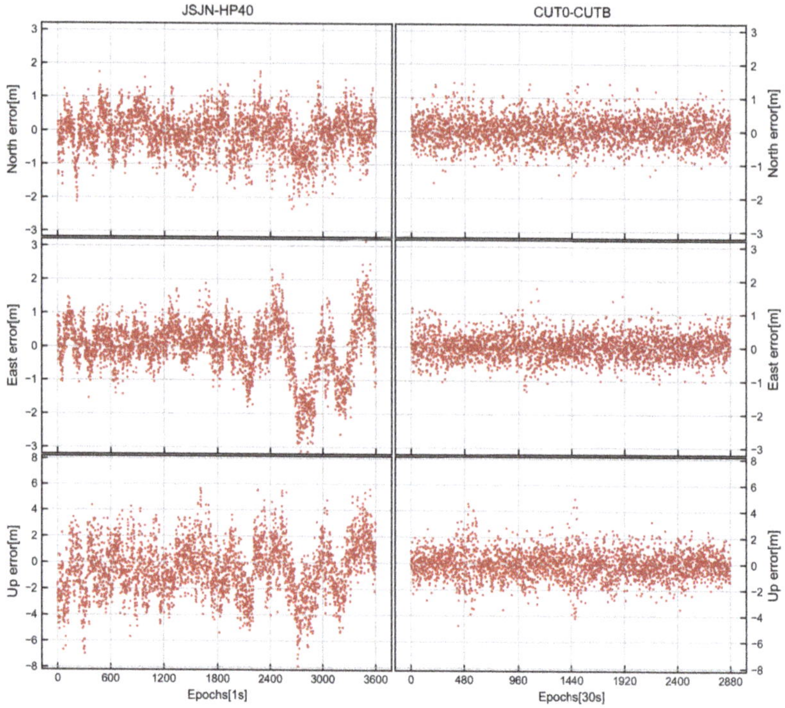

Figure 5. The time series of DD code positioning errors for JSJN−HP40 (**left**) and CUT0−CUTB (**right**).

3.3. Temporal Properties of Phase DISB

After analyzing receiver channel-dependent bias, the stability of phase DISB of the CDMA signals, including overlapping frequencies (GPS L1/Galileo E1, GPS L1/QZSS L1, BDS-2 B1I/BDS-3 B1I) and non-overlapping frequencies (GPS L1/BDS-2 B1I), is investigated in the part. The BDS-3 inherits the B1I frequency of BDS-2, and GPS, QZSS and Galileo share the same L1 frequencies, so it is important to research whether DISB is shared between these frequencies to achieve an inter-system model. Furthermore, if the DISB between GPS L1/BDS-2 B1I is also stable, then an inter-system model like that of the geodetic receivers can be realized. The CUT0-CUTB is used as a control group here.

Figures 6–8 display the time series, distribution and quantile–quantile (QQ) plot of phase DISB for overlapping frequencies. The results obtained show that the phase DISB of GPS L1/QZSS L1 and BDS-2 B1I/BDS-3 B1I is close to zero, and the STD is within 0.005 cycles for the JSJN-HP40, which is similar to CUT0-CUTB. For GPS L1-Galileo E1, although the phase DISB was nonzero, the STD was within 0.005 cycles, and during the test, it also showed the stable characteristics over time. Furthermore, the distribution of DISB is in accord with the theories perfectly, and the points on the QQ plot are also approximately near a straight line, which shows that the phase DISB is coincident with normal distribution. Table 2 shows the corresponding statistical values and critical value of DISB in the observation period; it can be seen from the table that the phase DISB meets $T < \chi^2_{\alpha}(n-1, 0)$, suggesting that the DISB is stable in the observation period.

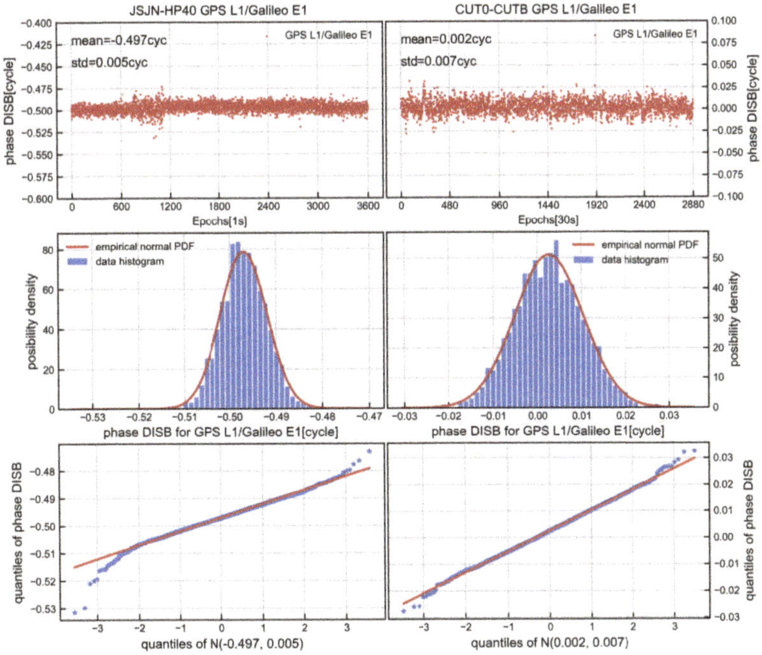

Figure 6. Time series, distribution and QQ plot of GPS L1/Galileo E1 phase DISB for JSJN−HP40 (**left**) and CUT0−CUTB (**right**).

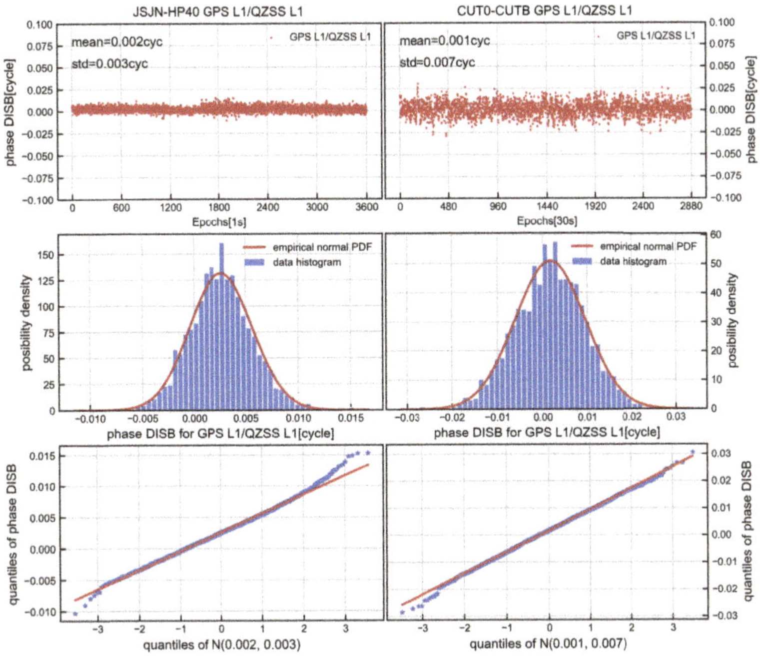

Figure 7. Time series, distribution and QQ plot of GPS L1/QZSS L1 phase DISB for JSJN−HP40 (**left**) and CUT0−CUTB (**right**).

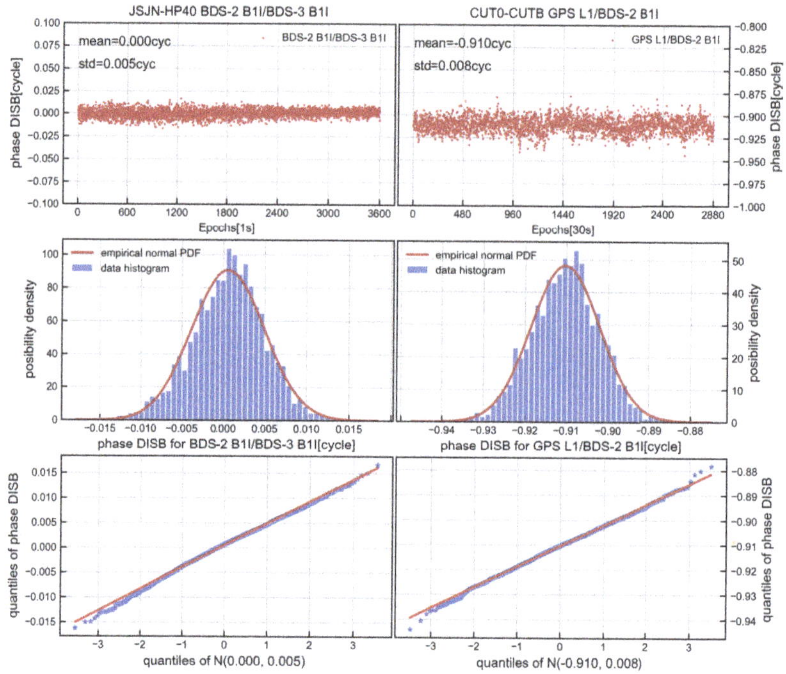

Figure 8. Time series, distribution and QQ plot of BDS−2 B1I/BDS−3 B1I phase DISB for JSJN−HP40 (**left**) and GPS L1/BDS−2 B1I phase DISB for CUT0−CUTB (**right**).

Table 2. Test statistics and critical value for phase DISB.

Baseline	Phase Combination	T	$\chi^2_{\alpha}(n-1,0)$
JSJN-HP40	GPS L1/Galileo E1	471.1	3740.7
	GPS L1/QZSS L1	102.8	3740.7
	BDS-2 B1I/BDS-3 B1I	106.9	3740.7
CUT0-CUTB	GPS L1/Galileo E1	368.4	3006.0
	GPS L1/QZSS L1	261.4	3006.0
	GPS L1/BDS-2 B1I	745.1	3006.0

Figure 9a,b display the integral part and the fractional part of the phase DISB of GPS L1/BDS-2 B1I estimated by the proposed method. It can be seen that the phase DISB of GPS L1/BDS-2 B1I is not stable as the integral part and the fractional part vary for each epoch. To illustrate this, the time series of DD phase measurements for inter-system differencing between GPS L1 and BDS B1I were calculated, and the results of G31 and C09 are shown in Figure 9c,d (both satellites can be continuously observed during the test period, and the G25 was chosen as the reference satellite). In the zero-baseline JSJN-HP40, the DD phase measurements only contain the ambiguity and phase observation noise for GPS intra-system differencing, while GPS L1/ BDS B1I inter-system differencing still includes phase DISB. It can be seen from Figure 9c that the phase data for the GPS is stable as normal, while the DD phase measurements of C09 show the same trends with the phase DISB of GPS L1/BDS B1I; this may be attributable to the GNSS chip hardware design of the HP40. According to previous research for geodetic receivers, the phase DISB for non-overlapping frequencies combination is stable over time. Obviously, the smartphone HP40 does not comply with this law. As a result, the inter-system model is no longer applicable to the HP40 GPS L1/BDS B1I data processing.

Based on the above analysis, for HP40, the inter-system model of GPS L1/QZSS L1 and BDS-2 B1I/BDS-3 B1I can be constructed without considering DISB; the phase DISB for GPS L1/Galileo E1 should be estimated in real time; and the non-overlapping combination of GPS L1/BDS B1I cannot be applied to build an inter-system model.

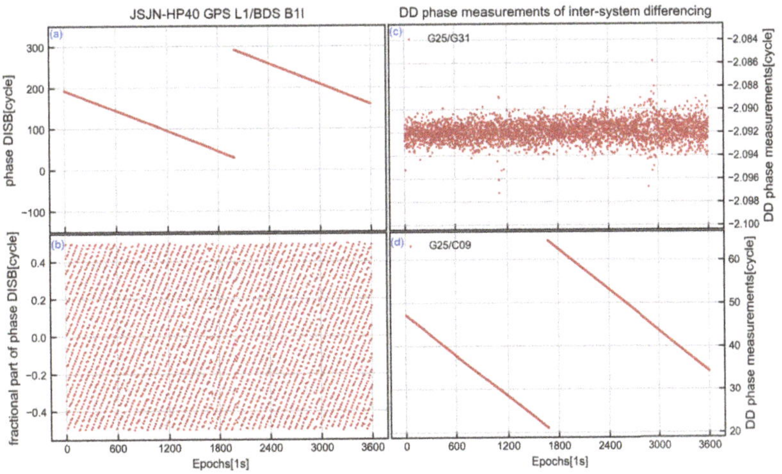

Figure 9. Time series of integral part (**a**) and the fractional part (**b**) of the phase DISB of GPS L1/BDS−2 B1I, and the DD phase measurements of G25/G31 (**c**) and G25/C09 (**d**).

4. Impact of DISB on Kinematic RTK Positioning

The characteristics of the receiver channel-dependent bias and DISB have been researched for the HP40 by using static data. However, in real application, most smartphones are in motion. As conducted in a previous study, the purpose of studying the stability of the DISB is to increase the strength of the positioning model so as to improve the positioning accuracy in real complex situations. In this part, we use a set of kinematics positioning experiments to evaluate the positioning accuracy of the intra-system model and inter-system model.

The kinematic dataset was collected on the lawn in front of the library of Southeast University, as shown in Figure 10, similar to the previous static experiment, and the JSJN station was set as the base station. The HP40 smartphone was used as the rover station and placed in the RF shielding box. The GNSS signals received by the external SinoGNSS AT340 antenna (one of the geodetic antennas of the SinoGNSS company) were sent to the HP40 by a sending antenna, and a CHCNAV i90 GNSS receiver (one of the high-performance IMU-RTK GNSS receivers of the CHC Navigation company) was placed in the RF shielding box for comparison purposes.

In the following positioning experiment, the ambiguity-fixed solutions from the CHCNAV i90 GNSS receiver are used as the true values to evaluate the solutions of the smartphone. A remote-control kart with a maximum speed of 6 km/h is used as a carrier to place these devices, and Figure 11 shows the real-time motion speed of the kart. Table 3 presents the experiment details; the dynamic baseline length was 0.4~0.6 km, and 25 min of data were collected with a sampling rate of 1s. Note that the L1/E1 frequency is used by GPS and Galileo and that there is a phase DISB between them; so, in the following experiment, we estimated the phase DISB in real time to build the inter-system model. In order to compare the positioning performance of the intra-system and inter-system models in a complex environment, we evaluated the positioning results of the two models when the cut-off angle was set to 35°.

At first, only the GPS + 2 Galileo is used, and then the Galileo satellites are gradually added according to the elevation angle. Table 4 presents the AR results in terms of the empirical success rates (the proportion of the number of ambiguity-fixed epochs relative to the number of total epochs) and the TTFF (time to first fix). Table 5 presents the statistics of the positioning results for the intra-system model and inter-system model. Figure 12(top,bottom) show the horizontal and vertical position error distributions using the intra-system and the inter-system model for two different combinations (GPS + 3 Galileo and GPS + 4 Galileo) with a cut-off angle of 35°.

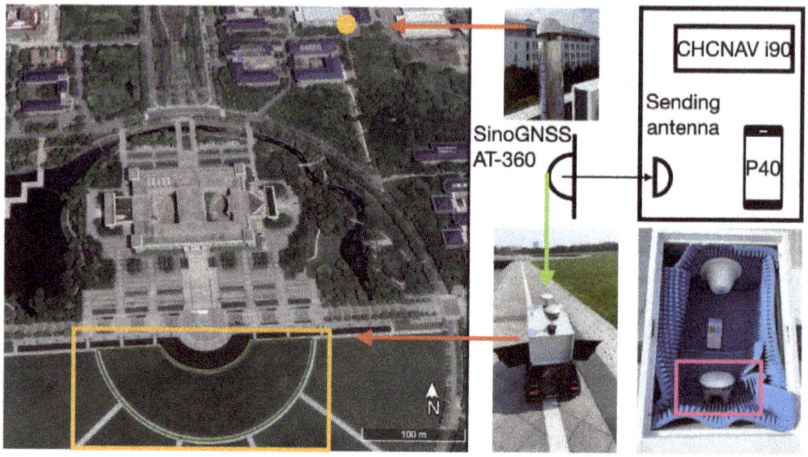

Figure 10. Layout of the smartphone and receiver for collecting kinematic datasets.

Table 3. Observation information for kinematic test.

Base Receiver	Rover Device	Antenna	Baseline Length	Duration (UTC Time)
JSJN	HP40	SinoGNSS AT340	0.4–0.6 km	25 min (03:35–04:00, 13 December 2022)
	CHCNAV i90			

Table 4. Empirical IAR success rates and TTFF for kinematic experiments.

Case	Fix Rate (%)		TTFF (s)	
	Intra	Inter	Intra	Inter
GPS + 2 Galileo	82.9	90.4	45	25
GPS + 3 Galileo	95.3	99.5	16	11
GPS + 4 Galileo	99.2	99.6	12	5
GPS + 5 Galileo	99.3	99.6	9	5

Table 5. Positioning comparison for the intra-system model and the inter-system model with different cases.

Case	Positioning Accuracy (cm)						Improvement (%)		
	Intra			Inter					
	N	E	U	N	E	U	N	E	U
GPS + 2 Galileo	0.37	0.37	1.80	0.33	0.26	1.10	8.9	30.6	38.9
GPS + 3 Galileo	0.34	0.28	1.17	0.32	0.25	1.05	5.5	10.3	10.2
GPS + 4 Galileo	0.31	0.25	1.03	0.30	0.24	0.94	4.7	6.5	9.3
GPS + 5 Galileo	0.31	0.24	0.89	0.30	0.24	0.85	3.3	2.5	4.9

The results show that the inter-system model can improve the positioning accuracy when only small satellites are used for positioning. For the case of GPS + two Galileo, the success rate increases to 82.9% for the intra-system model compared with 90.4% for the inter-system model, and the TTFF of the inter-system model is also shorter than the intra-system model. In addition, the positioning accuracies can increase about 8.9%, 30.6% and 38.9% in three directions. Furthermore, from the red box of Figure 12(top), we can see that only the inter-system model can obtain the fixed solution. With the ambiguities fixed, the positioning accuracy for the intra-system model and inter-system model are smaller than 5 cm, which means that the HP40 can achieve centimeter-level positioning, and when all the GPS and five Galileo satellites are used, both the inter-system model and intra-system model can achieve a high positioning performance. In conclusion, after correcting the phase DISB, the positioning performance of the inter-system model improved significantly, showing that centimeter-level positioning can be achieved in a complex environment for the HP40 with an external antenna. Therefore, using the inter-system model to improve the GNSS positioning performance in a complex observation environment for smartphones has important theoretical and practical significance. This approach will provide important support for the informatization of emerging industries, such as smart cities and intelligent transportation.

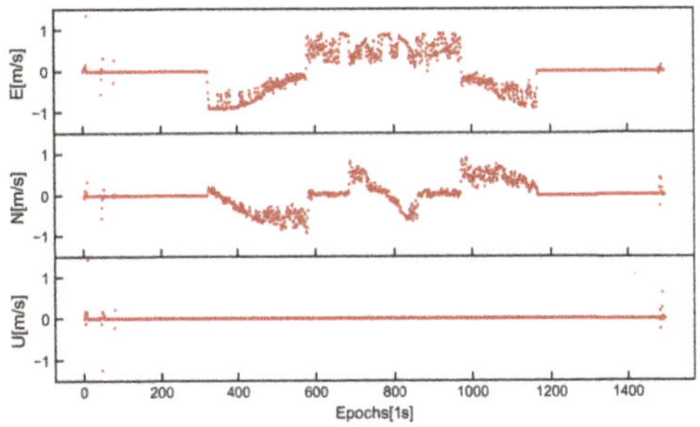

Figure 11. The velocity of remote−control kart for kinematic experiments.

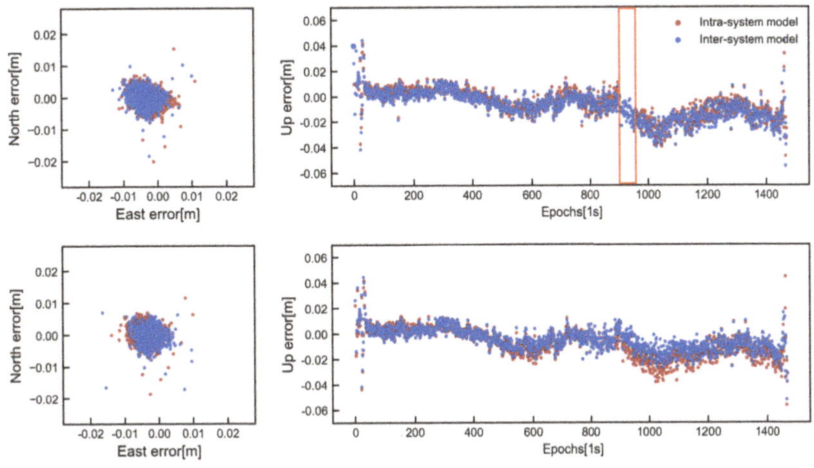

Figure 12. Horizontal position scatter and corresponding vertical time series for the intra−system model (red dots) and the inter−system model (blue dots) for the case of GPS + 3 Galileo (**top**) and GPS + 4 Galileo (**bottom**).

5. Discussion

The opening of the GNSS data interface of smartphones and the development of chips provide great opportunities for high-precision navigation for the public, but at the same time, the high-precision positioning of smartphones in urban environments still faces great challenges. In a relatively open observation environment, smartphones can also achieve a high positioning accuracy similar to a geodetic receiver. However, in the complex observation environment (such as the urban canyon with densely packed buildings), due to the characteristics of the GNSS signal transmission, the accuracy and reliability of the GNSS positioning still face great challenges, which are mainly reflected in the insufficient number of visible satellites, the weak intensity of the satellite geometric observation and the poor quality of the observation signals. The compatibility and interoperability of a GNSS multi-system can improve the positioning accuracy in complex environments to a certain extent, especially when there are few observation satellites.

For smartphone GNSS positioning, the receiver channel-dependent bias should be corrected in advance to achieve inter-system differencing positioning. After that, as shown in previous sections, for the HP40, the phase DISB between GPS L1/QZSS L1 and BDS-2 B1I/BDS-3 B1I is zero, and for GPS L1/Galileo E1, although the DISB is nonzero, it is stable. These stable DISB values can be used to strengthen the positioning model. However, we should also note that the phase DISB of GPS L1/BDS B1I is unstable, which is different with geodetic receivers. In view of the above problem, it is necessary to further study a more universal parameter estimation method so as to establish the inter-system processing method suitable for non-overlapping frequencies observations. The kinematic positioning experiment demonstrates that the inter-system model for a smartphone can effectively improve the positioning performance under complex environments. Although the inter-system differencing experiments performed were based on static or low dynamics, we think they can also be similarly carried out in a real-time, high-dynamic kinematic case, as long as the DISB is verified to be stable.

6. Conclusions

In this paper, we investigated the inter-system differencing between CDMA systems for smartphones. The DD receiver channel-dependent bias and DISB were assessed by using static datasets, and a real kinematic experiment was designed to assess the positioning performance of the inter-system model. The research findings of this contribution are as follows.

(1) For the HP40, the receiver channel-dependent phase biases are close to zero and show time-domain stability, and the STD of the receiver channel-dependent phase bias of all the satellites is less than 0.002 cycles, which provides the possibility of phase DISB estimation between frequency signals. However, there is an obvious systematic fluctuation for the DD code biases of the JSJN-HP40, which means that the HP40 is affected by the receiver channel-dependent code bias. We thus suggested setting a large initial STD value for code observations in the positioning to avoid the effect of the receiver channel-dependent code bias.

(2) The phase DISB of GPS L1/QZSS L1 and BDS-2 B1I/BDS-3 B1I is close to zero, and the STD is within 0.003 cycles for the HP40. For GPS L1/Galileo E1, although the phase DISB is nonzero, the STD is within 0.005 cycles, and during the test, it also showed stable characteristics over time. However, the phase DISB of GPS L1/BDS B1I is not stable.

(3) After introducing the stable phase DISB into the inter-system model, as expected, the kinematic positioning performance of the multi-GNSS RTK resulted in a 3–38.9% improvement in the positioning accuracies for complex environments, compared with the intra-system model, and the success rate was also improved and the TTFF was shortened.

Author Contributions: Conceptualization, R.S. and C.G.; methodology, R.S.; software, R.S. and R.Z.; validation, R.S.; formal analysis, R.S.; investigation, R.S.; resources, R.S. and L.G.; data curation, C.G. and R.S.; writing—original draft preparation, R.S; writing—review and editing, C.G., R.S., X.M. and W.G.; visualization, R.S., L.G. and R.Z.; supervision, C.G. and R.S.; project administration, C.G., R.S. and W.G.; funding acquisition, C.G., R.S. and X.M. All authors have read and agreed to the published version of the manuscript.

Funding: This research was funded by the Outstanding Postdoctoral Program of Jiangsu Province in 2022 (2022ZB113), the National Natural Science Foundation of China (41904022, 41974030) and the Ministry of Education-China Mobile Research Fund (MCM20200J01).

Data Availability Statement: The data that support the findings of this study are available on request from the corresponding author, upon reasonable request.

Acknowledgments: The authors gratefully acknowledge the GNSS Research Centre of Curtin University for providing the multi-GNSS observation data. Thanks also go to Li Guangcai for his support in the data acquisition. Valuable comments from the anonymous reviewers are acknowledged as well.

Conflicts of Interest: The authors declare no conflict of interest.

References

1. Zhang, B.; Hou, P.; Zha, J.; Liu, T. PPP–RTK functional models formulated with undifferenced and uncombined GNSS observations. *Satell. Navig.* **2022**, *3*, 3. [CrossRef]
2. Zhang, Z.; Yuan, H.; He, X.; Li, B.; Geng, J. Best Integer Equivariant Estimation With Quality Control in GNSS RTK for Canyon Environments. *IEEE Trans. Aerosp. Electron. Syst.* **2023**, 1–15. [CrossRef]
3. Ye, J. *Research on Pedestrian Navigation Algorithm Based on Multi-Sensor Fusion*; Chang'an University: Xi'an, China, 2020.
4. Ge, Y.; Cao, X.; Lyu, D.; He, Z.; Ye, F.; Xiao, G.; Shen, F. An investigation of PPP time transfer via BDS-3 PPP-B2b service. *GPS Solut.* **2023**, *27*, 61. [CrossRef]
5. Sikirica, N.; Malić, E.; Rumora, I.; Filjar, R. Exploitation of Google GNSS measurement API for risk assessment of GNSS applications. In Proceedings of the 2017 25th Telecommunication Forum (TELFOR), Belgrade, Serbia, 21–22 November 2017; pp. 1–3.
6. Banville, S.; Van Diggelen, F. Precise positioning using raw GPS measurements from Android smartphones. *GPS World* **2016**, *27*, 43–48.
7. Chen, B.; Gao, C.; Liu, Y.; Sun, P. Real-time precise point positioning with a Xiaomi MI 8 android smartphone. *Sensors* **2019**, *19*, 2835. [CrossRef]
8. Bahadur, B. A study on the real-time code-based GNSS positioning with Android smartphones. *Measurement* **2022**, *194*, 111078. [CrossRef]
9. Li, Z.; Wang, L.; Wang, N.; Li, R.; Liu, A. Real-time GNSS precise point positioning with smartphones for vehicle navigation. *Satell. Navig.* **2022**, *3*, 19. [CrossRef]
10. Li, G.; Geng, J. Characteristics of raw multi-GNSS measurement error from Google Android smart devices. *GPS Solut.* **2019**, *23*, 90. [CrossRef]
11. Zhang, Z.; Li, B.; Gao, Y.; Shen, Y. Real-time carrier phase multipath detection based on dual-frequency C/N0 data. *GPS Solut.* **2019**, *23*, 7. [CrossRef]
12. Geng, J.; Li, G. On the feasibility of resolving Android GNSS carrier-phase ambiguities. *J. Geod.* **2019**, *93*, 2621–2635. [CrossRef]
13. Darugna, F.; Wübbena, J.; Ito, A.; Wübbena, T.; Wübbena, G.; Schmitz, M. RTK and PPP-RTK using smartphones: From short-baseline to long-baseline applications. In Proceedings of the 32nd International Technical Meeting of the Satellite Division of The Institute of Navigation (ION GNSS+ 2019), Miami, FL, USA, 16–20 September 2019; pp. 3932–3945.
14. Bochkati, M.; Sharma, H.; Lichtenberger, C.A.; Pany, T. Demonstration of fused RTK (fixed)+ inertial positioning using Android smartphone sensors only. In Proceedings of the 2020 IEEE/ION Position, Location and Navigation Symposium (PLANS), Portland, OR, USA, 20–23 April 2020; pp. 1140–1154.
15. Sun, R.; Wang, J.; Cheng, Q.; Mao, Y.; Ochieng, W.Y. A new IMU-aided multiple GNSS fault detection and exclusion algorithm for integrated navigation in urban environments. *GPS Solut.* **2021**, *25*, 147. [CrossRef]
16. Liu, Q.; Gao, C.; Shang, R.; Peng, Z.; Zhang, R.; Gan, L.; Gao, W. NLOS signal detection and correction for smartphone using convolutional neural network and variational mode decomposition in urban environment. *GPS Solut.* **2023**, *27*, 31. [CrossRef]
17. Xia, Y.; Pan, S.; Meng, X.; Gao, W.; Ye, F.; Zhao, Q.; Zhao, X. Anomaly detection for urban vehicle GNSS observation with a hybrid machine learning system. *Remote Sens.* **2020**, *12*, 971. [CrossRef]
18. Xia, Y.; Meng, X.; Yang, Y.; Pan, S.; Zhao, Q.; Gao, W. First results of BDS positioning for LBS applications in the UK. *Satell. Navig.* **2021**, *2*, 8. [CrossRef]
19. Humphreys, T.E.; Murrian, M.; Van Diggelen, F.; Podshivalov, S.; Pesyna, K.M. On the feasibility of cm-accurate positioning via a smartphone's antenna and GNSS chip. In Proceedings of the 2016 IEEE/ION Position, Location and Navigation Symposium (PLANS), Savannah, GR, USA, 11–14 April 2016; pp. 232–242.
20. Riley, S.; Lentz, W.; Clare, A. On the path to precision-observations with android GNSS observables. In Proceedings of the 30th International Technical Meeting of The Satellite Division of The Institute of Navigation (ION GNSS+ 2017), Portland, OR, USA, 25–29 September 2017; pp. 116–129.
21. Paziewski, J.; Sieradzki, R.; Baryla, R. Signal characterization and assessment of code GNSS positioning with low-power consumption smartphones. *GPS Solut.* **2019**, *23*, 98. [CrossRef]

22. Paziewski, J. Recent advances and perspectives for positioning and applications with smartphone GNSS observations. *Meas. Sci. Technol.* **2020**, *31*, 091001. [CrossRef]
23. Paziewski, J.; Fortunato, M.; Mazzoni, A.; Odolinski, R. An analysis of multi-GNSS observations tracked by recent Android smartphones and smartphone-only relative positioning results. *Measurement* **2021**, *175*, 109162. [CrossRef]
24. Li, G.; Geng, J. Android multi-GNSS ambiguity resolution in the case of receiver channel-dependent phase biases. *J. Geod.* **2022**, *96*, 72. [CrossRef]
25. Li, B.; Miao, W.; Chen, G.; Li, Z. Ambiguity resolution for smartphone GNSS precise positioning: Effect factors and performance. *J. Geod.* **2022**, *96*, 63. [CrossRef]
26. Julien, O.; Alves, P.; Cannon, M.E.; Zhang, W. A tightly coupled GPS/GALILEO combination for improved ambiguity resolution. In Proceedings of the European Navigation Conference (ENC-GNSS'03), Calgary, AB, Canada, 9 September 2003; pp. 1–14.
27. Zhang, B.; Teunissen, P.J.; Yuan, Y. On the short-term temporal variations of GNSS receiver differential phase biases. *J. Geod.* **2017**, *91*, 563–572. [CrossRef]
28. Montenbruck, O.; Hauschild, A.; Steigenberger, P. Differential code bias estimation using multi-GNSS observations and global ionosphere maps. *Navig. J. Inst. Navig.* **2014**, *61*, 191–201. [CrossRef]
29. Odijk, D.; Teunissen, P.J. Characterization of between-receiver GPS-Galileo inter-system biases and their effect on mixed ambiguity resolution. *GPS Solut.* **2013**, *17*, 521–533. [CrossRef]
30. Paziewski, J.; Wielgosz, P. Accounting for Galileo–GPS inter-system biases in precise satellite positioning. *J. Geod.* **2015**, *89*, 81–93. [CrossRef]
31. Nadarajah, N.; Khodabandeh, A.; Teunissen, P.J. Assessing the IRNSS L5-signal in combination with GPS, Galileo, and QZSS L5/E5a-signals for positioning and navigation. *GPS Solut.* **2016**, *20*, 289–297. [CrossRef]
32. Odijk, D.; Nadarajah, N.; Zaminpardaz, S.; Teunissen, P.J. GPS, Galileo, QZSS and IRNSS differential ISBs: Estimation and application. *GPS Solut.* **2017**, *21*, 439–450. [CrossRef]
33. Odolinski, R.; Teunissen, P.J. Single-frequency, dual-GNSS versus dual-frequency, single-GNSS: A low-cost and high-grade receivers GPS-BDS RTK analysis. *J. Geod.* **2016**, *90*, 1255–1278. [CrossRef]
34. Wu, M.; Liu, W.; Wang, W.; Zhang, X. Differential inter-system biases estimation and initial assessment of instantaneous tightly combined RTK with BDS-3, GPS, and Galileo. *Remote Sens.* **2019**, *11*, 1430. [CrossRef]
35. Yuan, H.; Zhang, Z.; He, X.; Zeng, J. Tight integration of BDS-3/BDS-2/GPS/Galileo observations considering the new overlapping DISBs and its application in obstructed environments. *Adv. Space Res.* **2022**, *71*, 2879–2891. [CrossRef]
36. Gao, W.; Meng, X.; Gao, C.; Pan, S.; Wang, D. Combined GPS and BDS for single-frequency continuous RTK positioning through real-time estimation of differential inter-system biases. *GPS Solut.* **2018**, *22*, 20. [CrossRef]
37. Mi, X.; Zhang, B.; Yuan, Y.; Luo, X. Characteristics of GPS, BDS2, BDS3 and Galileo inter-system biases and their influence on RTK positioning. *Meas. Sci. Technol.* **2019**, *31*, 015009. [CrossRef]
38. Mi, X.; Zhang, B.; Yuan, Y. Multi-GNSS inter-system biases: Estimability analysis and impact on RTK positioning. *GPS Solut.* **2019**, *23*, 81. [CrossRef]
39. Chen, H.; Jiang, W.; Li, J. Multi-GNSS relative positioning with fixed inter-system ambiguity. *Remote Sens.* **2019**, *11*, 454. [CrossRef]
40. Shang, R.; Meng, X.; Gao, C.; Pan, S.; Gao, W.; Peng, Z. Particle filter-based inter-system positioning model for non-overlapping frequency code division multiple access systems. *J. Navig.* **2020**, *73*, 953–970. [CrossRef]
41. Liu, J.; Tu, R.; Han, J.; Zhang, R.; Zhang, P.; Fan, L. Estimability analysis of differential inter-system biases and differential inter-frequency biases for dual-frequency GPS and BDS combined RTK. *Meas. Sci. Technol.* **2019**, *31*, 025009. [CrossRef]
42. Zhang, B.; Teunissen, P. Zero-baseline Analysis of GPS/BeiDou/Galileo Between-Receiver Differential Code Biases (BR-DCBs): Time-wise Retrieval and Preliminary Characterization. *Navig. J. Inst. Navig.* **2016**, *63*, 181–191. [CrossRef]
43. Xiaohong, Z.; Mingkui, W.; Wanke, L. Model and Performance Analysis of Tightly Combined BeiDou B2 and Galileo E5b Relative Positioning for Short Baseline. *Acta Geod. Et Cartogr. Sin.* **2017**, *45*, 1.
44. Zhang, Z.; Li, Y.; He, X.; Chen, W.; Li, B. A composite stochastic model considering the terrain topography for real-time GNSS monitoring in canyon environments. *J. Geod.* **2022**, *96*, 79. [CrossRef]
45. Sui, M.; Gong, C.; Shen, F. An optimized stochastic model for smartphone GNSS positioning. *Front. Earth Sci.* **2022**, *10*, 1018420. [CrossRef]
46. Yuan, H.; Zhang, Z.; He, X.; Li, G.; Wang, S. Stochastic model assessment of low-cost devices considering the impacts of multipath effects and atmospheric delays. *Measurement* **2022**, *188*, 110619. [CrossRef]
47. Cai, C.; Gao, Y. Modeling and assessment of combined GPS/GLONASS precise point positioning. *GPS Solut.* **2013**, *17*, 223–236. [CrossRef]

MDPI
St. Alban-Anlage 66
4052 Basel
Switzerland
Tel. +41 61 683 77 34
Fax +41 61 302 89 18
www.mdpi.com

Remote Sensing Editorial Office
E-mail: remotesensing@mdpi.com
www.mdpi.com/journal/remotesensing

www.ingramcontent.com/pod-product-compliance
Lightning Source LLC
Chambersburg PA
CBHW040804250926
51850CB00007BA/1077
9783036576220